AQA A2

Biology

Steve Potter

A2

AQA Biology

Steve Potter

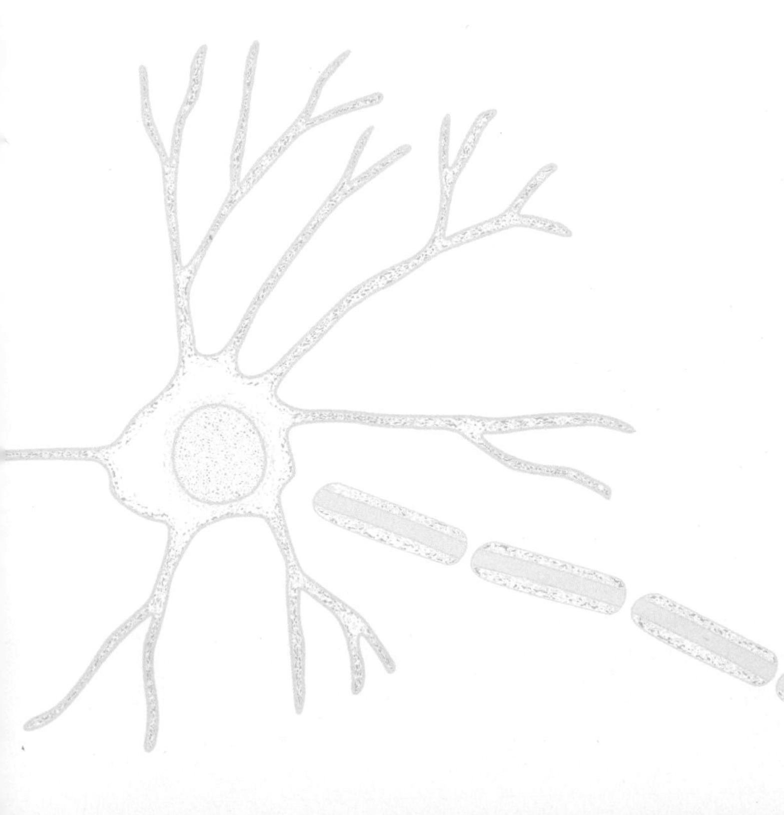

Philip Allan Updates, an imprint of Hodder Education, an Hachette UK company, Market Place, Deddington, Oxfordshire OX15 0SE

Orders
Bookpoint Ltd, 130 Milton Park, Abingdon, Oxfordshire OX14 4SB
tel: 01235 827720
fax: 01235 400454
e-mail: uk.orders@bookpoint.co.uk

Lines are open 9.00 a.m.–5.00 p.m., Monday to Saturday, with a 24-hour message answering service. You can also order through the Philip Allan Updates website: www.philipallan.co.uk

ISBN 978-1-84489-218-1

Impression number 5 4 3 2
Year 2013 2012 2011 2010 2009

Printed in Italy.

P01397

Contents

Introduction

About this book

This textbook is written specifically for students following the AQA A2 Biology course. The topics are covered in broadly the same order as in the specification, however some items have been reordered to give a more coherent structure.

Chapters 1–6 deal with the content of Module 4. The first four chapters cover:

- populations and the environment
- energy transduction by photosynthesis and respiration
- the transfer of energy through ecosystems
- cycling of elements through ecosystems

These topics are all concerned with the interaction between energy and matter in ecosystems.

Chapters 5 and 6 cover the topics of inheritance, selection and speciation. Some of the patterns of inheritance that occur as genes are passed from one generation to the next are explained, as is the effect of natural selection on gene frequencies in populations. Finally, Chapter 6 explains how this can lead to the evolution of new species.

Chapters 7–11 cover the content of Module 5. Chapters 7, 8 and 9 deal with how organisms detect and respond to changes in their internal and external environments. This includes explanations of:

- how sense cells detect changes and initiate nerve impulses
- the nature of nerve impulses and how they are transmitted
- how hormones bring about responses in specific cells
- how specific effectors bring about responses that either cause changes in the body or help to maintain a constant internal environment

Chapter 10 returns to the topic of protein synthesis, outlined in the AS course. Here it is covered in more detail, with more emphasis on the way in which the synthesis of proteins is regulated.

Finally, Chapter 11 looks at the ways in which genes can be cloned and the ways in which cloned genes can be used.

Each chapter begins with a chapter outline and a brief introduction. Margin comments accompany the text. Some provide extra information to help clarify a point without interrupting the flow of text. These are identified by the symbol ◀. Others are examiner hints. These are identified by the symbol ⓔ.

Feature boxes are included to provide extra detail or to give information about applications of a topic. Some of these features relate to How Science Works and may show how:

- the work of one research team often depends on previous research findings
- scientists can communicate their research findings
- correlational evidence does not constitute proof
- the way in which careful design of an investigation allows cause and effect to be inferred, if not proved
- the way in which research is refined to show cause and effect

These boxes are identified with the following symbol:

The main content of each chapter is followed by a comprehensive summary; this would be a good place to start your revision of that topic.

Each chapter ends with questions designed to test your understanding of that topic. Multiple-choice questions are followed by longer, structured examination-style questions. Some of these test aspects of How Science Works.

The unit tests

Terms used in the unit tests

You will be asked precise questions in the examination, so you can save a lot of valuable time as well as ensuring you score as many marks as possible by knowing what is expected. Terms most commonly used are explained below.

Describe

This means exactly what it says — 'tell me about…' — and you should not need to explain why.

Explain

Here you must give biological reasons for why or how something is happening.

Complete

You must finish off a diagram, graph, flow chart or table.

Draw/plot

This means that you must construct some type of graph. For this, make sure that:

- you choose a scale that makes good use of the graph paper (if a scale is not given) and does not leave all the plots tucked away in one corner
- you plot an appropriate type of graph — if both variables are continuous, then a line graph is usually the most appropriate; if one is a discrete variable, then a bar chart is appropriate
- you plot carefully using a sharp pencil and draw lines accurately

From the...

This means that you must use only information in the diagram/graph/photograph or other forms of data.

Name

This asks you to give the name of a structure/molecule/organism etc.

Suggest

This means 'give a plausible biological explanation for' — it is often used when testing understanding of concepts in an unfamiliar situation.

Compare

In this case, you have to give similarities *and* differences.

Calculate

This means add, subtract, multiply, divide (do some kind of sum!) and show how you got your answer — always show your working!

'Dos and don'ts'

When you finally open the test paper, it can be quite a stressful moment. For example, you may not recognise the diagram or graph used in question 1. It can be quite demoralising to attempt a question at the start of an examination if you do not feel confident about it. So:

- *do not* begin to write as soon as you open the paper
- *do not* answer question 1 first, just because it is printed first (the examiner did not sequence the questions with your particular favourites in mind)
- *do* scan all the questions before you begin to answer any
- *do* identify those questions about which you feel most confident
- *do* answer first those questions about which you feel most confident, regardless of order in the paper
- *do* read the questions carefully — if you are asked to explain, then explain, don't just describe
- *do* take notice of the mark allocation and don't supply the examiner with all your knowledge of osmosis if there is only 1 mark allocated (similarly, you will have to come up with four ideas if 4 marks are allocated)
- *do* try to stick to the point in your answer (it is easy to stray into related areas that will not score marks and will use up valuable time)
- *do* take care with:
  - drawings — you will not be asked to produce complex diagrams, but those you do produce must resemble the subject
  - labelling — label lines must touch the part you are required to identify; if they stop short or pass through the part you will lose marks
  - graphs — draw small points if you are asked to plot a graph and join the points with ruled lines or, if specifically asked for, a line or smooth curve of best fit through all the plots
- *do* try to answer all the questions

Chapter 1

Populations

This chapter covers:

- the nature of populations
- the factors that affect the size of populations
- competition
- methods of investigating populations
- the structure of human populations

Population is a much-used word and everyone has an understanding of what it means. We talk of the world population, the population of England, the population of Huddersfield and so on. How is it that we can we use the word population in all these different contexts?

What is a population?

In biology, the word 'population' has a specific meaning.

A population is all the individuals of a particular species in a particular habitat at a particular time.

A habitat is an area where a population lives and finds the nutrients, water, living space and other essentials it needs to survive.

e Try to avoid describing a habitat as 'an animal's home'!

So, the examples above of how we sometimes use the word are all correct. All that changes in the examples is the size of the particular habitat. The size of the particular habitat can be very different for different organisms.

There are several populations of cheetah throughout Africa (Figure 1.1). Each population occupies a large territory. The serrated wrack seaweeds on a beach form a population that occupies a smaller area. The tadpoles in a garden pond make up a small local population of young frogs.

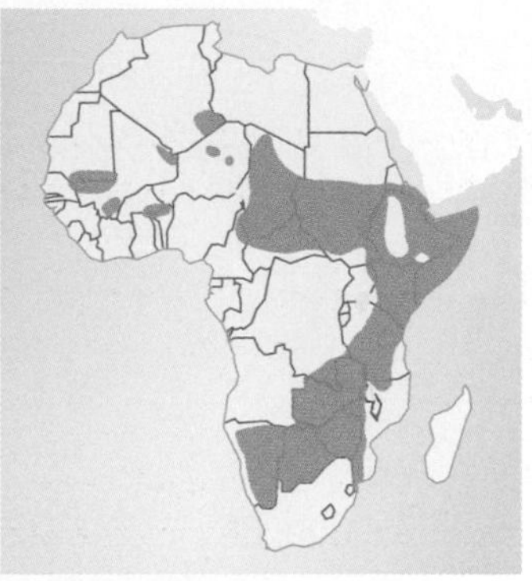

Figure 1.1 The distribution of cheetahs in Africa, showing different populations

We could describe the pond as the habitat of the tadpoles and tadpoles are, in fact, found in most areas of the pond. However, other organisms are found only in certain parts of the pond and so have a more restricted habitat (Figure 1.2). The floating plants are found only at the pond surface; the decomposers are found in the dead material (detritus).

Tadpoles in a pond. When these mature and leave the pond they will become part of a wider population of adult frogs.

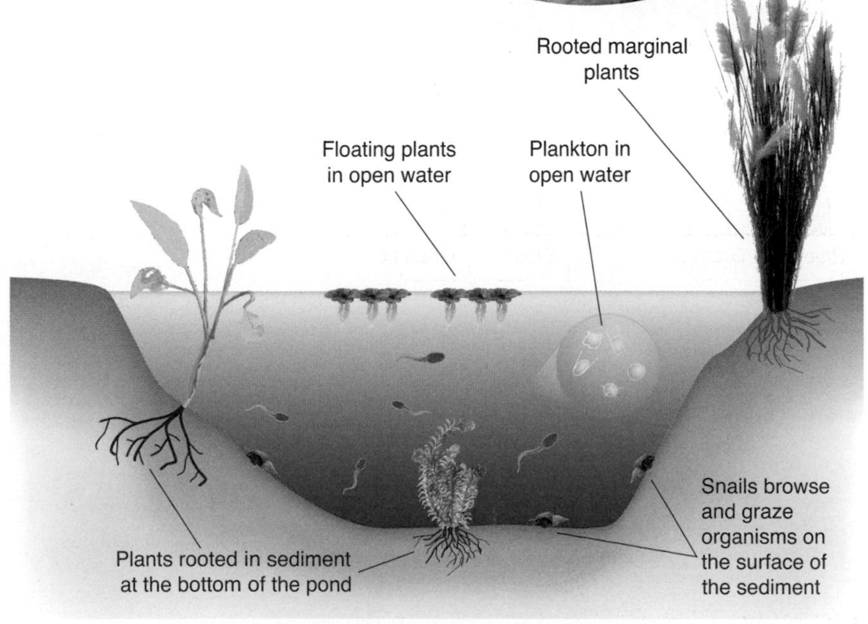

Figure 1.2 Diagram of a pond showing the habitats of some organisms

How can several populations live in the same area?

In a pond, there are populations of different types of organism. They are able to live in the same area because each exploits a different habitat. For example:

- plankton exploit the open-water regions
- decomposers inhabit the detritus at the bottom of the pond
- snails browse the surface of the sediment at the bottom of the pond and graze on small organisms

Some organisms share the same habitat. However, they make different demands on that habitat. This combination of habitat and the demands made is called the **ecological niche**.

The ecological niche of an organism describes its role within a habitat.

For example, floating plants and tadpoles are both found in the open-water habitat. The plants use sunlight, carbon dioxide, water and minerals from the water; the tadpoles feed on algae and other small organisms. The plants and tadpoles have different niches.

In another example, blue tits and great tits spend much of their time foraging in trees for insects and insect larvae. When one or the other species is present alone, each species forages at about the same height. However, if both species are present in the same trees, they forage at different heights and so avoid competing for the same niche (Figure 1.3).

Research has shown that two species cannot occupy the same niche. If two species are present initially, they compete for the available resources in the niche. One will be more successful and the other will be made locally extinct. This is called the competitive exclusion principle.

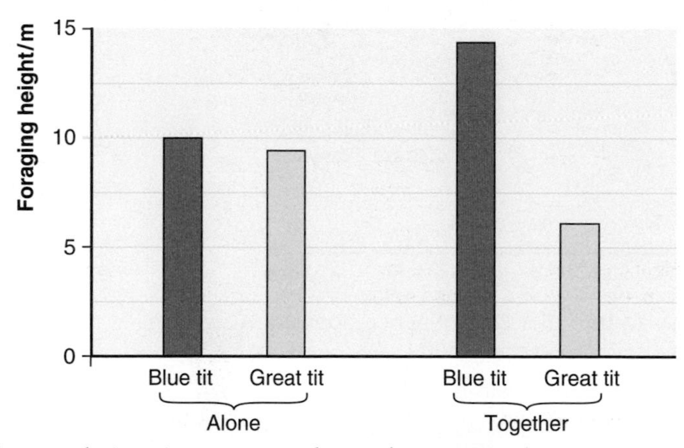

Figure 1.3 Foraging heights of blue tits and great tits

- All the populations in an area make up the **community**.
- All the habitats in an area make up the **physical environment** of that area.

Together, the community and the environment make up an **ecosystem**.

An ecosystem is a community of organisms that interact with each other and with the physical environment in which they are found.

Ecosystems vary enormously in size from, for example, a small garden pond through small fields to the Amazon rainforest.

What factors influence the size of populations?

A pond covered in *Lemna*

Duckweed (*Lemna minor*) is a common pond plant. Within weeks, a population of duckweed can increase from just a few plants to covering a pond.

So what allows the population to increase this much? It must be because the duckweed initially had an abundance of all or most of the factors needed for growth (light, carbon dioxide, water, mineral ions) and an absence (or near absence) of the factors that could limit the population (herbivores, disease-causing organisms).

However, this rapid population growth cannot continue indefinitely. Eventually, one of two things usually happens. Either:

- the duckweed exhausts its supply of mineral ions and the population crashes

or

- a level is reached that can be sustained by the available nutrients and the number of plants dying is matched by the number of new plants produced

These factors are illustrated in Figure 1.4. The second situation produces what is known as the **carrying capacity** for the duckweed in that particular environment.

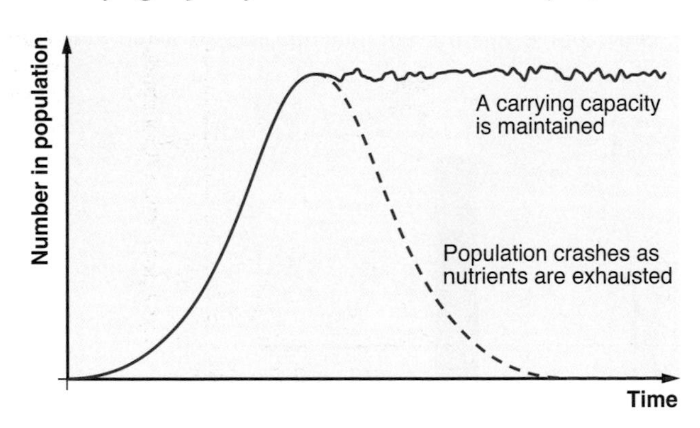

Figure 1.4 Population growth in *Lemna*

However, these are not the only factors that can influence population size. The population could be reduced by:

- a disease-causing organism entering the environment
- a predator (a herbivore in the case of duckweed) entering the environment

The factors that influence populations are grouped into two main categories:

- **biotic factors** — the effects of other organisms of the same or different species
- **abiotic factors** — the effects of factors in the physical environment (e.g. light, temperature, carbon dioxide concentration, oxygen concentration)

Biotic factors

Biotic factors have a number of effects on population size:

- **Predation** The presence of a predator (or herbivore in the case of plants) reduces the numbers.

- **Disease-causing organisms** If disease is widespread, then population growth is slowed
- **Intraspecific competition** Competition between members of the same species operates in two main ways:
  - Reducing the resources of all of the population can reduce fertility and so reduce population growth.
  - Reducing the resources of just some of the population (as others compete more effectively) means that these members of the population die while the others reproduce. However, population growth is still reduced.
- **Interspecific competition** When two species compete for the same resource, a complex situation can develop. Although the competitive exclusion principle states that two species cannot occupy the same niche, this is not absolute. The following *can* happen:
  - One of the two species out-competes the other, which may then die out.
  - Both species suffer a reduction in numbers because they have almost equal ability to 'harvest' the resource and the effects of intraspecific competition for a reduced resource come into play.
  - The species are able to coexist.

The predator–prey relationship

One of the best-known examples of the relationship between a predator and its prey is that between the lynx and the snowshoe hare (Figure 1.5).

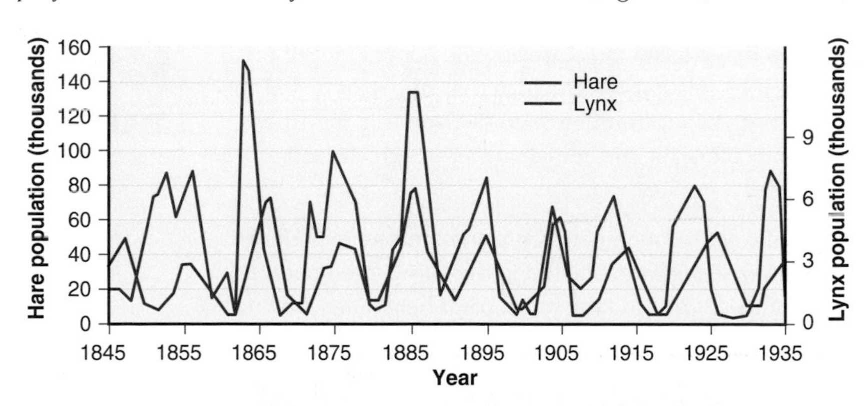

Figure 1.5 Changes in the population size of snowshoe hare and lynx between 1850 and 1935

A lynx with snowshoe hare prey

The population sizes of predator and prey are interdependent:

- An increase in the prey population means more food for the predator.
- The predator population increases.
- The increased numbers of predators kill more prey, so the prey population decreases.
- There is now less food for the predator, so the predator population decreases.
- The reduced numbers of predators kill fewer prey, so the prey numbers increase.

Notice that the change in numbers of the predator population always lags behind that of the prey population. This is because the predator is dependent on the prey for food. Notice also that the number of predators is always lower than the number of prey. This is because of the loss of energy along a food chain (see page 53).

Intraspecific competition

Intraspecific competition is competition between members of the *same* species for a resource (often food) in the same habitat.

For example, gypsy moth caterpillars infested much of southern New England (on the east coast of the USA) in the summer of 1980.

The density of the infestation was quite low, which allowed most of the caterpillars to metamorphose into adults. The adults mated and laid egg masses (each mass containing several hundred eggs) on almost every tree in the region. In the following spring, the eggs hatched and the caterpillars began feeding. As they fed and grew, they stripped the trees of their leaves.

Gypsy moth caterpillar

The millions of caterpillars were soon competing for a very limited resource — any remaining leaves! As a result, the population crashed and only a few caterpillars metamorphosed into adults.

Intraspecific competition is a major factor in controlling the populations of predators. In a typical predator–prey relationship (such as that of the lynx and the snowshoe hare), when the prey population begins to fall, there is intraspecific competition between the predators for the remaining prey, which leads to a population decline.

There can also be intraspecific competition between plants. In one experiment, different numbers of sunflower seeds were planted in the same-sized pots. The number of live sunflower plants per pot and the mean plant height are shown in Tables 1.1(a) and 1.1(b).

With increased numbers, there is increased intraspecific competition, which results in fewer, smaller plants surviving. This sort of experiment has applications in agriculture — for example, crop seeds are supplied with recommended sowing densities.

Table 1.1 (a) Number of live plants per pot

Time/ days	2 seeds per pot	4 seeds per pot	8 seeds per pot	16 seeds per pot	32 seeds per pot
7	2	3	6	11	24
14	2	3	6	11	24
21	2	3	4	10	22
28	2	3	3	8	20
35	2	3	3	7	17

Time/ days	2 seeds per pot	4 seeds per pot	8 seeds per pot	16 seeds per pot	32 seeds per pot
7	8.5	7.5	6.1	5.2	4.8
14	9.3	8.3	7.2	5.9	5.3
21	18.9	16.2	11.4	9.6	7.3
28	20.2	18.6	14.6	13.4	11.3
35	27.4	22.5	17.4	16.3	15.2

Table 1.1 (b) Mean height of plants per pot/cm

Interspecific competition

Interspecific competition occurs when two *different* species compete for the same resource in the same habitat. Although most organisms have their own ecological niche, there may be overlap between the niches of two species. This can be shown diagrammatically. Figure 1.6 represents the relationship between the niches of two species of bird that feed on slightly different-sized insects and which have slightly different temperature tolerances.

The range of conditions an organism *could* survive in is called its fundamental niche. The range of conditions in which it is *actually found* is called its realised niche. The realised niche is usually more restricted because of competition.

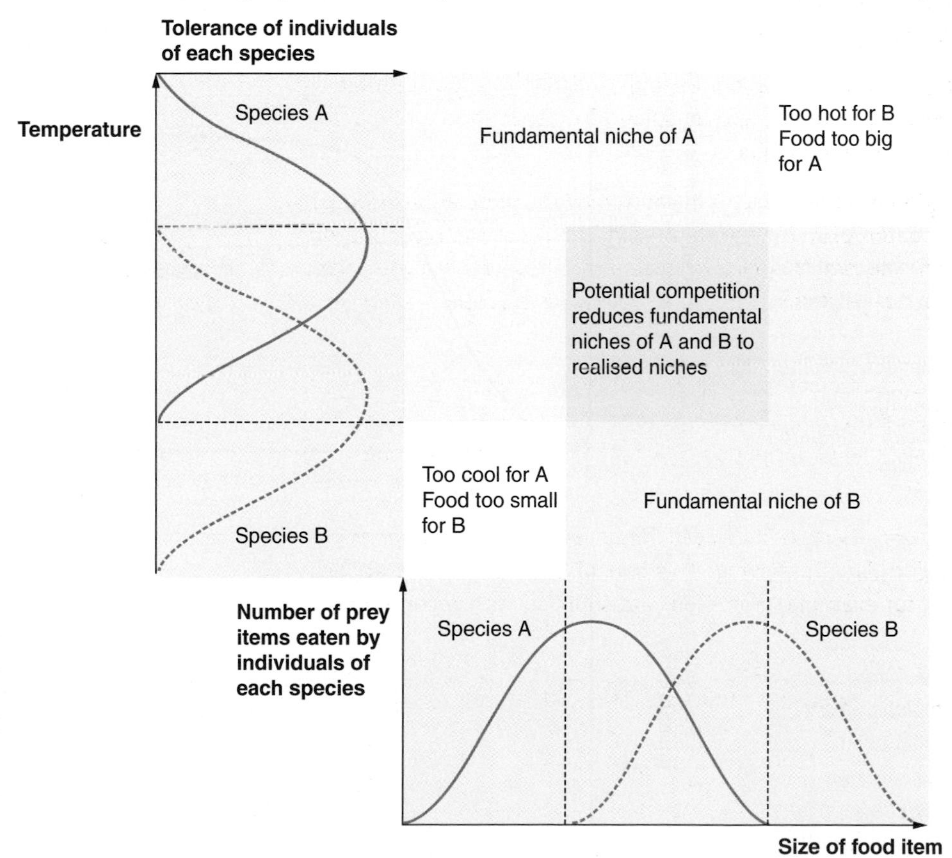

Figure 1.6 Diagram to represent the relationship between two overlapping niches

From Figure 1.6 it can be seen that there are conditions in which only one of the two species could survive and that there is one set of conditions in which the combined preferences of the two species overlap. Wherever these conditions are found, if both species are present, there will be interspecific competition between them. The outcome of this competition depends on how well each species competes for the resources in the conditions.

Box 1.1 An investigation into interspecific competition in *Paramecium*

In 1934, G. F. Gause grew three different species of *Paramecium* (a unicellular organism) in a culture medium of oatmeal and yeast. Initially, he grew each species in isolation and monitored the population densities (Figures 1.7 a–c). Subsequently, he grew combinations of species together and monitored the population densities (Figures 1.7 d–e).

Gause concluded that if the two species utilise the same resource in the same way, then one species out-competes the other totally, which becomes locally extinct. However, if the two species use the resource in a slightly different manner, then coexistence is possible.

All three species of *Paramecium* feed on yeast. Therefore, when cultured in pairs, it might be expected that one species would out-compete the other and make it extinct locally. This happened to *Paramecium caudatum* when it was cultured with *Paramecium aurelia*.

However, when *Paramecium caudatum* was cultured with *Paramecium bursaria*, both species survived; they were able to coexist. Gause found that this was because *Paramecium bursaria* contains a number of unicellular green algae that photosynthesise and produce oxygen. Therefore, *Paramecium bursaria* is able to exist near the bottom of the culture vessel whereas *Paramecium caudatum* must be near the surface to obtain oxygen.

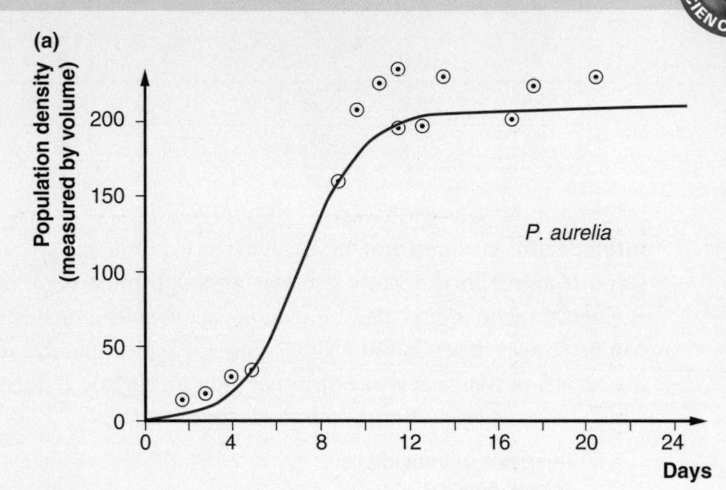

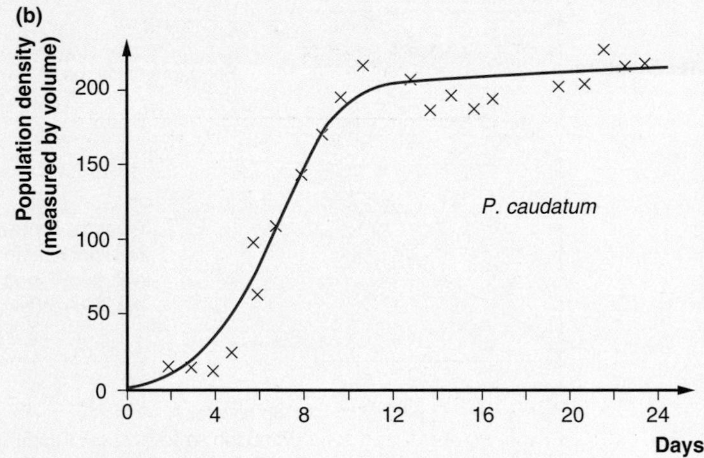

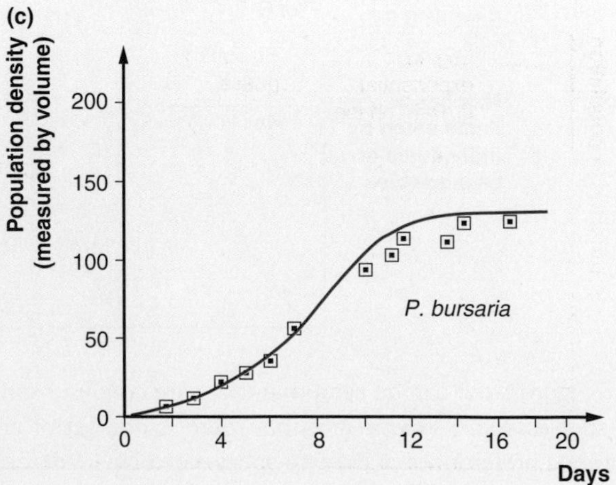

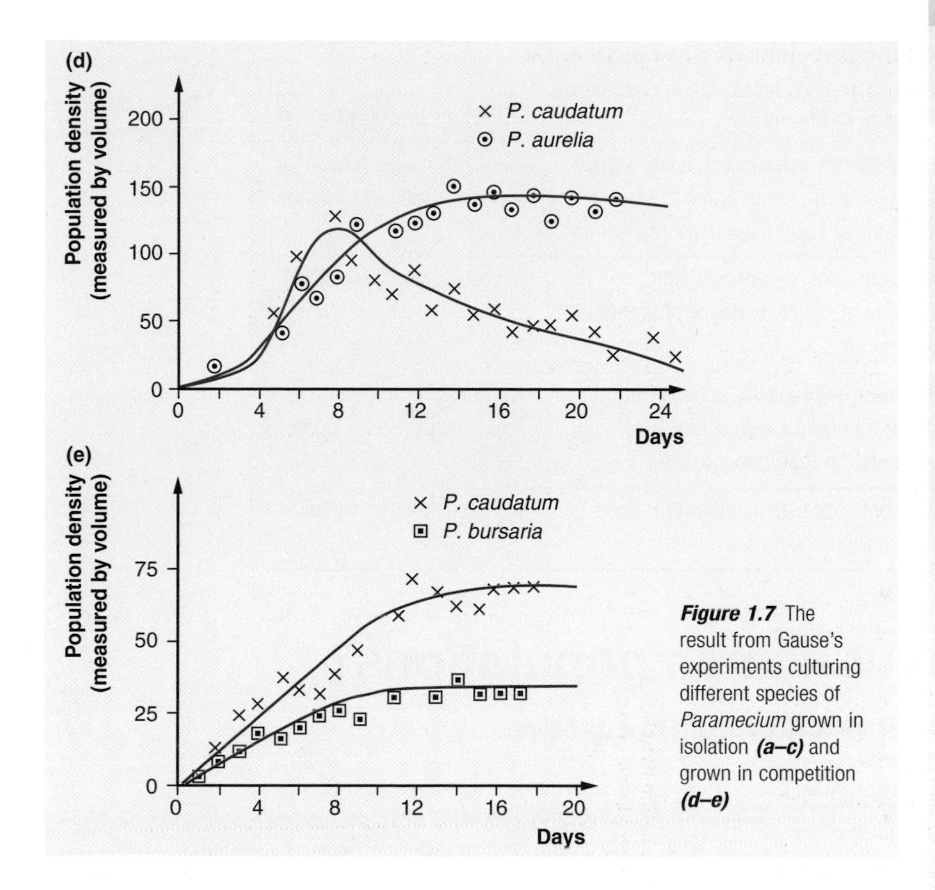

Figure 1.7 The result from Gause's experiments culturing different species of *Paramecium* grown in isolation ***(a–c)*** and grown in competition ***(d–e)***

How do the numbers in a population change over time?

As a result of the combined influence of some or all of the factors described above, most populations develop through four stages, illustrated in Figure 1.8.

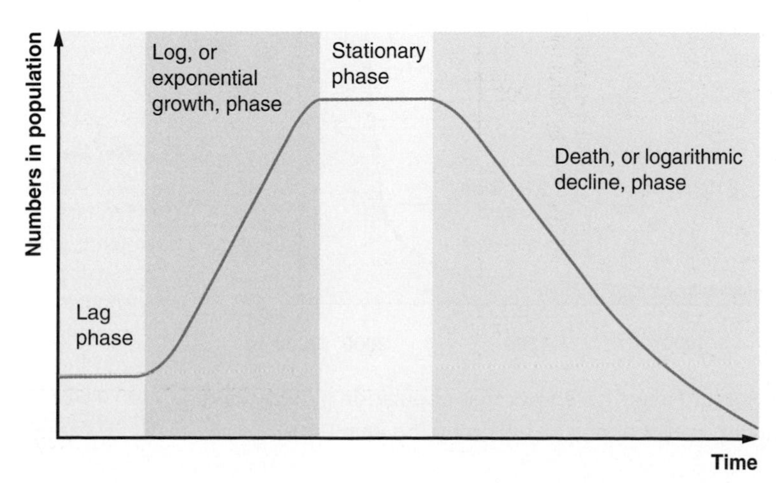

Figure 1.8 The growth phases of a population

The four phases of the curve are described in Table 1.2.

Phase	What is happening	Effect on population size
Lag	Population is establishing itself; some organisms are not adapted to the environment and die, others reproduce	Numbers either remain low and static or increase slowly
Log	All organisms are adapted and reproduce rapidly because of plentiful resources	Numbers increase rapidly
Stationary	The carrying capacity is reached; the same numbers are dying as are produced in reproduction	Numbers remain more or less constant; they fluctuate about a 'mean' level
Decline	Nutrients are exhausted and/or toxic excretory products accumulate	Numbers decline rapidly

Table 1.2 Phases in the development of a population

What about human populations?

The growth of the human population

The human population has been increasing ever more rapidly over the past 200 years. This is because, in general, the quality and quantity of food available has increased and the impact of disease-causing organisms has decreased because of improved sanitation and medical care.

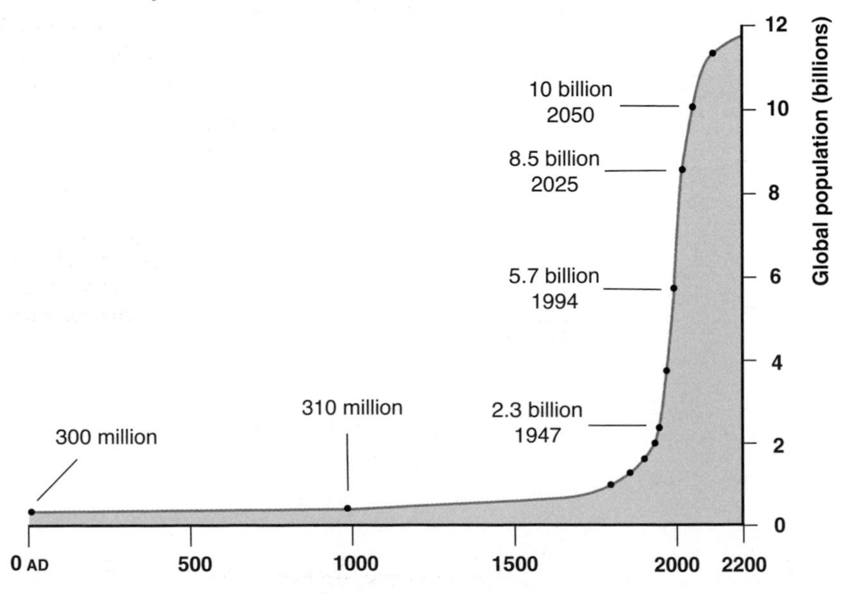

Figure 1.9 The growth of the human population over the past 2000 years together with the projected growth for the next 200 years

However, careful examination of Figure 1.9 shows that population growth should slow and the numbers should level at about 12 billion by the year 2200.

Will the levelling of the human population come too late? If the predictions about global warming are correct, there could be considerably less land suitable for agriculture and therefore less food available to support the increasing population.

The human population is subject to the same checks as other populations — biotic and abiotic factors influence population size. However, the human population is also subject to other factors that affect its development. These include the point at which the particular population develops agriculture and then becomes industrialised. These factors affect growth rate, death rate and life expectancy.

These changes are called the **demographic transition**. The four stages of the demographic transition are shown in Figure 1.10.

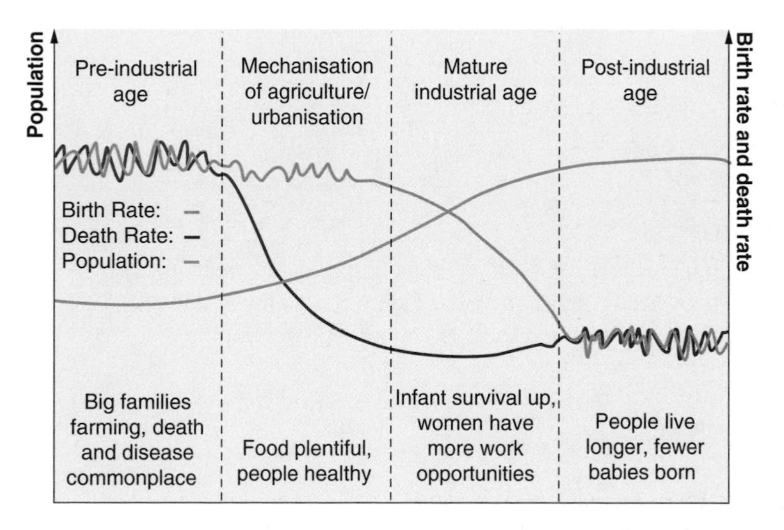

Figure 1.10 The stages of the demographic transition

Population growth is the actual increase of a population. So, for example, there may be 100 000 more people in a country at the end of a year. Growth rate is the increase (or decrease) in numbers expressed as a percentage of the original population. If a population of 10 million grew by 100 000 in 1 year, the growth rate would be $(100\,000/10\,000\,000) \times 100 = 1\%$.

Notice that in the second and third stages of the demographic transition, the death rate falls before the birth rate. This creates a period when the population is increasing. In the final stage, the birth rate and death rate are low and the population is stable.

Most developed countries are in this final stage of the demographic transition; developing countries are in one of the two middle stages. As a result, most of the population growth occurs in developing countries. This is shown in Figures 1.11 and 1.12.

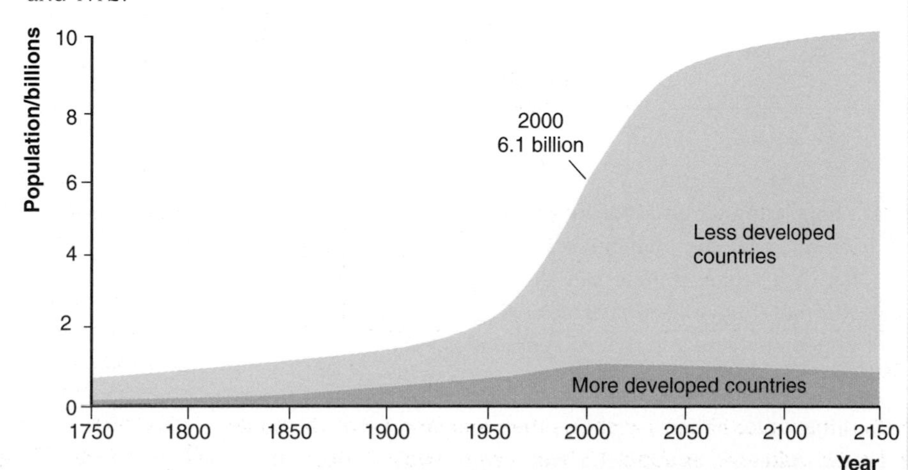

Figure 1.11 The change in population of developed and developing countries from 1750 projected to 2150

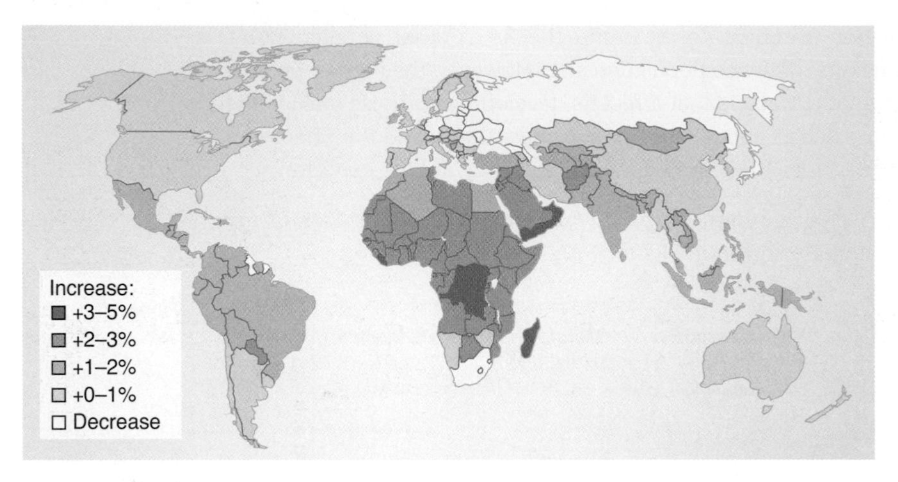

Fig. 1.12 Growth rates of different countries in 2007

Population structure

In the demographic transition, the relative number of young and old people changes. This is best shown by using age pyramids. Figure 1.13 shows age pyramids for France (a developed country) and India (a country still developing).

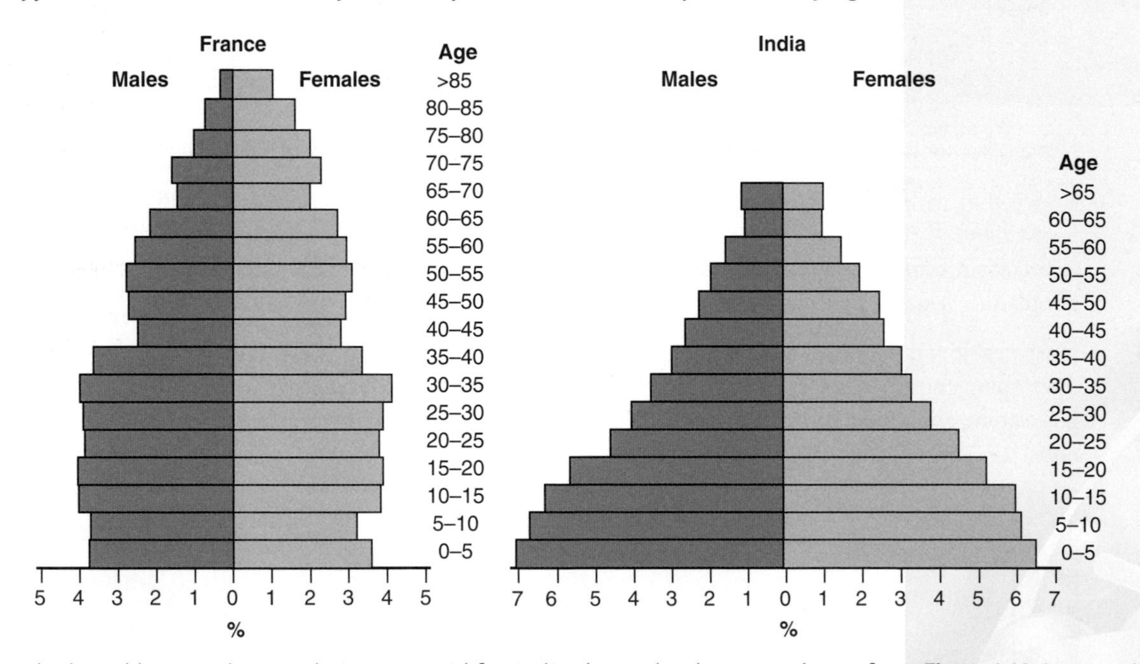

Figure 1.13 Age pyramids for France and India

The broad base to the population pyramid for India shows that large numbers of children are being born and that the population is increasing. The structure in France is different. There are similar numbers in all the age groups until old age is reached. This suggests that the number of babies being born is approximately equal to the number of people dying and that the population is stable.

Notice also that life expectancy is greater in France. This is true of most developed countries, including the UK. There are now more people aged 60 + in the UK than there are people under 18, as the BBC news extract from August 2008 in Box 1.2 shows.

Box 1.2 Children outnumbered by over-60s

People aged over 60 in the UK outnumber children for the first time, according to official figures.

The statistics point to the ageing nature of the UK's society

TopFoto

The Office for National Statistics revealed that 13 262 256 people were 60 or over in mid-2007 — up from 12 928 071 the previous year.

Meanwhile, the number of under-18s fell from 13 119 654 to 13 111 023 over the same time period.

Help the Aged said an older population would require social care reform and the end of 'arbitrary' retirement ages.

Those currently defined as pensioners — men aged above 65 and women aged above 60 — make up 19% of the population, compared with 18.9% for children below the age of 16.

Chichester has the highest concentration of pensioners in England and Wales, with more than 32%.

How can we investigate populations?

Biologists sometimes need to estimate:

- population numbers
- the way in which a population is distributed across an area

Although it is possible to count the numbers of large organisms, such as oak trees, in an area and to record where they are found, biologists often have to use sampling techniques. It would be impossible to count, for example, all the butter-cups in a field or all the lugworms on a sea shore to find the population size. Equally, it would be impossible to record exactly where each one is found in order to describe the distribution of the species in those habitats.

How do biologists investigate the numbers and distribution of small static organisms?

To make an estimate of the size of a population of a small static species, such as clover or buttercups, biologists:

- take a sample of the population
- record the numbers in the sample
- make an estimate of the total population, assuming that the sample is representative of the total population

> *e* If there is bias in a sample, then the results of an investigation will misrepresent the nature of the population. For example, if a biologist selects an area of buttercups to be the sample because 'it looks typical', then bias is introduced because the biologist's perception of what is typical may be misinformed.

For a sample to be representative of the population as a whole it must not be biased in any way; it must be a **random sample**.

To estimate numbers, biologists use **quadrats**. There are several styles of these; the simplest is a square metal frame. The size of quadrat chosen varies according to the area under investigation. However, a quadrat with an area of $0.25\,m^2$ is commonly used.

The procedure is:

- Divide the area into a grid.
- Use the random number generator on a calculator to produce a pair of coordinates — for example A5, B6 (this pair would define the top left-hand corner of the shaded square in Figure 1.14) and A2, B4.
- Place a quadrat with its top left-hand corner on the intersection of the coordinates.
- In each sample quadrat, count the number of organisms of the species under investigation This is the **frequency** of occurrence of the organism.
- Repeat this procedure according to the number of quadrats you intend to use.
- Calculate the mean number per quadrat.

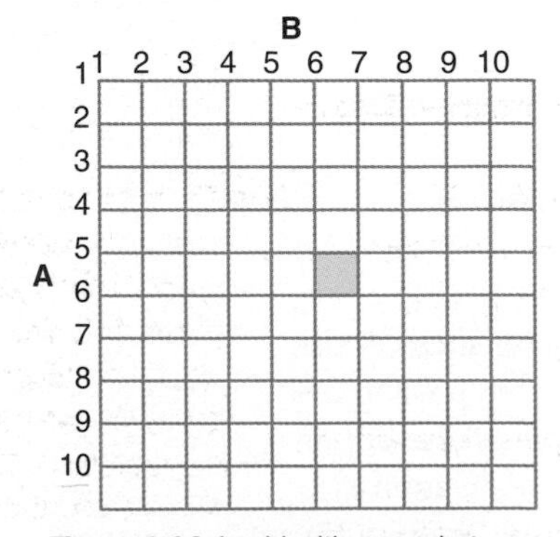

Figure 1.14 A grid with a quadrat positioned at A5, B6

This student is recording the numbers of daisies in a quadrat

- Find the area of the quadrat.
- Estimate the area of the field or sample area, making sure that both areas are in the same units (e.g. both in m^2 or both in cm^2).
- Estimate the population size using the formula:

$$\frac{\text{mean number of organisms per quadrat} \times \text{area of field}}{\text{area of quadrat}}$$

Sometimes, it is difficult to count the number of organisms — even in this small area. Can you imagine trying to count the number of grass plants in a quadrat? How many blades of grass are there on each plant? It is nearly impossible.

To get round this problem, biologists estimate the **percentage cover** of an organism. This can be done crudely by making an estimate of how much of the quadrat is covered by that organism. If the organism takes up half of the quadrat, then the percentage cover is 50%.

However, a better way of estimating percentage cover is to use a quadrat that is subdivided into smaller squares (Figure 1.15).

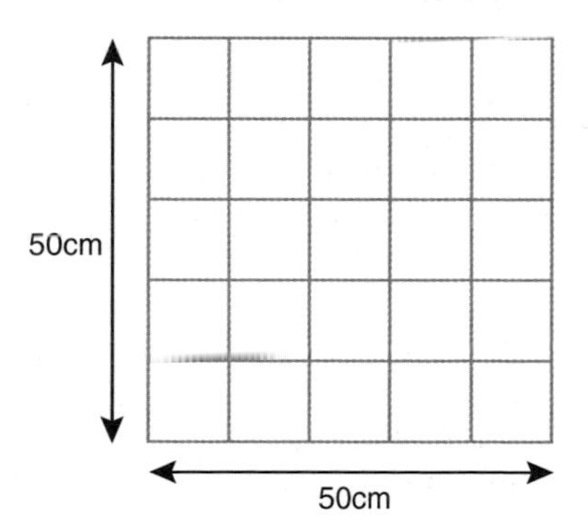

There are 25 small squares. Therefore, each small square represents 4% of the total area.

If the organism covers seven squares, the percentage cover is $7 \times 4\% = 28\%$.

If the organism covers eight squares and two part-squares, then the percentage cover is $9 \times 4 = 36\%$.

Figure 1.15 A gridded quadrat

Gridded quadrats can also be used to estimate frequency. If an organism occurs in 12 of the 25 squares (the actual number in each small square does not matter), the frequency of occurrence is:

$$\frac{12 \times 100}{25} = 48\%$$

How do biologists estimate the numbers of mobile species?

To estimate the size of a population of animals that move, the **mark–release–recapture** technique is used:

- Collect a sample of the animals from the area and count them (N_1).
- Put a small mark, using a harmless substance, in an unobtrusive place on each animal. This is to ensure that the animal is not harmed by the process and that the marking does not affect its chance of survival by making it more noticeable to predators.
- Release the marked animals and allow time for them to disperse among the population.
- Collect a second sample and note the total size of the sample (N_2) and the number that are marked (n).

There are a number of ways of collecting small animals. One of these is a pitfall trap. These are small containers that are buried in the ground and camouflaged. The animals fall into the traps and cannot escape.

If the number of marked individuals in the second sample (n) makes up the same proportion of the second sample (N_2) as the total number of marked individuals (N_1) does of the whole population (x), we can write this mathematically, as:

$$\frac{n}{N_2} = \frac{N_1}{x}$$

To calculate the population size, the formula is rearranged:

$$x = \frac{N_1 \times N_2}{n}$$

For example, if 50 (N_1) woodlice were caught originally, marked and released, and subsequently 40 (N_2) were caught of which 10 were marked (n), the population would be estimated at:

$$x = \frac{50 \times 40}{10} = 200$$

Dan Guravich/SPL

A marked butterfly ready for release

When using this technique it is assumed that:

- there are no migrations
- there is no reproduction
- there are no deaths
- marking does not affect behaviour
- on release, there is random mixing of the marked and unmarked individuals
- the second sample is representative of the population as a whole

How do biologists measure the distribution of organisms across an area?

This technique involves using a **transect**. Once the area under investigation has been defined, the method is as follows:

- Lay a tape measure across the sample area.
- At regular intervals (e.g. every 4 m) lay five quadrats to one side (always the same side) of the tape (Figure 1.16).

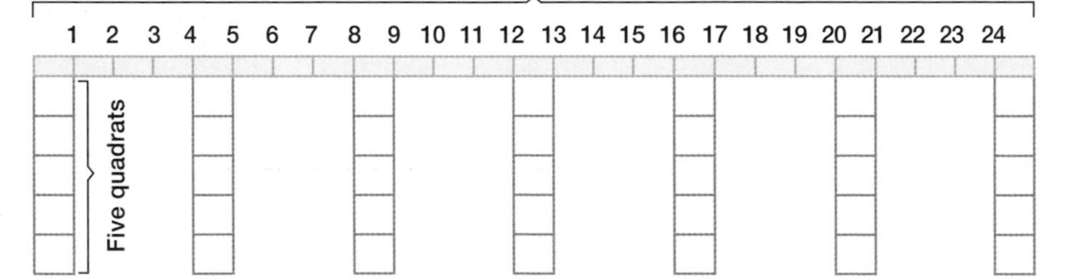

Figure 1.16 A belt transect

The abundance of the different organisms at each sampling point along the transect can be estimated by:

- simply recording presence or absence in each of the five quadrats and converting the number of occurrences to a percentage frequency (e.g. a species that occurs in four of the five quadrats has a frequency of 80%)

Students investigating the distribution of organisms along a transect

Martyn F. Chillmaid/SPL

- estimating the percentage of each quadrat covered by the species and taking the mean for the five quadrats to give the average percentage cover
- counting the numbers of each organism in each quadrat to obtain a mean number for each sampling point

> Notice that in investigating a belt transect, the quadrats are not positioned randomly, they are positioned at regular intervals. This is called systematic sampling. We use systematic sampling whenever we want to show the changes or trends that occur across an area. Random sampling could miss out large sections, whereas systematic sampling ensures that all sections of the area are represented in the sample (Figure 1.17).

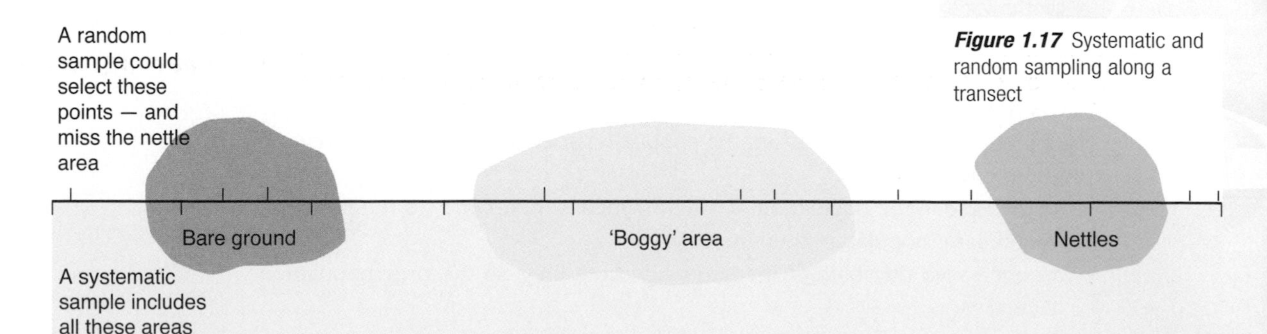

Figure 1.17 Systematic and random sampling along a transect

The results from a transect can be shown in a **kite diagram**. In a kite diagram, the abundance of a species at each sampling point is plotted and then joined to a similar plot at the next sampling point. Figure 1.18 shows a simple kite diagram for three species along a transect.

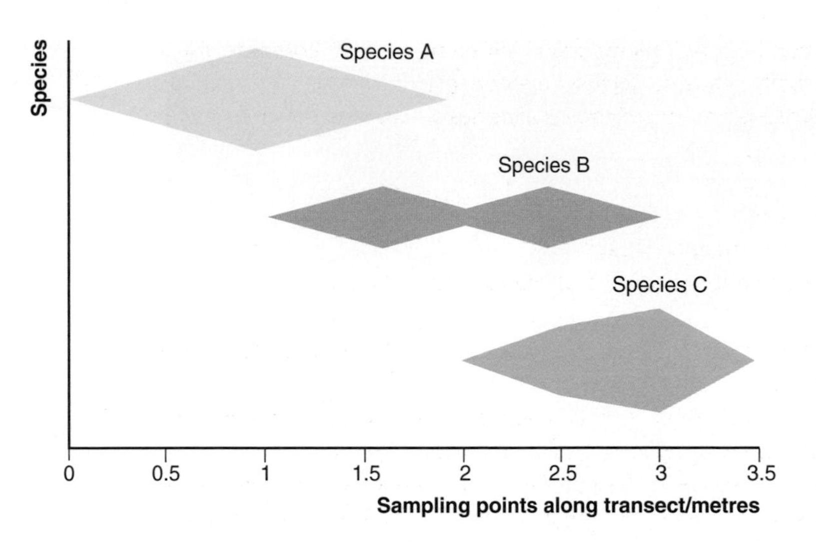

Figure 1.18 A kite diagram showing the relative abundance of three species along a 3.5 metre transect

Summary

Ecological factors

- A population is all the individuals of a particular species in a particular habitat at a particular time.
- A habitat is an area in which a population lives and finds nutrients, water, living space and other essentials it needs to survive.
- An ecological niche describes the role of an organism within a habitat.
- A community is all the organisms present in a particular environment at a particular time.
- An environment is the sum of the physical factors in an area that interact with the community in that area.
- An ecosystem is a community of organisms that interact with each other and with the physical environment in which they are found.
- The size of a population is affected by:
  - biotic factors — predation, disease-causing organisms, intraspecific or interspecific competition for a resource (often food)
  - abiotic factors — for example, temperature, light intensity, concentration of oxygen or carbon dioxide, availability of water
- In a predator–prey relationship, the population sizes of predator and prey are interdependent:
  - an increase in the population of the prey means more food for the predator
  - the predator population then increases
  - the increased numbers of predators kill more prey, so the prey population decreases
  - there is now less food for the predator, so the predator population decreases
  - the reduced numbers of predators kill fewer prey, so the prey numbers increase
- In intraspecific competition, members of the same species compete for a resource; overuse of the resource can lead to a population crash

- In interspecific competition, members of different species compete for a resource; this usually leads to local extinction of the weakest competitor, although coexistence is possible if the two species do not use the resource in the same way.

Population growth

- The four stages of a population growth curve are:
  - lag phase — population is low and stable while it adapts to the new surroundings
  - log phase — the adapted population increases rapidly because there are abundant resources
  - stationary phase — numbers are high and stable as deaths are matched by births; this is the carrying capacity for the population in that area
  - decline phase — overuse of resources or accumulation of toxic waste products causes a decline in the population; if the resources recover, the population may also recover, otherwise it may decline to zero
- The stages of development of the human population of a country are represented in the demographic transition as the country changes from a pre-agricultural society to an industrial society.
- Age pyramids show the percentage of males and females of each age group in a population.
- Expanding, stationary and contracting populations have different age pyramids (Figure 1.19).

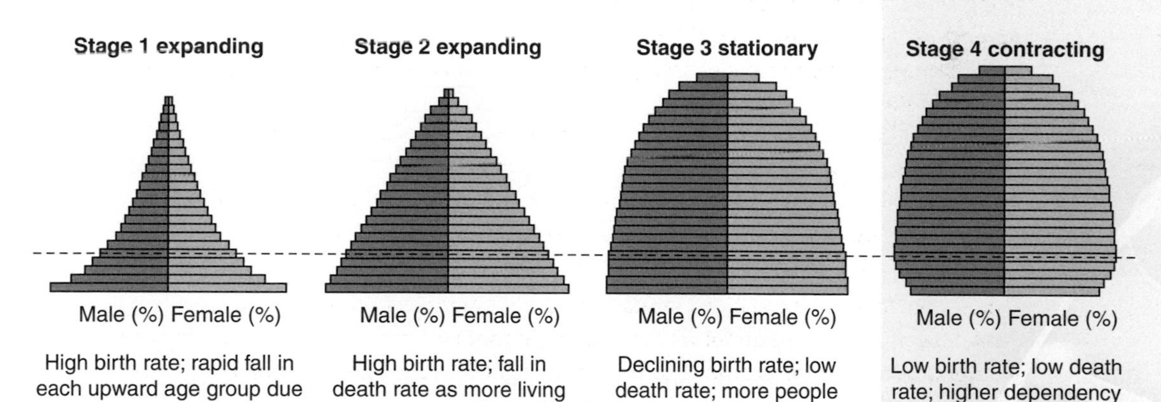

Figure 1.19 Age pyramids for expanding, stationary and contracting populations

Fieldwork

- Numbers and distributions of static organisms in an area are often investigated using quadrats.
- Random sampling is used to estimate numbers.
- Systematic sampling is used to show distribution across an area.
- The mark–recapture–release technique is used to investigate the numbers of a mobile species in an area.

Questions

Multiple-choice

1 A population is:

A a group of one species in the same place at the same time
B a group of different species in the same place at the same time
C all the organisms of one species in the same place at the same time
D all the organisms of different species in the same place at the same time

2 A niche is different from a habitat because:

A a habitat describes the physical surroundings in which an organism lives
B a niche describes the role of an organism in its habitat
C both A and B
D neither A nor B

3 An ecosystem is:

A the place where an organism lives
B a population interacting with its habitat
C a community interacting with its environment
D all of the above

4 The sequence of phases in a population growth curve is:

A log — lag — stationary — decline
B lag — log — stationary — decline
C lag — log — decline — stationary
D log — lag — decline — stationary

5

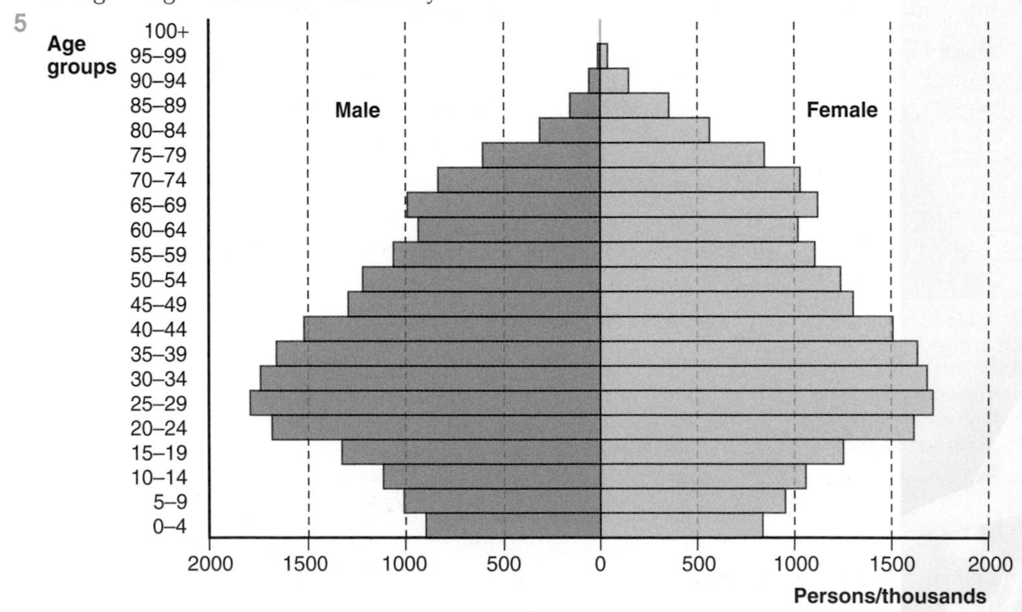

This age pyramid represents:

A an expanding population
B a static population
C a population in decline
D an unbalanced population

6 In a predator–prey relationship:
 A the numbers of the predator increase before those of the prey and are generally higher
 B the numbers of the predator increase after those of the prey and are generally higher
 C the numbers of the predator increase after those of the prey and are generally lower
 D the numbers of the predator increase before those of the prey and are generally lower

7 To investigate the distribution of an organism across an area, students should use:
 A a belt transect
 B random quadrats
 C the mark–release–recapture technique
 D none of the above

8 The main benefit of random sampling is that:
 A it provides a representative sample
 B it provides an unbiased sample
 C both A and B
 D neither A nor B

9 Interspecific competition is:
 A competition between members of different species in the same ecosystem
 B competition between members of the same species in the same ecosystem
 C competition between members of the same species in the same habitat
 D competition between members of different species in the same habitat

10 The best description of the abiotic factors of a pond ecosystem is:
 A the pond water and dissolved oxygen
 B the pond water and all the dissolved substances
 C the pond water, the dissolved substances and the mud at the bottom and sides of the pond
 D the pond water, the dissolved substances, the mud at the bottom and sides of the pond and the air above the pond

Examination-style

1 The diagram shows a typical population growth curve:
 (a) Explain why the population does not increase:
  (i) during phase A.
  (ii) during phase C *(3 marks)*
 (b) Give two reasons for the decline during phase D. *(2 marks)*
 (c) Which phase corresponds most closely to the current rate of change of the human population? Explain why this is the case. *(3 marks)*

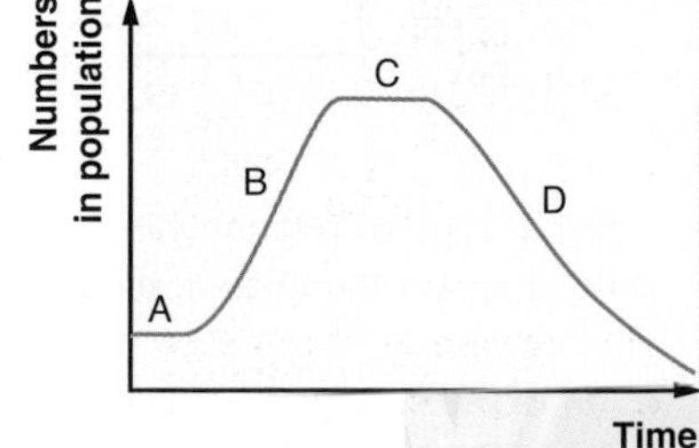

Total: 8 marks

2 The diagram shows the age pyramid of the population of Ohio, USA in 2001.

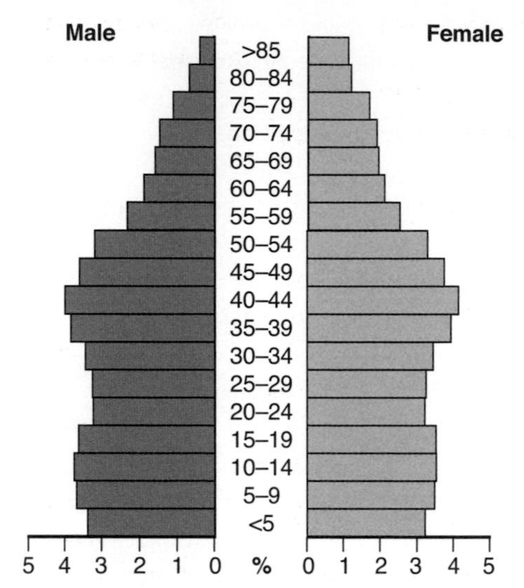

(a) Describe the evidence in the pyramid that suggests that women have a longer life expectancy than men. *(2 marks)*

(b) Is this an expanding, static or shrinking population? Use evidence from the diagram to support your answer. *(3 marks)*

(c) State three ways in which the population structure of a developing area would be different from the population in the developed area of Ohio. *(3 marks)*

Total: 8 marks

3 Foxes kill and eat rabbits. The graph shows the numbers of foxes and rabbits in an area over 50 years.

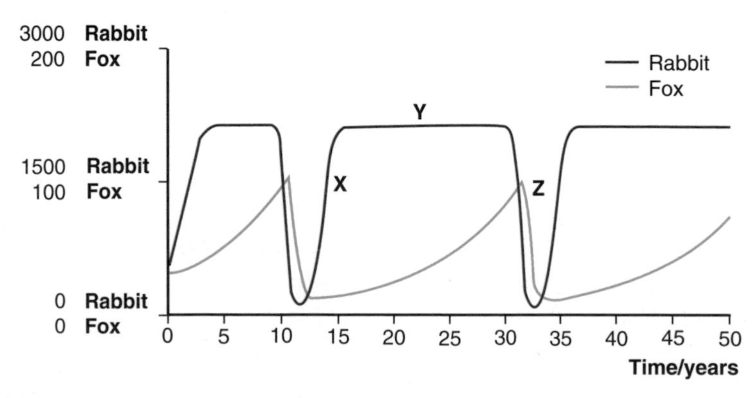

(a) What type of relationship is shown in the graph? *(1 mark)*

(b) (i) Explain the change in the number of rabbits at **X** on the graph. *(2 marks)*

(ii) Suggest why the number of rabbits remains constant at **Y** on the graph. Explain your answer. *(3 marks)*

(iii) Explain the change in the number of foxes at **Z** on the graph. *(2 marks)*

(c) (i) Suggest which technique the researchers used to estimate the number of rabbits and foxes. *(1 mark)*

(ii) Describe how the technique is carried out *(4 marks)*

Total: 13 marks

4 In an investigation into the number of woodlice in an area, students caught a sample of 35 woodlice, marked them with a dye and then released them. The following day, they caught a second sample of 42 woodlice, of which seven were marked.

(a) Explain:

(i) two features of the procedure of marking the woodlice *(2 marks)*

(ii) why the students waited a day before catching the second sample of woodlice *(2 marks)*

(b) Using the woodlice numbers given in the question, calculate the number of woodlice in the area under investigation. *(2 marks)*

(c) State three assumptions made in using this technique to estimate population size. *(3 marks)*

Total: 9 marks

5 The diagram shows the distribution of two species of barnacle on a rocky shore in Scotland. Adult barnacles are immobile (static) animals.

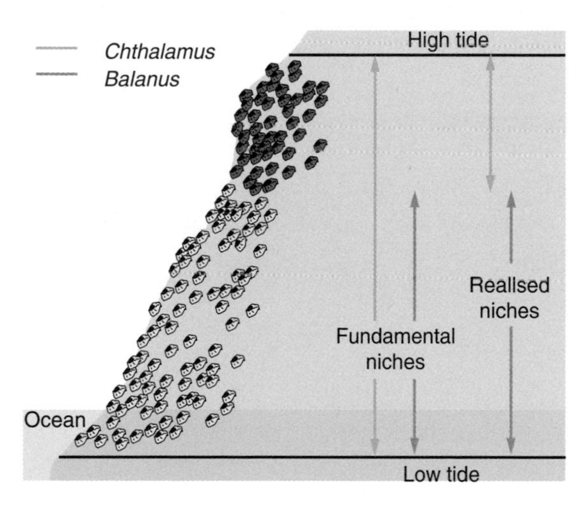

(a) Describe how the researchers could have obtained the data about the distribution of the barnacles. *(4 marks)*

(b) Use information from the diagram, together with your own knowledge, to explain what is meant by the terms 'fundamental niche' and 'realised niche'. *(4 marks)*

(c) (i) Suggest why the fundamental and realised niches of *Balanus* are the same whereas those of *Chthalamus* are different. Explain your answer. *(4 marks)*

(ii) Suggest how the distribution of *Chthalamus* would differ if *Balanus* were absent. *(3 marks)*

Total: 15 marks

Chapter 2

How do cells obtain useful chemical energy to drive their metabolic processes?

This chapter covers:

- the suitability of the ATP molecule as an immediate energy source in cells
- the role of photosynthesis in transducing light energy into useful chemical energy, including:
  - the light-dependent reactions
  - the light-independent reactions
  - the relationship between the structure of a chloroplast and its function
- the factors that affect the rate of photosynthesis, including the principle of limiting factors
- the application of the principle of limiting factors to crop production in greenhouses
- the role of aerobic respiration in releasing energy from organic molecules to synthesise molecules of ATP, including:
  - glycolysis
  - the link reaction
  - the Krebs cycle
  - the electron transport chain and oxidative phosphorylation
  - the relationship between structure and function of the mitochondrion
  - the role of anaerobic respiration in releasing energy from organic molecules to synthesise molecules of ATP

◀ Energy transduction is the transfer of energy from one system to another.

Photosynthesis and respiration are arguably *the* most important biological processes. Without them, life as we know it today would not exist. In photosynthesis, light energy is transduced by a light-dependent system into chemical energy. In a light-independent system, carbohydrates are synthesised using this chemical energy. Both these systems are located in chloroplasts (Figure 2.1).

As well as being the basis for the synthesis of other organic molecules, carbohydrates are also used in respiration to transfer the energy held in them into molecules of ATP. Some of the reactions of respiration take place in the cytoplasm; most take place in the mitochondria. Energy held in molecules of ATP is used to drive all the energy-requiring processes that take place in cells.

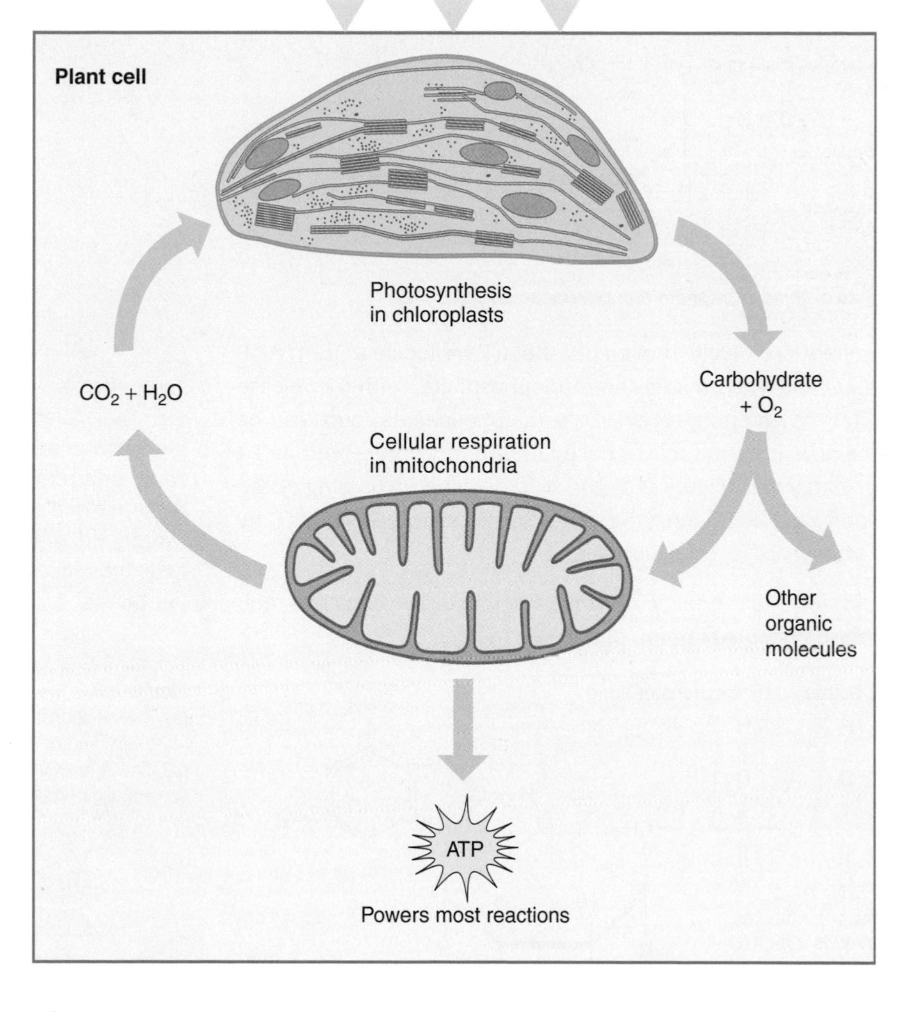

Figure 2.1 An overview of energy transfer in a plant cell

Photosynthesis in plants is discussed this chapter. However, remember that algae and some bacteria can also photosynthesise.

Of course, animals and many other organisms cannot photosynthesise and so rely on the organic molecules produced by plants. The plants are eaten and the organic molecules digested, absorbed and circulated to the cells where they are respired to release energy.

What is ATP?

ATP is short for **a**denosine **trip**hosphate. The 'adenosine' part of the molecule is itself made from two parts:

- the base adenine, which is found in DNA and RNA
- ribose, the sugar found in RNA

In ATP, three phosphate groups are attached to the adenosine.

Figure 2.2 A molecule of ATP

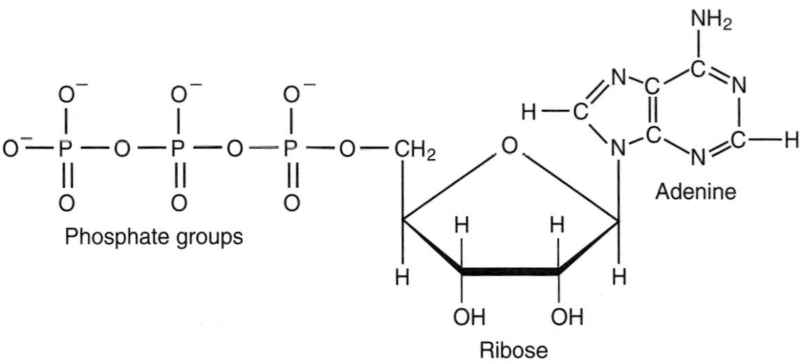

Phosphate groups

Ribose

Adenine

ATP consists of three phosphate groups, ribose and adenine

Phosphate groups can be successively 'broken off' the ATP molecule to form ADP (**a**denosine **di**phosphate) and AMP (**a**denosine **m**ono**p**hosphate) with the release of energy (Figure 2.3 (a)) When energy is available, the phosphate groups can be reattached. In practice, it is only the third phosphate group (represented as P_i) that is detached and reattached (Figure 2.3 (b)). It is detached in a single-step hydrolysis reaction controlled by the enzyme ATPase. The resynthesis of ATP by reattaching the third phosphate is more complex.

There are about 50 g or 10^{23} molecules of ATP in the average adult human body. On average, each of these molecules is hydrolysed and then resynthesised three times per minute.

All organisms depend for their energy on the continued breaking off and reattachment of this third phosphate group of the ATP molecule.

Figure 2.3
(a) The relationship between AMP, ADP and ATP
(b) The hydrolysis and resynthesis of ATP

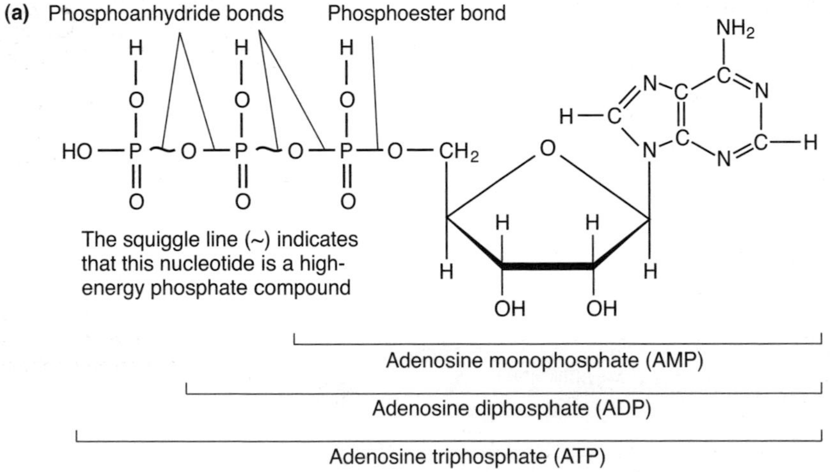

(a) Phosphoanhydride bonds Phosphoester bond

The squiggle line (~) indicates that this nucleotide is a high-energy phosphate compound

Adenosine monophosphate (AMP)

Adenosine diphosphate (ADP)

Adenosine triphosphate (ATP)

(b)

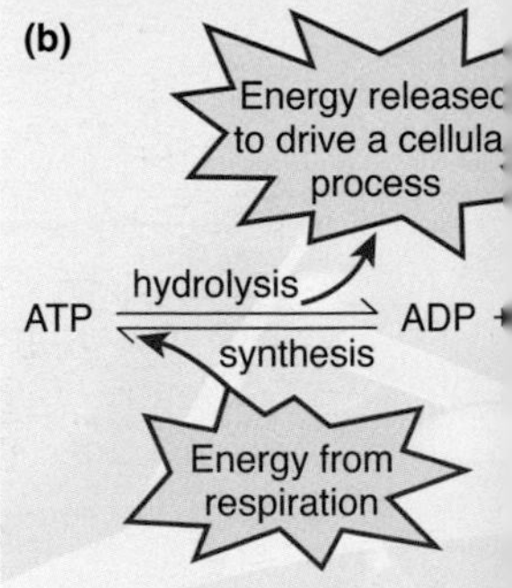

ATP is the molecule that releases energy to drive biological processes. It is said to be **coupled** to these processes. It is an ideal molecule for this function because:

- the energy is released from the molecule quickly, in a single-step hydrolysis reaction
- the energy is released in small amounts that are closely matched to the amounts needed for cellular reactions
- the molecule is easily moved around within the cell but cannot leave the cell

The energy released from ATP is used in:

- the synthesis of macromolecules, such as proteins
- active transport
- muscle contraction
- the initial reactions of respiration

How is light energy transduced in photosynthesis?

This takes place in a series of reactions called the **light-dependent reactions**. Light energy is absorbed by special photosensitive pigments such as **chlorophyll** in the **chloroplasts**.

How is the structure of a chloroplast suited to its function?

The chlorophyll molecules are arranged in '**photosystems**' and are linked to chains of molecules that transport electrons from one molecule to another. These are called **electron transport chains** (ETCs). The molecules of the photosystems and the electron transport chains are fixed in the membranes of the thylakoids (Figure 2.4). This makes the processes much more efficient than if the molecules were just floating around in a liquid.

Figure 2.4
(a) A three-dimensional representation of the structure of a chloroplast

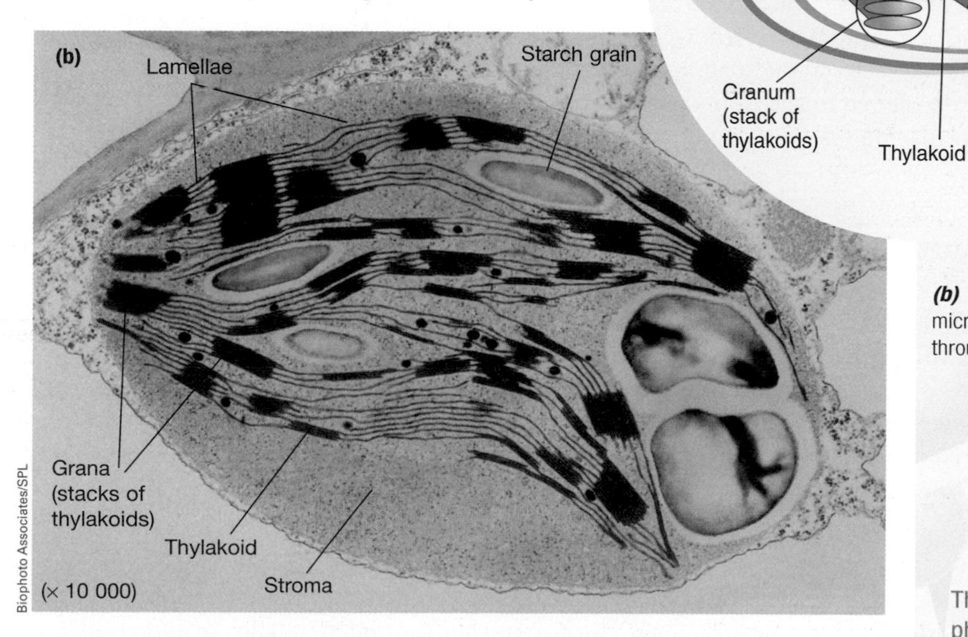

(b) A transmission electron micrograph of a section through a chloroplast

There are two types of photosystem: **photosystem I** and **photosystem II**. They are sensitive to light of different wavelengths and are linked to different electron transport chains.

The numbering of the photosystems relates only to the order in which their structures were discovered.

The light-dependent reactions take place in the membranes of the **thylakoids**.

The liquid **stroma** is the site of the light-independent reactions, in which carbohydrates are synthesised. Chemical reactions such as these take place more effectively in solution, than if they were fixed in membranes.

How do the light-dependent reactions take place?

The light-dependent reactions use light energy to synthesise two molecules that 'drive' the light-independent reactions. The two molecules are:

- ATP, which provides the energy for the reactions
- NADPH, which provides the hydrogen ions for a key reduction reaction

NADPH is usually called reduced NADP (**n**icotinamide **a**denine **d**inucleotide **p**hosphate) because NADP has been reduced by the addition of a hydrogen atom (H).

When light energy enters chloroplasts, several processes happen simultaneously (Figure 2.5). These processes involve transfer of electrons and create conditions in which many water molecules decompose into oxygen, hydrogen ions and electrons:

$$2H_20 \rightarrow 0_2 + 4H^+ + e^-$$

This light-dependent splitting of water is called **photolysis**.

Note that light does not split the water molecule directly; it creates conditions in which the water molecule splits up.

The following processes then occur.

Process **A**, leading to an accumulation of protons inside the thylakoid:

- Electrons (e^-) in chlorophyll molecules in photosystem II are excited — they become more energetic.
- This extra energy enables them to escape from the chlorophyll and pass to the first molecule (Pq) in an electron transport chain. (These electrons are replaced by electrons formed in the photolysis of water molecules.)
- The electrons pass from one carrier to the next, losing energy as they do so.
- One of the molecules in the chain is a proton (hydrogen ion) pump. As electrons are transferred to and from this molecule, the energy they lose powers the pump that moves protons from the stroma into the space inside the thylakoid.
- This leads to an accumulation of protons inside the thylakoid.

Process **B**, leading to the production of reduced NADP:

- Electrons in chlorophyll molecules in photosystem I are excited and escape from the molecule. (These electrons are replaced by the electrons that have passed down the electron transport chain from photosystem II.)
- The electrons then pass along a second electron transport chain that 'presents' them on the outside of the thylakoid membrane. Here they react with protons (hydrogen ions) and NADP in the stroma of the chloroplast to form NADPH (reduced NADP).

Process **C**, leading to the production of ATP:

- The protons inside the thylakoid move down a concentration gradient through a membrane-bound enzyme called **ATP synthase**.
- As they pass through the enzyme, they cause it to 'spin' and this spinning provides the energy to synthesise ATP from ADP and P_i. The ATP is formed in the stroma.

A useful analogy is to think of water turning a water wheel; the energy of the spinning wheel can be used to generate electricity. The process of creating a concentration gradient of protons (hydrogen ions) across the thylakoid membrane and then allowing them to pass through ATP synthase is called **chemiosmosis**.

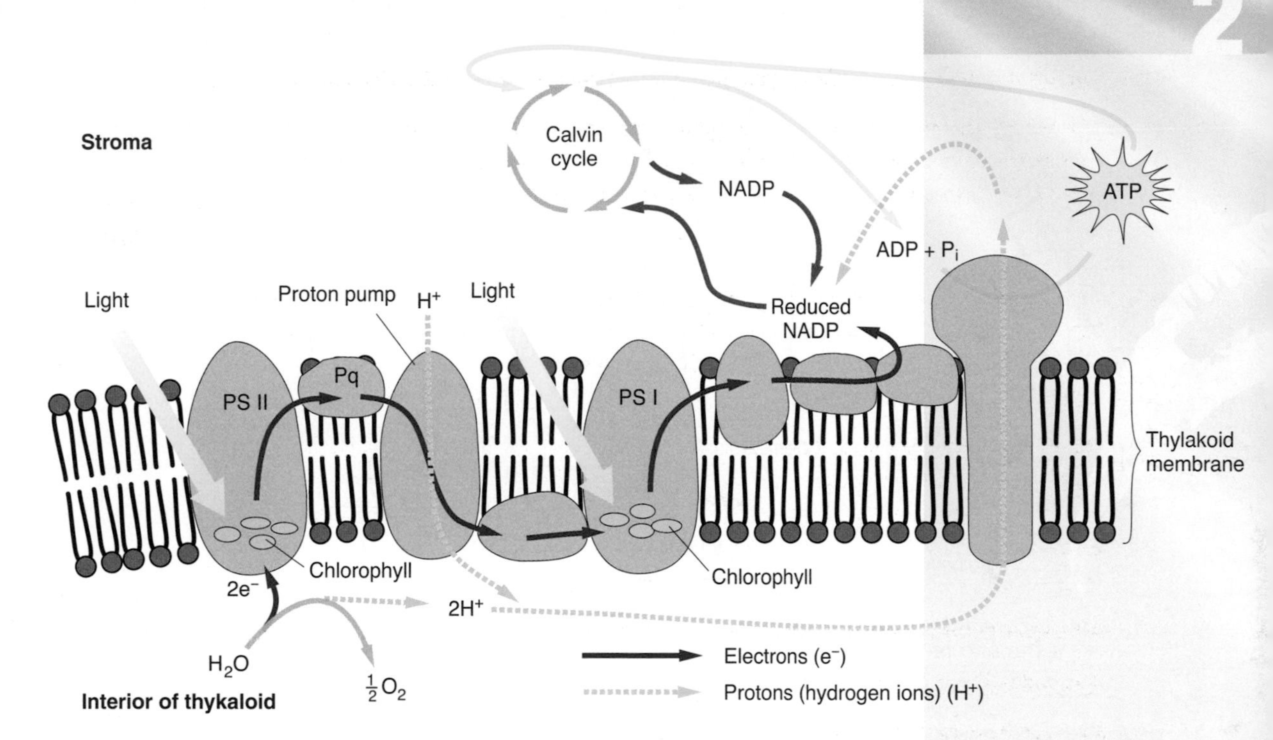

Figure 2.5 The arrangement of molecules involved in the light-dependent reactions in the thylakoid membrane

In summary, light energy is used to excite electrons:

- The electrons cause the transfer of protons to the inside of the thylakoid membrane as they pass along the first electron transport chain; this eventually leads to the formation of ATP.
- The electrons react with hydrogen ions and NADP at the end of the second electron transport chain to form reduced NADP; this reaction can only happen because of the extra energy possessed by the electrons.

The ATP and reduced NADP produced are used to drive the synthesis of carbohydrates in the light-independent reactions of photosynthesis.

The above description outlines how the various molecules are arranged in the thylakoid membranes and how electrons and protons (hydrogen ions) are moved from one molecule to another.

The process can also be represented as a kind of energy–time graph for the electrons. This does not show the physical relationship between the molecules; it shows when and where the electrons gain and lose energy throughout the light-dependent reactions.

This model shows the following:

- Energised electrons escape from chlorophyll in photosystem II to an electron acceptor and, from there, move down an electron transport chain.
- As the electrons pass down the transport chain, they lose energy. This energy is used to drive the synthesis of ATP from ADP and P_i.
- The electrons are accepted by chlorophyll in photosystem I.
- They are energised once more and pass along a different electron transport chain.

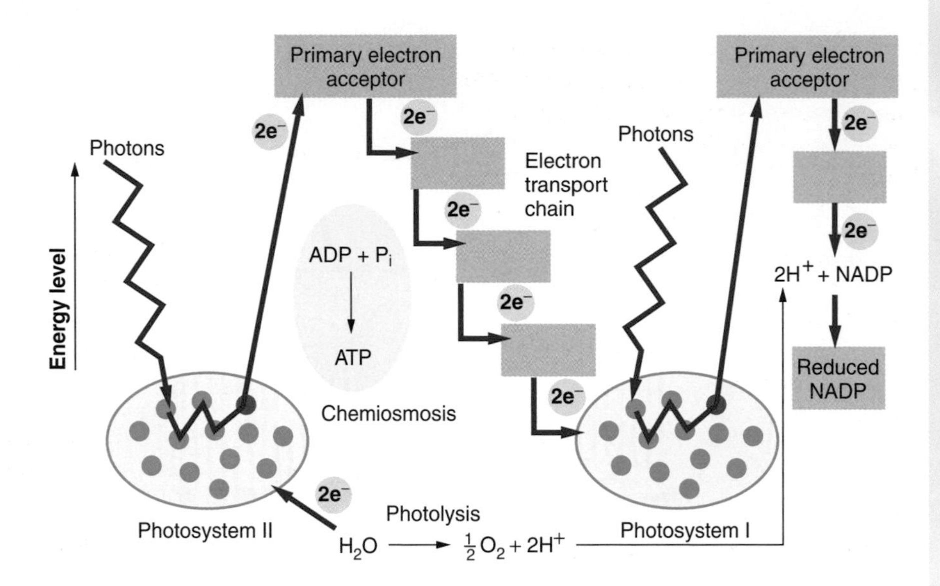

Figure 2.6 The gain and loss of energy by electrons during the light-dependent reactions

- At the end of this chain, the electrons combine with hydrogen ions (from the photolysis of water) and NADP to form reduced NADP (Figure 2.6).

Although this model shows when and where the electrons gain and lose energy, it does not show:

- the mechanisms by which ATP and reduced NADP (NADPH) are formed
- the physical relationship between the molecules in the thylakoid membrane

How is carbohydrate synthesised in the light-independent reactions?

The light-independent reactions of photosynthesis occur in the stroma of chloroplasts. They comprise a complex cycle of reactions that involves the addition of carbon dioxide to a pre-existing five-carbon molecule (a molecule containing five carbon atoms). The resulting molecules are modified to regenerate the original molecule and, at the same time, glucose is synthesised. This sequence of reactions was discovered by Melvin Calvin, an American biologist. The light-independent reactions are sometimes known as the Calvin cycle (Figure 2.7).

Box 2.1 How was the sequence of the light-independent reactions discovered?

In the 1950s, Melvin Calvin and his associates, working at Berkeley University in the USA, experimented with the unicellular alga *Chlorella* by exposing it to radioactive carbon dioxide.

The algae were contained in a 'lollipop'-like structure through which light was shone. Radioactive carbon dioxide was then passed through the apparatus for a known period of time.

After certain periods of time, the algae were killed and the chemicals inside them that contained radioactive carbon (which could only have come from the carbon dioxide) were identified using two-dimensional chromatography.

As time passed, more compounds were formed that contained the radioactive carbon. By refining the experiment and using shorter and shorter time intervals, Calvin identified the first stable compound formed as **glycerate phosphate (GP)**, a compound that contains three carbon atoms.

The main stages of the light-independent reactions are as follows:

- Carbon dioxide reacts with **ribulose biphosphate** (**RuBP**) (a five-carbon compound) in the stroma. The reaction is catalysed by the enzyme **Rubisco**.
- Two molecules of the three-carbon compound glycerate phosphate (**GP**) are formed by this reaction.
- Each molecule of GP is converted to **triose phosphate** (**TP**), another three-carbon compound). This is a reduction reaction that uses hydrogen ions from reduced NADP and energy from ATP.
- Some of the TP formed is used to regenerate RuBP (ATP is required); some is used to form glucose and other useful organic compounds.

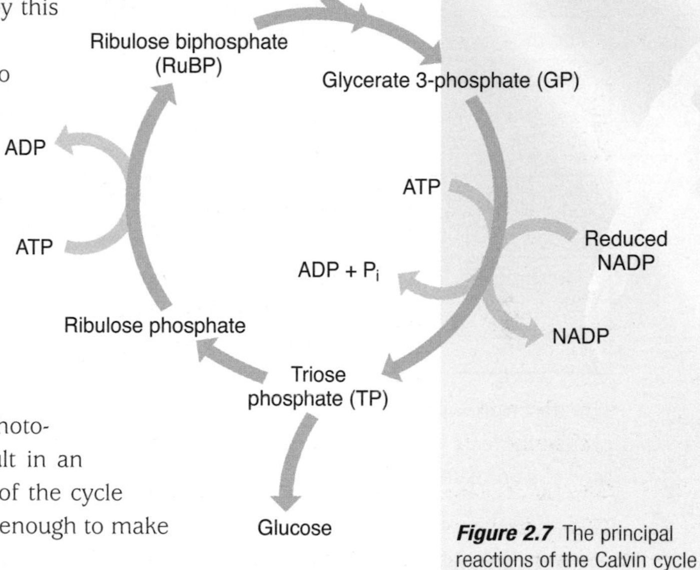

Figure 2.7 The principal reactions of the Calvin cycle

In the light-independent reactions of photosynthesis, three 'turns of the cycle' result in an output of one molecule of TP. Six turns of the cycle give an output of two molecules of TP — enough to make one molecule of glucose.

During the light-independent reactions, reduced NADP is oxidised to NADP and ATP is hydrolysed to ADP and P_i. These are then re-used in the light-dependent reactions to regenerate ATP and reduced NADP. This is illustrated in Figure 2.8.

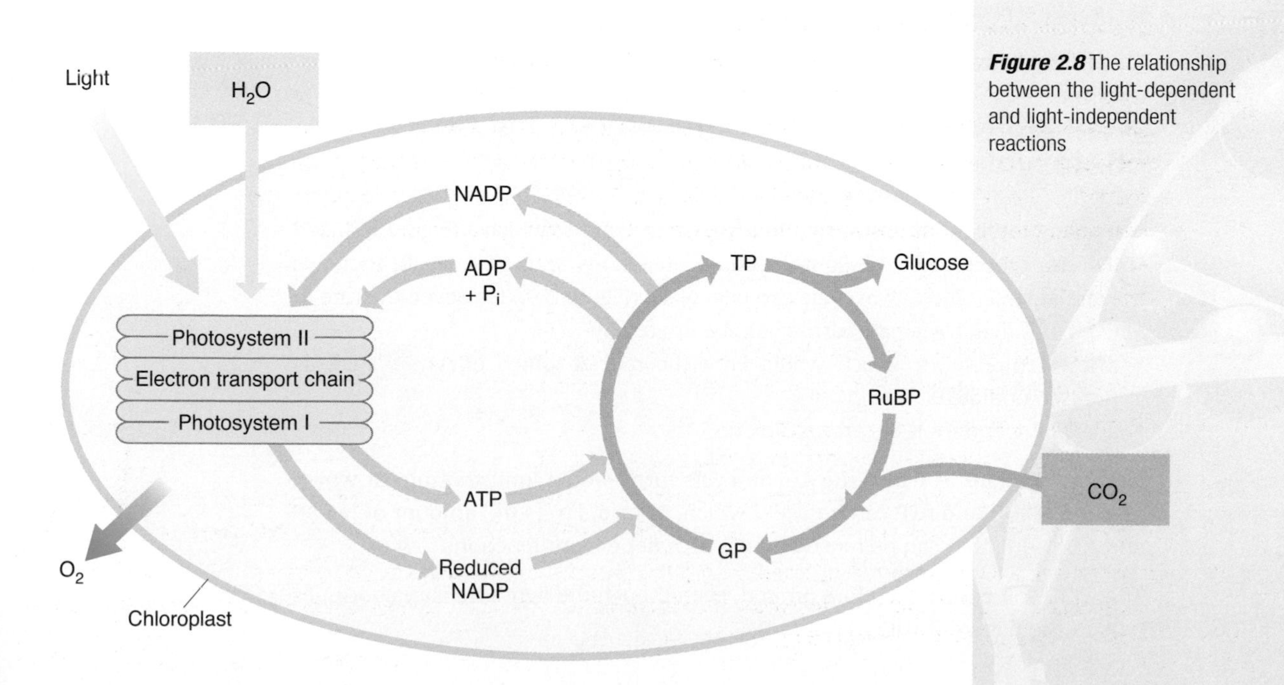

Figure 2.8 The relationship between the light-dependent and light-independent reactions

What factors affect the rate of photosynthesis?

Photosynthesis is dependent on a number of factors (Table 2.1).

Table 2.1 Factors that affect the rate of photosynthesis

Factor	Effect on photosynthesis
Light intensity	Low light intensity can limit the light-dependent reactions by reducing the number of electrons in chlorophyll molecules that become photo-excited
Carbon dioxide concentration	Low concentrations can limit the light-independent reactions by reducing the rate of the initial reaction with RuBP
Temperature	Low temperatures can limit the rate of enzyme activity — for example, of ATP synthase (light-dependent reactions) and Rubisco (light-independent reactions)

◀ The optimum temperature for the activity of the enzymes involved in photosynthesis varies with the geographical location. The enzymes of plants that live within the Arctic Circle have much lower optimum temperatures than those of plants found in the tropics.

The descriptions in Table 2.1, make reference to the factors 'limiting' the rate of photosynthesis when they are in short supply. But which factor actually limits the rate of photosynthesis?

The answer to this question differs on different days. On a cold bright day in winter, temperature is likely to hold back the rate of photosynthesis. On a warm cloudy day in summer, light intensity is likely to limit the rate. On a warm, sunny day in summer, it could be the concentration of carbon dioxide that limits the rate.

In general terms we can say that:

The rate of photosynthesis is limited by the factor that is present in a limiting quantity.

This is known as the principle of limiting factors.

e Think of a number of cyclists who want to cycle as a group. The speed if the group is determined by the speed of the slowest cyclist.

What is the effect of light intensity on the rate of photosynthesis?

Increasing the light intensity should increase the rate at which ATP and reduced NADP are produced in the light-dependent reactions and, as a result, increase the rate at which the Calvin cycle can take place (Figure 2.9). However, the rate at which the Calvin cycle can 'turn' could be limited by:

- a low temperature, which would limit the rate at which enzymes such as Rubisco operate
- a low concentration of carbon dioxide

Limiting the rate at which the Calvin cycle turns would limit the rate at which reduced NADP and ATP can be used, which, in turn, limits the amount of NADP and ADP + P_i that can be reused by the light-dependent reactions.

The whole process is therefore limited, even though the light intensity continues to increase (Figure 2.10).

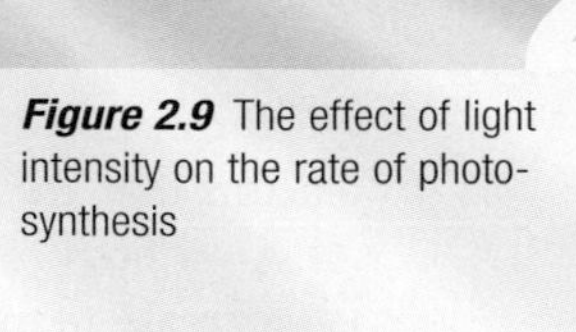

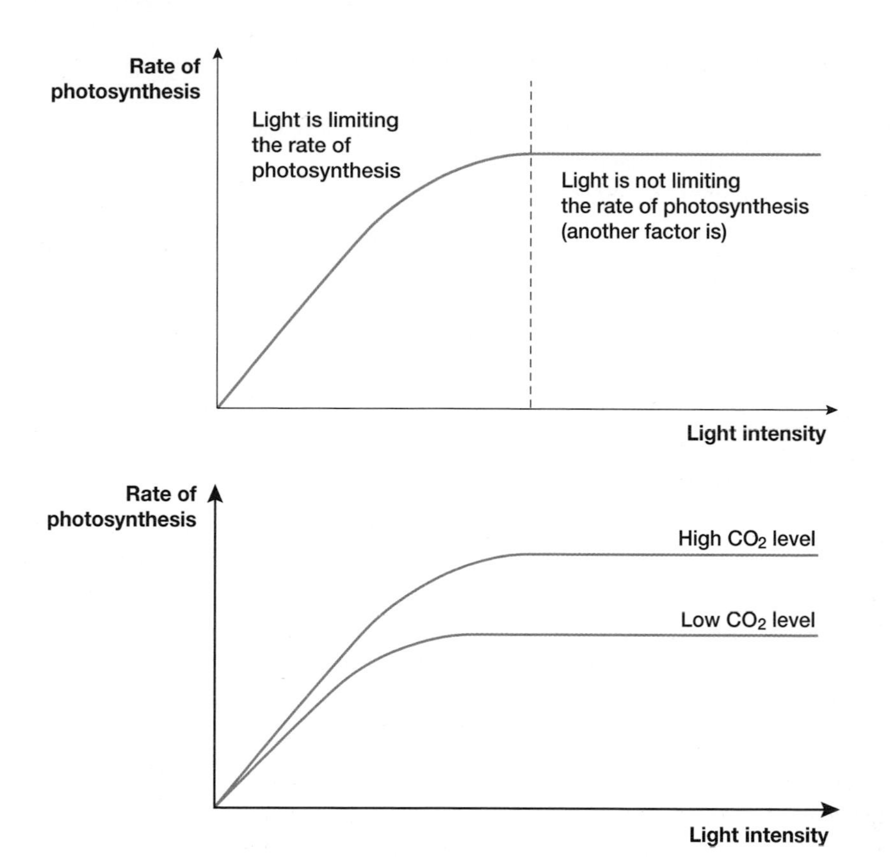

Figure 2.9 The effect of light intensity on the rate of photosynthesis

Figure 2.10 Increasing the concentration of the limiting factor allows the rate of photosynthesis to increase further with increasing light intensity

As well as the major factors discussed here, a number of other factors influence the rate of photosynthesis. These include:

- the wavelength of the light; photosynthesis takes place faster in the presence of red and blue light because these wavelengths are absorbed more efficiently than others; leaves appear green because 'green' wavelengths are reflected
- the amount of chlorophyll present

Increasing the factor that is limiting the Calvin cycle increases the rate at which ATP and reduced NADP can be used and, therefore, the rate at which NADP and ADP + P_i are recycled to the light-dependent reactions. The whole process proceeds faster with increasing light intensity — until something again limits the Calvin cycle.

The rate of photosynthesis can be limited by temperature as well as by the concentration of carbon dioxide. Figure 2.11 shows how different combinations of these two factors can limit the rate of photosynthesis as light intensity increases.

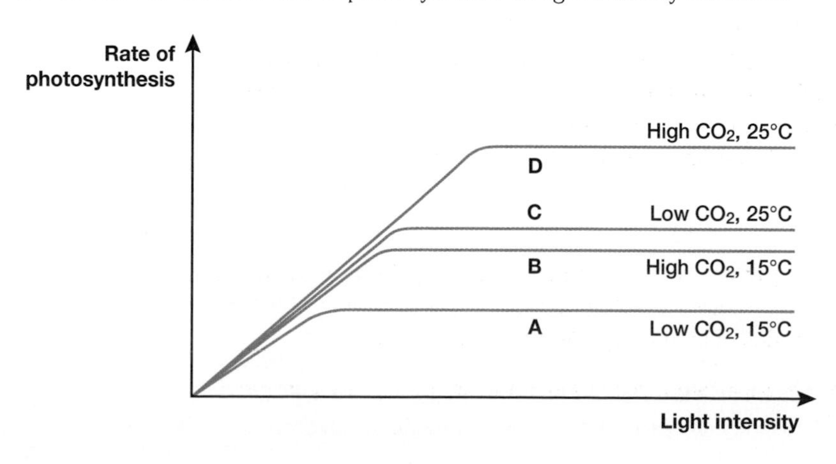

Figure 2.11 The effects of different temperatures and carbon dioxide concentrations on the rate of photosynthesis as light intensity increases

In the region of the graphs where light is non-limiting (horizontal lines), the factors that are limiting are:

- **A** — both temperature and carbon dioxide; increasing either produces an increase in the rate of photosynthesis to level **B** or **C**
- **B** — temperature (the factor that has not been increased from **A**); increasing the temperature increases the rate to level **D**
- **C** — carbon dioxide (the factor that hasn't been increased from **A**); increasing the carbon dioxide concentration increases the rate to level **D**

What is the effect of temperature on the rate of photosynthesis?

Increasing the temperature increases the kinetic energy of particles. This means that they move faster and collide with each other more frequently. Therefore, increasing the temperature increases the rate of chemical reactions. However, in biological systems (including photosynthesis), many reactions are controlled by enzymes and the effect of a change in temperature is more complex.

In photosynthesis, the reaction between RuBP and CO_2 to form GP (as part of the light-independent reactions) is catalysed by the enzyme Rubisco. This is a key reaction that affects many subsequent reactions. Any change in the rate of this reaction affects the overall rate of photosynthesis.

Increasing the temperature up to the optimum for enzyme action increases the rate of the formation of enzyme–substrate complexes and increases the rate of reaction.

Increasing the temperature beyond the optimum for the enzyme causes denaturation and a rapid loss of enzyme function (Figure 2.12).

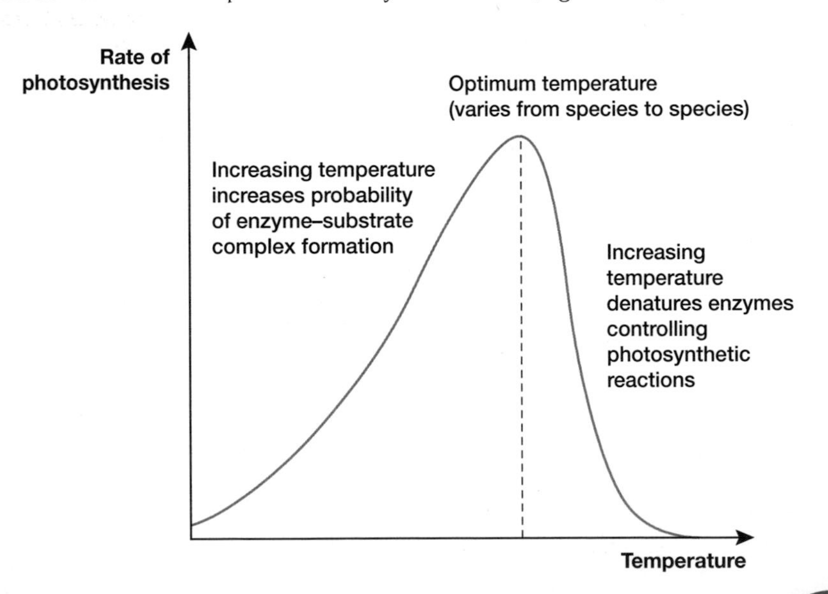

Figure 2.12 The effect of temperature on the rate of photosynthesis

Box 2.2 Enhancing photosynthesis in crop production

HOW SCIENCE WORKS

Crops are often grown in large glass greenhouses or in even larger polytunnels.

In both cases, the crops are grown in an indoor, controlled environment covered by a transparent material that allows light to penetrate. Growers can apply knowledge of the principle of limiting factors to enhance photosynthesis. Increasing the carbon dioxide concentration and increasing the temperature (up to a point) can both increase the rate of photosynthesis and, therefore, the yield of the crop.

Many crops are grown in large glass greenhouses or in polytunnels

Simply enclosing the plants in a greenhouse or polytunnel increases the temperature (because of the 'greenhouse effect'), without any extra heating costs (Figure 2.13). However, this only happens during daylight hours. At night, the greenhouse cools down and growth processes slow down.

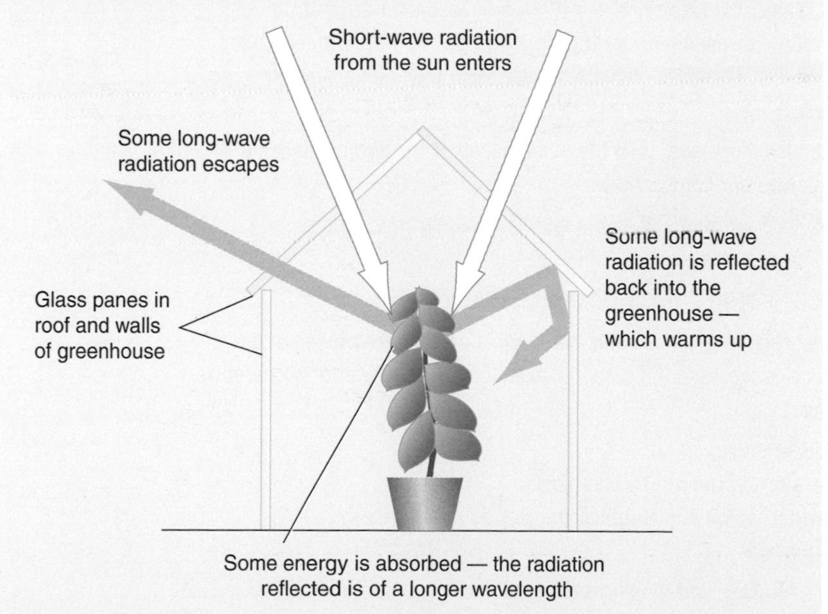

Figure 2.13 The greenhouse effect

Short-wave radiation from the sun enters the greenhouse, but the long-wave radiation created cannot all escape. The greenhouse heats up.

However, before investing in any equipment to maintain increased temperatures and carbon dioxide concentrations, the grower needs to be aware of the likely gains:

- What will be the extra yield from increasing the concentration of carbon dioxide?
- What will be the extra yield from increasing the temperature?

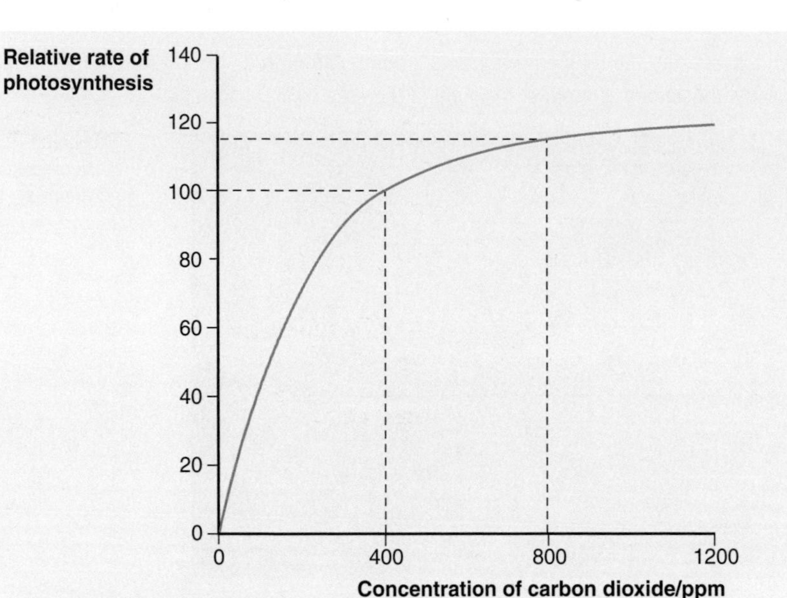

The concentration of carbon dioxide in the air is just under 400 ppm (parts per million). This results in a relative rate of photosynthesis of 100.

Figure 2.14 The effect of carbon dioxide concentration on photosynthesis

As Figure 2.14 shows, doubling the concentration of carbon dioxide (without altering any other factor) increases the rate of photosynthesis by less than 20%. If the concentration of carbon dioxide is increased further, there is very little increase in the rate of photosynthesis. Therefore, something else must be limiting the rate — probably temperature.

If the temperature is also increased, then the combination of the two produces a larger increase in the rate of photosynthesis.

Burning a fossil fuel, such as methane, in a controlled manner produces carbon dioxide and releases heat energy.

$$CH_4 + 2O_2 \rightarrow CO_2 + 2H_2O + \text{release of energy as heat}$$

Some growers do just this; they place sufficient burners at different places in the polytunnel/greenhouse to increase the carbon dioxide concentration and temperature in the whole enclosure.

Larger growers frequently use:

- a dedicated generator to produce carbon dioxide
- a boiler to heat water which is distributed in pipes to heat several polytunnels/greenhouses

The costs of buying, installing and running such equipment must be set against the gains from increased productivity. The effects of increasing the temperature and carbon dioxide concentration on a specific crop must also be studied carefully (Figure 2.15).

It might be too costly to heat the greenhouse to 25°C and provide the extra carbon dioxide necessary to achieve productivity level **D**. Productivity levels **B** and **C** are similar, but one may be less costly than the other. The grower must take all these factors into consideration.

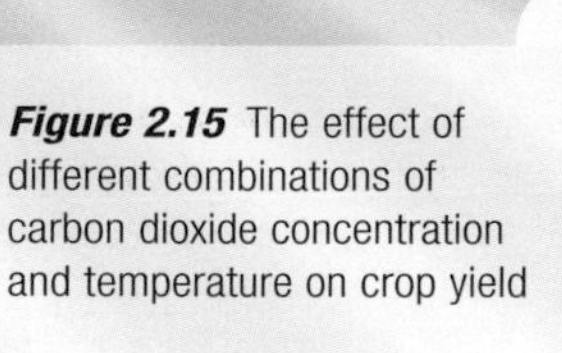

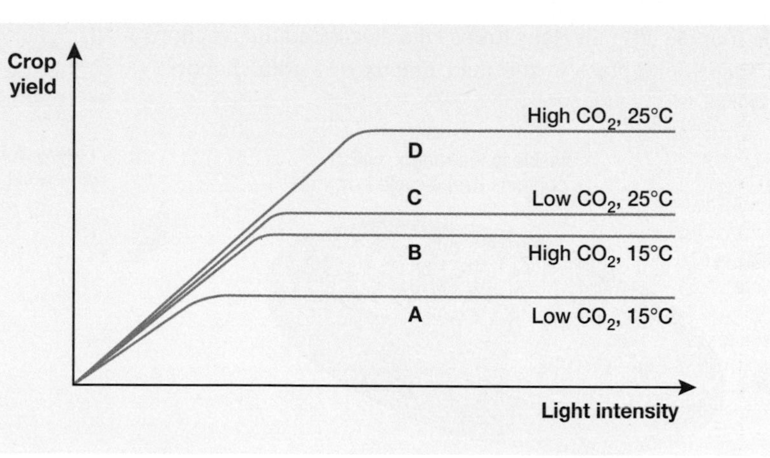

Figure 2.15 The effect of different combinations of carbon dioxide concentration and temperature on crop yield

How is ATP produced in respiration?

Respiration proceeds by two main pathways:

- **aerobic respiration**, which requires the presence of oxygen
- **anaerobic respiration**, which can take place in the absence of oxygen

Aerobic respiration

Aerobic respiration is a complex process comprising many reactions. Most of the ATP produced in aerobic respiration is produced in the mitochondria by moving protons through an ATP synthase enzyme, in a similar way to ATP production in photosynthesis.

Some ATP is produced in respiration by a process called **substrate-level phosphorylation**. In this process, a phosphate group is transferred from a phosphorylated substance (XP in the equation below) to ADP, producing a molecule of ATP:

$$\text{XP} + \text{ADP} \xrightarrow{\text{Substrate-level phosphorylation}} \text{X} + \text{ATP}$$

The reaction is controlled by an enzyme and can only occur if there is sufficient energy in the phosphate group to form the bond with ADP.

Glucose cannot enter mitochondria and so must be converted to a molecule that can. This molecule is the three-carbon compound, **pyruvate**. This first stage of respiration is called **glycolysis**, which means 'glucose splitting'.

Pyruvate is then converted into a two-carbon compound that enters a cycle of reactions, which produces hydrogen carriers. The conversion to the two-carbon compound is called the **link reaction** and the

Transmission electron micrograph of a mitochondrion

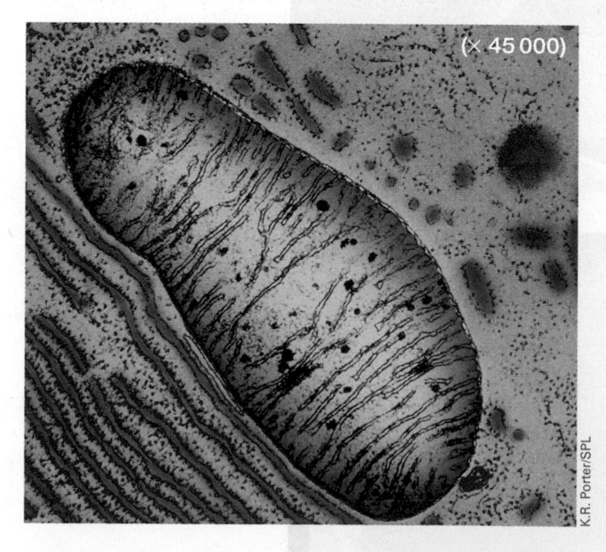

cycle is the **Krebs cycle** (named after Sir Hans Krebs who elucidated the reactions involved). Both these stages take place in the fluid **matrix** of a mitochondrion (Figures 2.16 and 2.17).

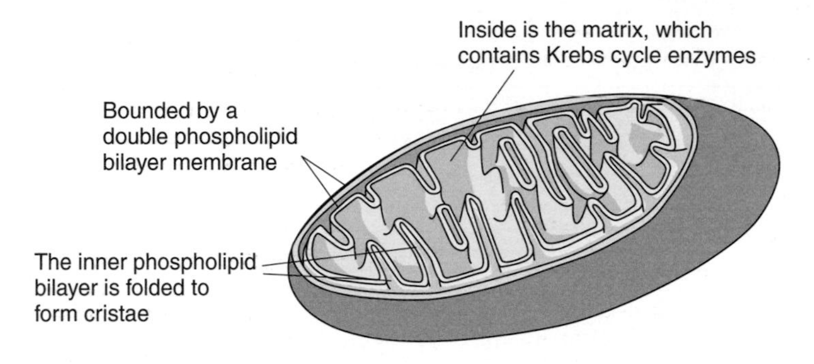

Figure 2.16 Structure of a mitochondrion

A mitochondrion is bounded by two membranes. The outer membrane is freely permeable to most substances; the inner membrane is selectively permeable. Most substances can only pass through the inner membrane via appropriate carrier molecules. It is also the site of the molecules of an electron transport chain and membrane-bound ATP synthase enzymes. The membrane is folded into cristae, which increases the area of the membrane, allowing increased electron transport and ATP synthesis.

In all three stages, hydrogen atoms are transferred to the coenzyme **NAD** to produce **reduced NAD**. These molecules later release their hydrogen atoms as protons (hydrogen ions) and electrons.

Notice that the hydrogen carrier is NAD *not* NADP, which is the hydrogen carrier in photosynthesis.

The protons are used in the chemiosmotic synthesis of ATP as the electrons pass along a transport chain. The molecules of the electron transport chain and the ATP synthase enzymes are located in the inner membrane of the mitochondrion, which is folded into **cristae** to increase the available membrane area.

At the end of the electron transport chain, the electrons combine with protons (hydrogen ions) and oxygen atoms to form water. Oxygen is needed for the electron transport chain to operate, so the formation of ATP in this way is called **oxidative phosphorylation**.

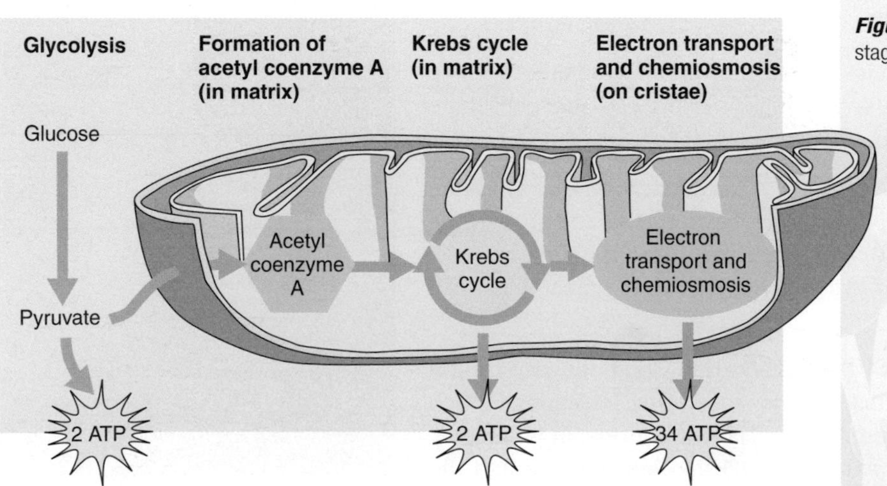

Figure 2.17 Outline of the stages of aerobic respiration

The link reaction, Krebs cycle and the reactions of the electron transport chain all depend on the presence of oxygen. None of these occurs in anaerobic respiration. Glycolysis can take place in the absence of oxygen and is the only energy-releasing process in anaerobic respiration.

What happens in the main stages of aerobic respiration?

Glycolysis

The reactions of glycolysis take place in the cytoplasm. Two molecules of ATP are used initially to phosphorylate each molecule of glucose. In its non-phosphorylated state, glucose is quite unreactive; phosphorylation makes it more reactive. Once phosphorylated, it is converted to another six-carbon sugar (fructose 1,6 bisphosphate) before being split into two molecules of the three-carbon sugar GP. This is then converted into pyruvate, with the production of ATP (by substrate-level phosphorylation) and reduced NAD (NADH). The main reactions of glycolysis are shown in Figure 2.18.

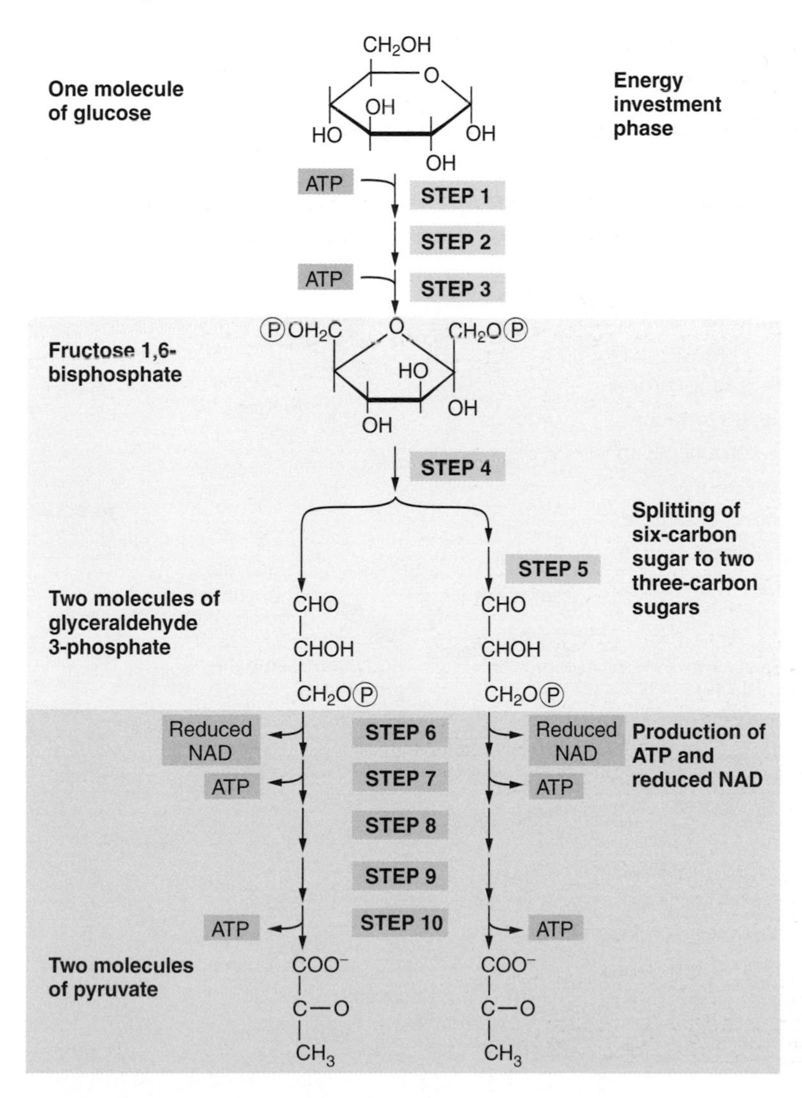

Figure 2.18 The main reactions of glycolysis

In Figure 2.18, the detail of most of the reactions has been omitted. Note also that you do not need to know the names of any of the intermediate compounds in the formation of pyruvate from glucose. However, you *must* know that:

- two molecules of ATP are used
- four molecules of ATP are formed
- two molecules of reduced NAD are formed per molecule of glucose

In glycolysis, there is a net gain of two ATP molecules per molecule of glucose (two molecules are used initially and then four are produced). Two molecules of reduced NAD are also produced. The molecules of pyruvate pass into the mitochondria through carrier molecules in the mitochondrial membrane.

The link reaction and Krebs cycle

Both of these stages take place in the fluid matrix of the mitochondrion.

In the link reaction (Figure 2.19), the following take place:

- Pyruvate reacts with a molecule of coenzyme A to form a molecule of acetyl coenzyme A.
- Hydrogen atoms are lost in the process and are used to form reduced NAD from NAD.
- Carbon dioxide is produced.

The removal of a carbon atom to form carbon dioxide is called **decarboxylation**.

Pyruvate (3C)
(from glycolysis,
two molecules per glucose)
CO_2
NAD
CoA
Reduced
NAD
Acetyl CoA (2C)
CoA
To Krebs
cycle

Figure 2.19 Stages in the link reaction

Details of the link reaction are often omitted. For simplicity, acetyl coenzyme A is referred to as a two-carbon compound.

In the Krebs' cycle (Figure 2.20), the following reactions occur:

- The two-carbon group from acetyl coenzyme A reacts with a four-carbon compound called **oxaloacetate** to form a six-carbon compound called **citrate.** The original coenzyme A is regenerated.
- Citrate then loses a carbon atom (is decarboxylated) to form a five-carbon compound and CO_2 is produced.
- The five-carbon compound is also decarboxylated to form a four-carbon compound and CO_2 is again produced. A molecule of ATP is produced by substrate-level phosphorylation.
- The four-carbon compound undergoes several molecular transformations to regenerate the original four-carbon compound, oxaloacetate, and the cycle can begin again.
- In several reactions in the cycle, reduced NAD is produced; in just one reaction, reduced FAD is produced. This is a similar molecule to reduced NAD.

Reduced NAD and reduced FAD are 'vehicles' for carrying protons and electrons released from glucose molecules to the electron transport chain and proton pumps in the cristae of the inner membrane of the mitochondrion.

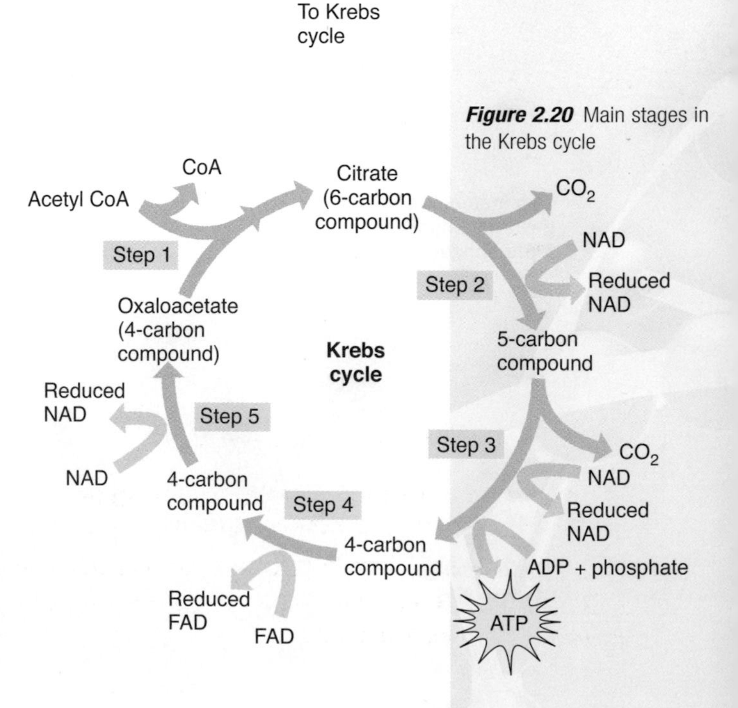

Figure 2.20 Main stages in the Krebs cycle

The electron transport chain and the chemiosmotic production of ATP

The hydrogen atoms carried by reduced NAD and reduced FAD are released at the cristae and split into protons and electrons.

The electrons pass along the transport chain, losing energy as they pass from one carrier to the next. Three of the electron carriers are proton pumps that move protons from the matrix of the mitochondrion to the intermembrane space.

As the electrons are transferred through these three proton pumps, the energy they lose powers the pumps that move the protons. Electrons from reduced NAD make this happen at all three pumps (Figure 2.21).

At the end of the electron transport chain, the electrons combine with protons and oxygen to form molecules of water. This is why oxygen is known as the terminal electron acceptor.

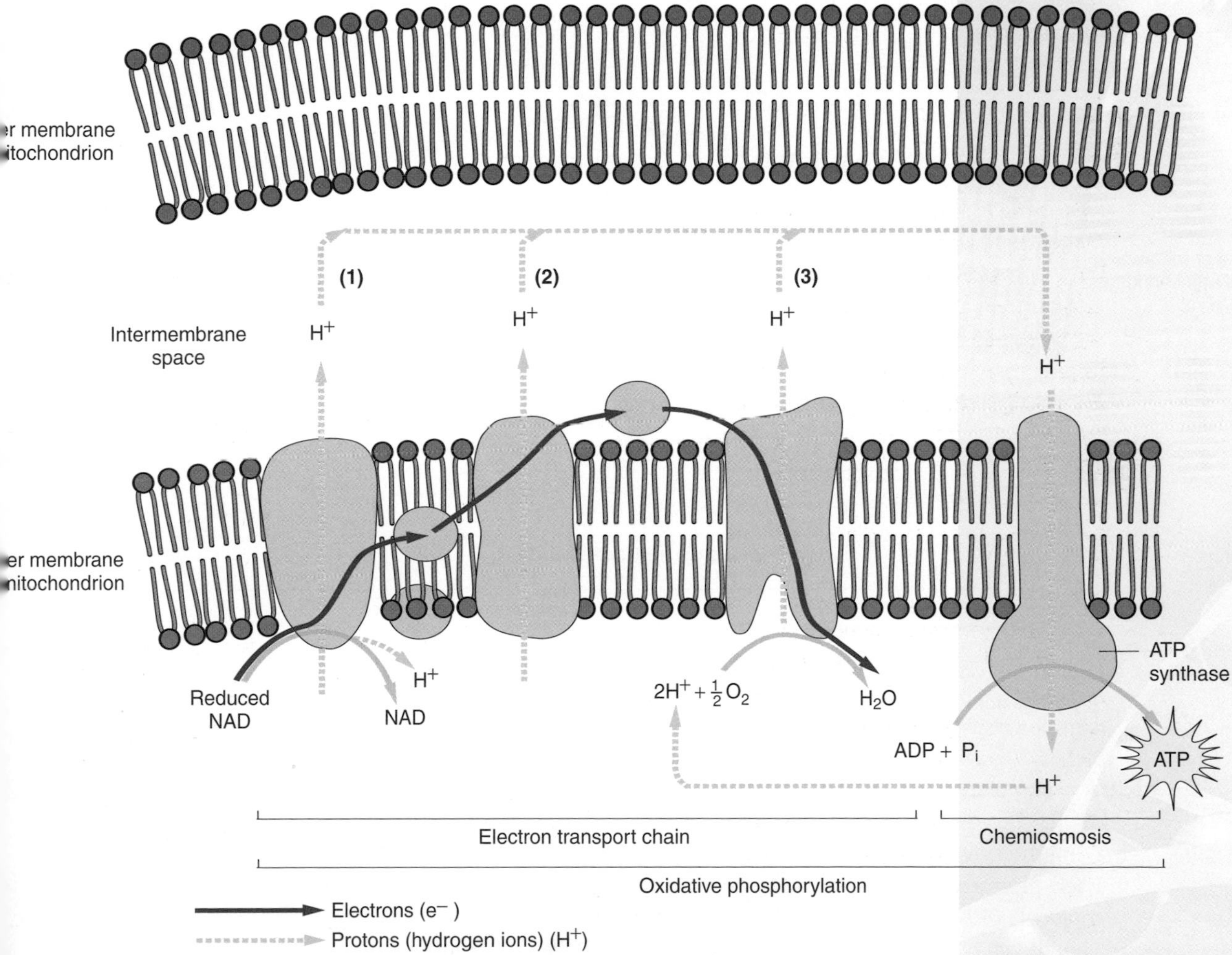

Electrons from reduced NAD pass through all the carriers in the electron transport chain and drive all three proton pumps

Figure 2.21 Electron transport for reduced NAD

Electrons from reduced FAD have less energy and join the transport chain at a different point. They cause only two protons to be pumped into the intermembrane space (Figure 2.22).

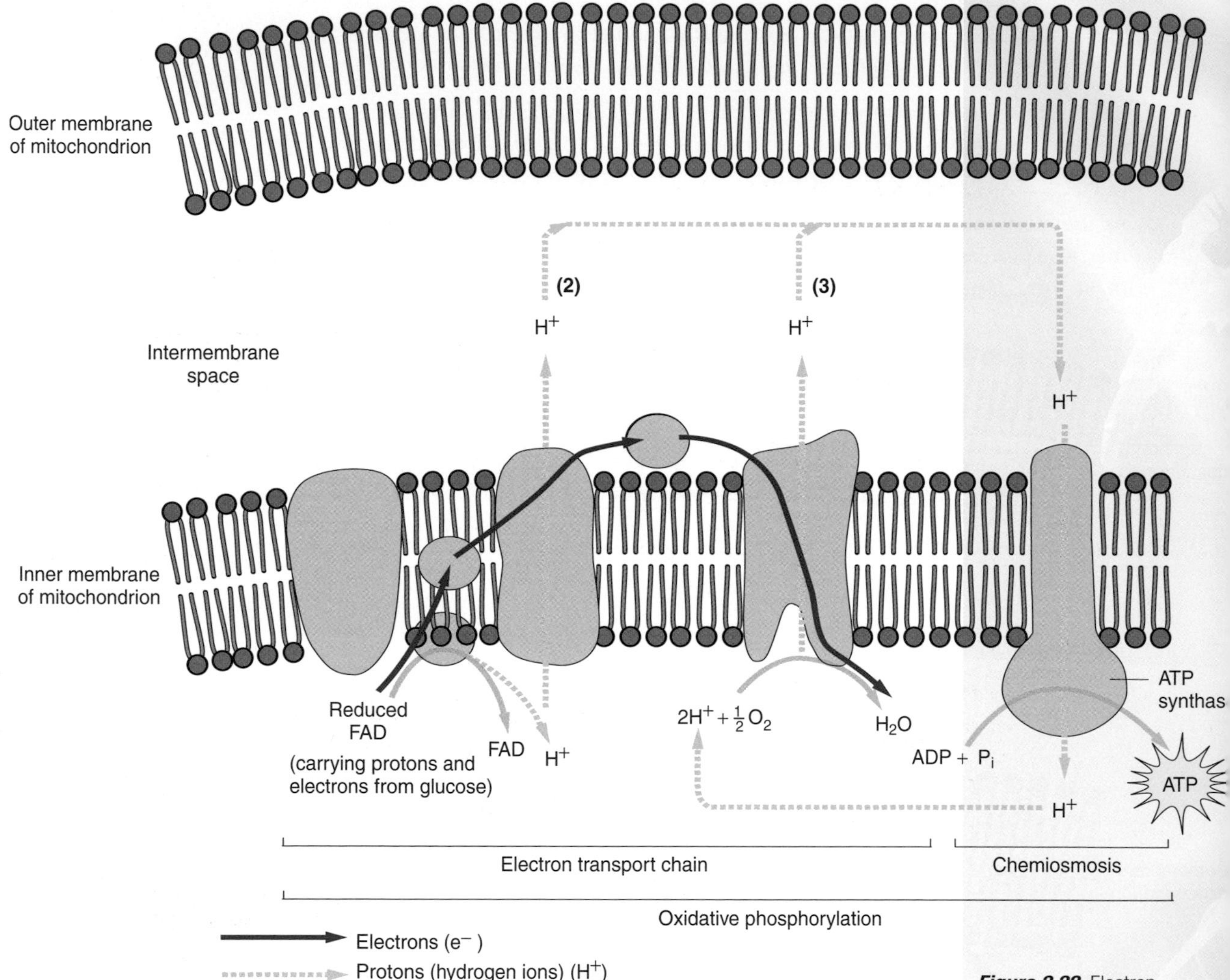

Figure 2.22 Electron-transport for reduced FAD

The accumulation of protons in the intermembrane space creates a proton gradient. Protons pass down the gradient through ATP synthase molecules, making the synthase 'spin' and produce ATP from ADP and P_i:

- Each proton that passes through ATP synthase causes the synthesis of one molecule of ATP.
- The oxidation of one molecule of reduced NAD results in three protons passing through ATP synthase and so leads to the synthesis of three molecules of ATP.
- The oxidation of one molecule of reduced FAD results in two protons passing through ATP synthase and so leads to the synthesis of just two molecules of ATP.

The electron transport chain and chemiosmosis together make up the process of oxidative phosphorylation.

2

Box 2.3 A summary of aerobic respiration

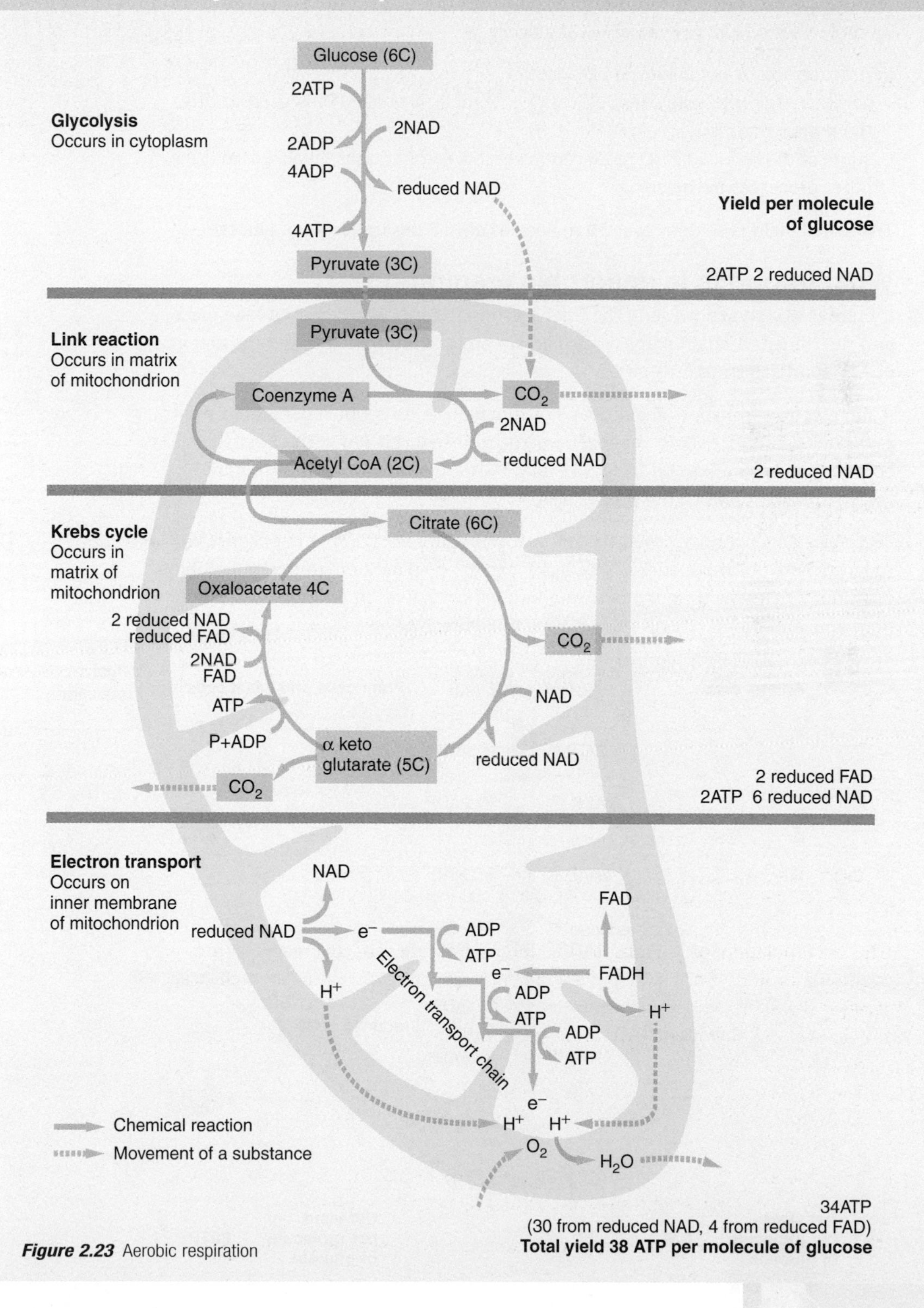

Figure 2.23 Aerobic respiration

By adding up the number of molecules of ATP produced, the model of aerobic respiration we have discussed predicts that there will be a net yield of 38 molecules of ATP per molecule of glucose.

In practice, this is not achieved because:

- some energy (the equivalent of just over 2 molecules of ATP) is used to drive the proton pumps
- some of the reduced NAD produced in aerobic respiration is 'hijacked' to drive other processes in the cell.

The actual yield is seldom over 30 molecules of ATP per molecule of glucose.

What happens in anaerobic respiration?

If there is no oxygen present, the final reaction to form water cannot take place and the transport chain comes to a halt. No protons are pumped and the action of ATP synthase stops.

If the electron transport chain does not function, NAD is not regenerated from reduced NAD and FAD is not regenerated from reduced FAD. The Krebs cycle comes quickly to a halt because NAD and FAD are required. NAD is required in the link reaction and so this ceases also.

Glycolysis can continue, even though it too requires NAD. This is because the reduced NAD formed during glycolysis can be regenerated under anaerobic conditions by converting the pyruvate into either lactate (in animals) or ethanol (in plants and fungi). This is summarised in Figure 2.24.

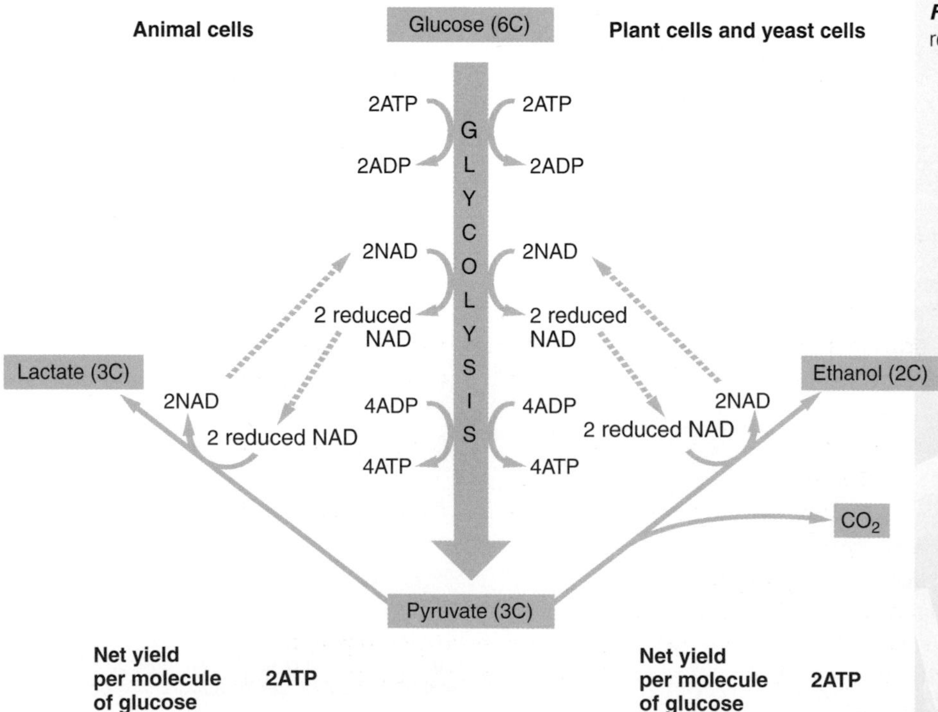

Figure 2.24 Anaerobic respiration

The production of ATP by the anaerobic pathway is only a temporary solution for most organisms. Anaerobic respiration produces a net yield of only 2 molecules of ATP per molecule of glucose (rather than the predicted 38 and achieved 30 in aerobic respiration). The product of anaerobic respiration (lactate or ethanol) is toxic to the cell in which it is produced.

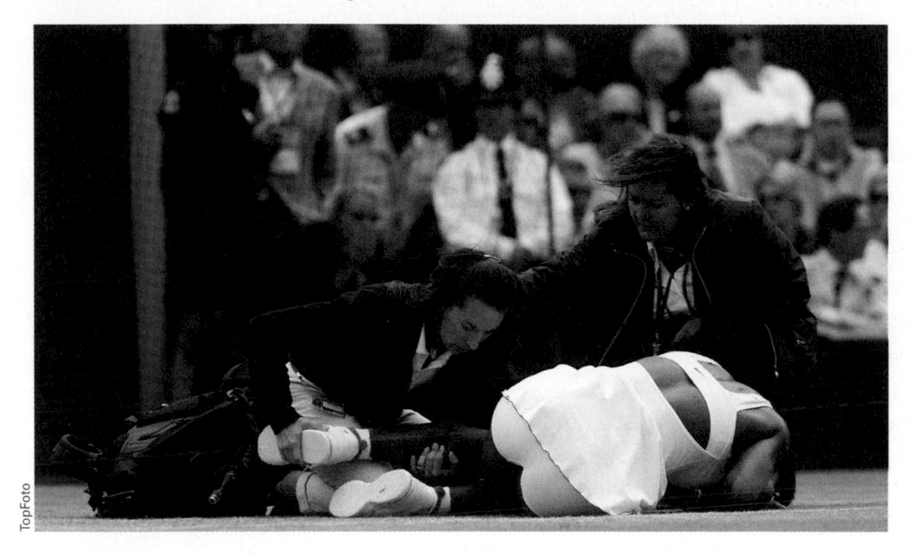
TopFoto

The lactate built up in this tennis player's muscle cells is causing considerable pain

Summary

ATP

- ATP is an ideal energy-storage molecule in a cell because:
  - energy is released from the molecule quickly, in a single-step hydrolysis reaction
  - energy is released in small amounts, which are matched closely to the amounts needed for cellular reactions
  - the molecule is moved around easily within the cell but cannot leave the cell

Photosynthesis

- Chloroplasts are well adapted to carry out photosynthesis because:
  - the grana provide a large surface area for the arrangement of chlorophyll molecules and the associated electron transport systems of the light-dependent reactions
  - the stroma provides a fluid medium for the reactions of the light-independent reactions
- The light-dependent reactions produce ATP and reduced NADP, both of which are needed in the light-independent reactions.
- In the light-independent reactions:
  - CO_2 combines with RuBP (5C) to form two molecules of GP
  - GP is reduced to TP; reduced NADP supplies the hydrogen ions and ATP supplies the energy; the NADP and ADP + P_i are recycled to the light-dependent reactions

- some TP (1/6) is used to synthesise useful carbohydrates — for example glucose
- most TP (5/6) is used to regenerate RuBP so that the cycle of reactions can begin again

- The rate of photosynthesis is influenced by light intensity, concentration of carbon dioxide and temperature; the factor present in the least quantity limits the rate of photosynthesis.

Respiration

- The main stages of aerobic respiration are glycolysis, the link reaction, Krebs cycle, the electron transport chain and the chemiosmotic synthesis of ATP. In the descriptions below, all figures are per molecule of glucose:
  - In glycolysis, glucose (6C) is converted to pyruvate (3C) with the net gain of two molecules of ATP and two molecules of reduced NAD.
  - In the link reaction, pyruvate is converted to actetyl coenzyme A (effectively 2C) with the loss of carbon dioxide and the production of two molecules of reduced NAD.
  - In the Krebs cycle, acetyl coenzyme A combines with oxaloacetate (4C) to form citrate (6C) which is then decarboxylated to a 5C compound. This is decarboxylated to a 4C compound, which is then converted into oxaloacetate. The cycle produces six molecules of reduced NAD, two molecules of reduced FAD and two molecules of ATP (by substrate-level phosphorylation)
  - As electrons from reduced NAD and reduced FAD pass along the electron transport chain, they lose energy which is used to pump protons from the matrix to the intermembrane space.
  - Protons then pass down an electrochemical gradient back into the mitochondrion through molecules of ATP synthase. Each proton that passes through the enzyme causes one molecule of ATP to be synthesised.
- In anaerobic respiration, the reactions of the electron transport chain, Krebs cycle and the link reaction cannot occur as, without oxygen as the terminal electron acceptor, NAD and FAD cannot be regenerated from reduced NAD and reduced FAD.
- Glycolysis still occurs in anaerobic conditions because the NAD needed is regenerated from reduced NAD by the reduction of pyruvate to lactate (animal cells) or ethanol (plant cells and yeast cells).

Questions

Multiple-choice

1 In the light-dependent reactions of photosynthesis:

A NAD is reduced
B ADP is produced
C both of the above
D neither of the above

The graph shows the effect of light intensity on the rate of photosynthesis at different concentrations of carbon dioxide and at different temperatures. Questions 2 and 3 relate to this graph.

Rate of photosynthesis
Light intensity
S Z
Y
R X
Q W
P

2 Which of lines W, X, Y, and Z represents conditions of a high concentration of carbon dioxide and a high temperature?

A W
B X
C Y
D Z

3 Which of the regions P, Q, R and S, represents conditions in which light intensity is limiting the rate of photosynthesis?

A P
B Q
C R
D S

4 In the light-independent reactions of photosynthesis:

A ATP is used to convert GP into TP
B reduced NADP is used to convert GP into TP
C both A and B
D neither A nor B

5 Which of the following takes place during the Krebs cycle?

A oxidative phosphorylation
B substrate-level phosphorylation
C electron transport
D the link reaction

6 In anaerobic respiration:

A oxidative phosphorylation does not take place
B substrate-level phosphorylation does take place
C NAD is reduced in glycolysis
D all of the above

7 Mitochondria and chloroplasts:

A are both enclosed by a double membrane
B both produce ATP
C both A and B
D neither A nor B

The diagram shows the structure of a chloroplast. Questions 8 and 9 relate to this diagram.

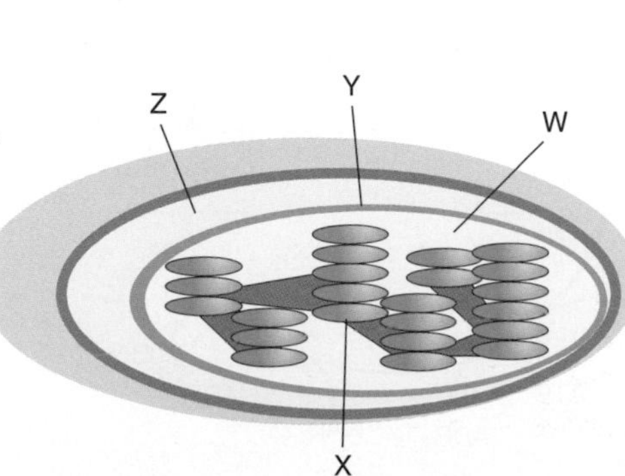

8 During the light-dependent reactions of photosynthesis, ATP is produced in the regions labelled:

A W

B X

C Y

D Z

9 NADP moves from:

A W to X

B X to W

C X to Y

D Y to W

10 ATP is an ideal molecule to act as a short-term energy store because it:

A releases energy in small amounts

B releases energy in a one-step reaction

C is easily transferred within a cell but cannot leave the cell

D all of the above

Examination-style

1 The diagram shows some of the stages in respiration.

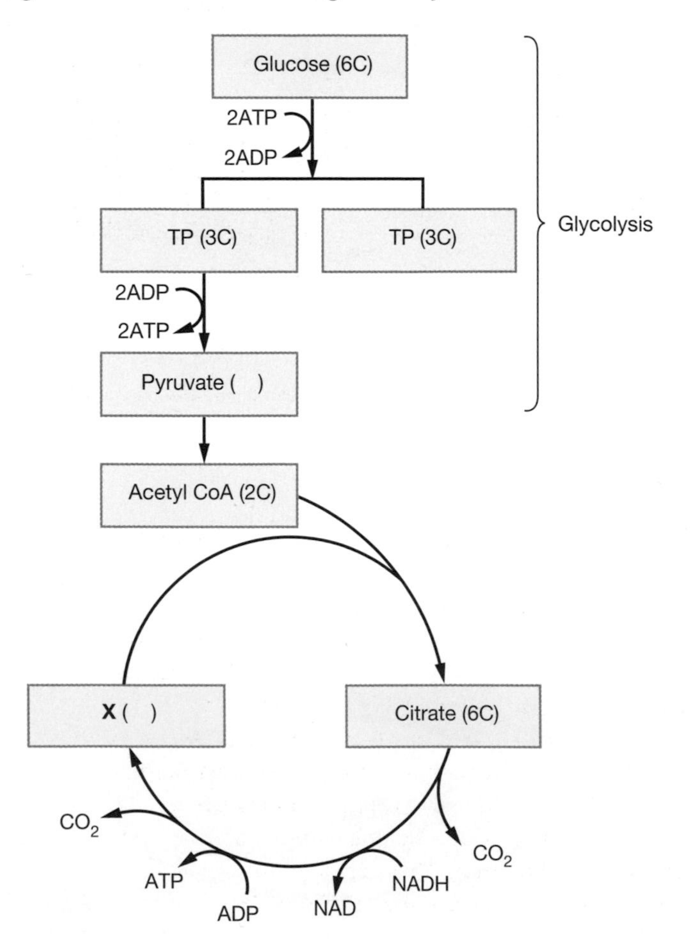

(a)(i) How many carbon atoms are present in pyruvate and in compound **X**? *(1 mark)*

(ii) How many ATP molecules are generated during glycolysis? Explain your answer. *(2 marks)*

(b) Identify **one** stage in the diagram where substrate-level phosphorylation takes place. *(1 mark)*

(c) What becomes of the reduced NAD generated in glycolysis and the Krebs cycle? *(2 marks)*

Total: 6 marks

2 The diagram shows the location in the thylakoid membrane of the molecules involved in the light-dependent reactions of photosynthesis.

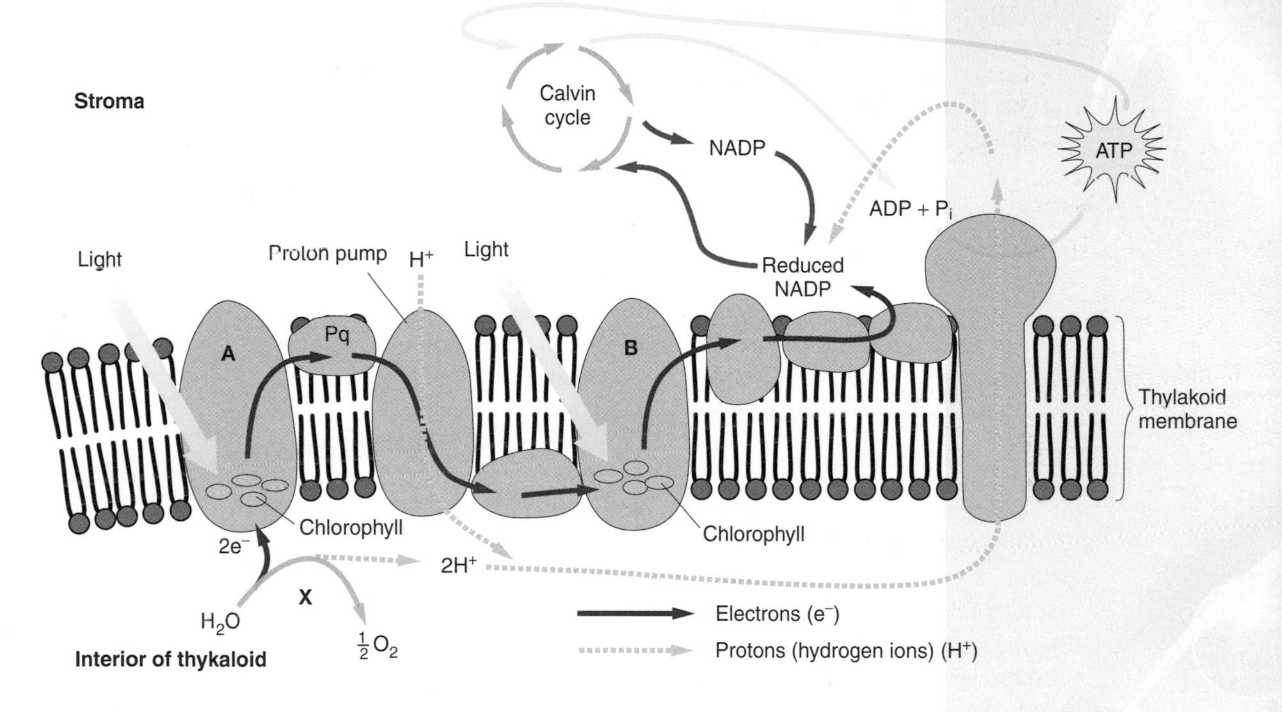

(a) Name the structures labelled **A** and **B** *(2 marks)*

(b) Name the reaction labelled **X**. What is the importance of this reaction in the light-independent reactions? *(3 marks)*

(c) Use your knowledge of the structure of a chloroplast to explain why light striking a chloroplast is not all absorbed by chlorophyll. *(3 marks)*

Total: 8 marks

3 The diagram shows apparatus that was used to investigate respiration in yeast. Before starting the experiment, the students carrying out the investigation ensured that the yeast was respiring actively. They measured the relative position of the bubble of liquid in the capillary tubing every minute. A '+' indicated movement away from the boiling tube; a '–' indicated movement towards the boiling tube. They repeated the experiment five times. Their mean results are shown in the table.

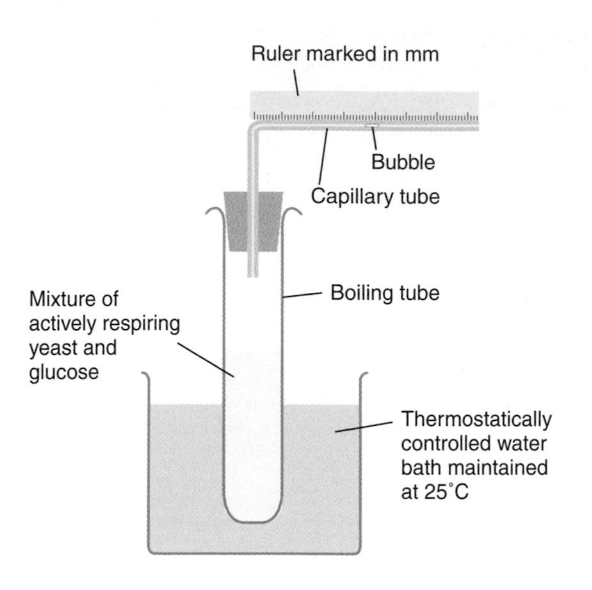

Time/ minutes	Total movement of bubble/mm
0	0
1	0
2	0
3	+2
4	+5
5	+9
6	+13
7	+16
8	+21
9	+25
10	+28

(a) Explain why the students stood the boiling tube in a water bath. *(3 marks)*

(b) (i) In terms of pressure, explain why the bubble moved away from the boiling tube between the third and tenth minutes. *(2 marks)*

(ii) Use your knowledge of aerobic and anaerobic respiration to explain why the bubble did not move for the first three minutes and then moved away from the boiling tube. *(5 marks)*

Total: 10 marks

4 The diagram summarises some of the reactions of aerobic respiration.

(a) Name the processes taking place at X, Y and Z. *(3 marks)*

(b) In the absence of oxygen, the Krebs cycle cannot take place, even though its reactions do not use oxygen. Explain why. *(3 marks)*

(c) Reduced NAD is produced in glycolysis. Explain what becomes of this reduced NAD in aerobic respiration and in anaerobic respiration. *(4 marks)*

Total: 10 marks

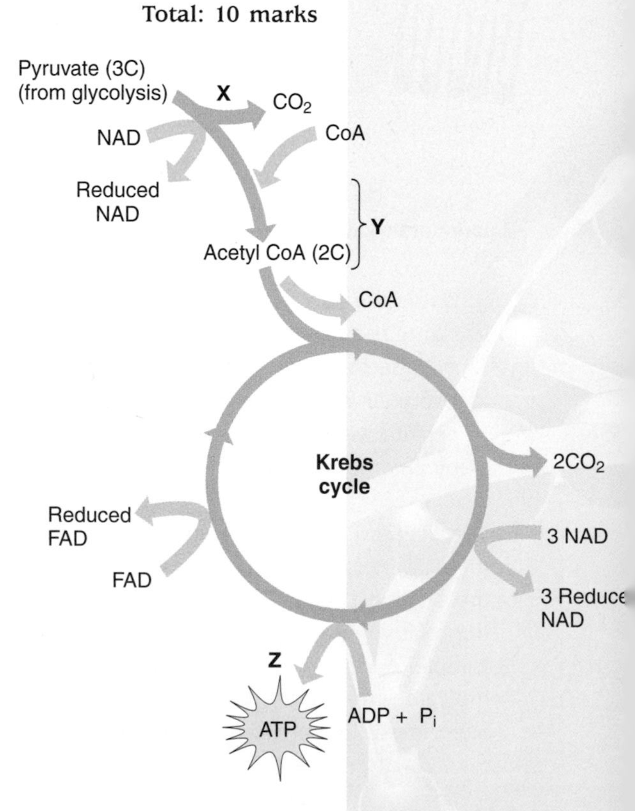

5 **(a)** The flow chart summarises the fate of light energy striking a leaf.

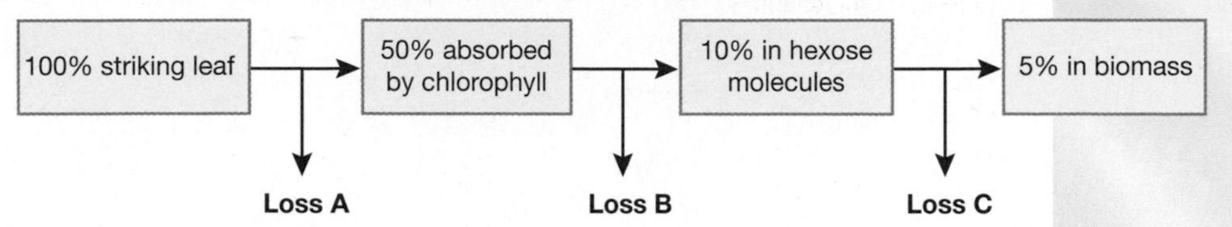

(i) Suggest *two* processes that could contribute to **loss A**. *(2 marks)*

(ii) What does **loss B** tell you about the percentage efficiency of the process of photosynthesis? Explain your answer. *(2 marks)*

(iii) In this example, what percentage of light energy becomes part of the net primary production? Explain your answer. *(3 marks)*

(b) The diagram below summarises the main reactions of the light-independent stage of photosynthesis. The graph shows changes in the levels of RuBP and GP in a chloroplast when the light source is removed.

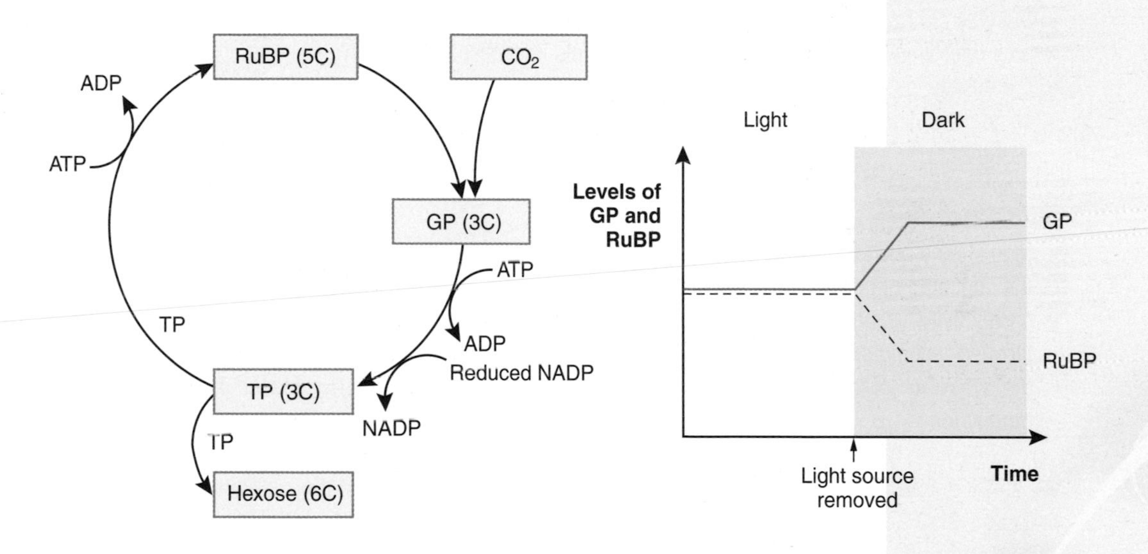

(i) Give *two* possible fates of the hexose produced. *(2 marks)*

(ii) Use the diagram to explain the changes in the levels of RuBP and TP in the chloroplast when the light source is removed. *(6 marks)*

Total: 15 marks

Chapter 3

How is energy transferred through ecosystems?

This chapter covers:
- the way energy enters ecosystems
- how energy is transferred between trophic levels in ecosystems
- the energy losses associated with energy transfer
- ecological pyramids, including pyramids of:
  - number
  - biomass
  - energy
- food chains and food webs
- the concepts of gross productivity and net productivity
- productivity of natural ecosystems and ecosystems based on modern intensive farming techniques
- the ways in which modern farming methods seek to reduce energy losses in production

We saw in the last chapter that all organisms depend for their existence on the continual hydrolysis of ATP to release energy. This energy then drives all their metabolic processes. ATP is synthesised in respiration and hydrolysed to release energy as it is needed.

We know from Chapter 1 and AS Unit 2 that an ecosystem is 'a self-sustaining system in which organisms interact with each other and with their physical environment'. In order to sustain itself, an ecosystem needs:
- a regular input of energy to drive the processes
- the necessary materials for organisms to produce new biomass

In this chapter we shall look at how energy enters an ecosystem, is transferred through it and finally leaves the ecosystem. We shall consider the role of plants, animals and decomposers in transferring energy. But we must establish two important principles at the outset:
- Energy cannot be created or destroyed, it can only be transferred and, in the transfer, converted from one form to another.
- The amount of usable energy in a system decreases as energy is transferred.

These two principles are non-mathematical approximations to the first and second laws of thermodynamics. The second law is particularly important and explains why we must eat continually. When we move, energy held in molecules of ATP is used to contract muscle fibres. However, the energy cannot be recycled. It is lost as heat as the muscle contracts; more ATP must be synthesised by respiring more food molecules for the next contraction of the muscle.

What processes are involved in transferring energy through an ecosystem?

All organisms in an ecosystem release the energy they need from organic molecules in respiration. Photosynthetic organisms synthesise their organic molecules from inorganic ones, but all other organisms must take in organic molecules. They must feed on either living or dead organisms. This allows us to classify organisms in an ecosystem into:

- **producers** — organisms that produce their own organic molecules
- **consumers** — organisms that feed on other living organisms
- **decomposers** — organisms that feed on dead organisms

The simplest way of considering the transfer of energy is to look at a simple **food chain**, for example:

Grass → Gazelle → Cheetah

Using this simple food chain as an example, the processes involved in the transfer of energy are as follows:

- Light energy enters the grass (the **producer**).
- In photosynthesis, the light energy is transduced to chemical energy in organic molecules, such as glucose.

A cheetah chasing a gazelle

- Some of this chemical energy is used in growth to build new cells (assimilation) and is retained within the grass.
- Some is used in respiration and, once used, is lost as heat.
- The gazelle (the **primary consumer**) eats and digests some of the grass, transferring chemical energy to its cells.
- Some of this chemical energy is used in growth to build new cells (assimilation) and is retained within the gazelle.
- Some is used in respiration and, once used, is lost as heat.
- The cheetah (the **secondary consumer**) eats and digests some parts of the gazelle, transferring chemical energy to its cells.
- Some of this chemical energy is used in growth to build new cells (assimilation) and is retained within the cheetah.
- Some is used in respiration and, once used, is lost as heat.
- At each stage in the food chain, organisms that are not eaten eventually die and the energy contained in their cells is transferred during decay to decomposers. When the decomposers respire, this energy is lost as heat.
- The different feeding levels (producer, primary consumer and so on) are called **trophic levels**.

The flow of energy through this simple food chain is represented is shown in Figure 3.1.

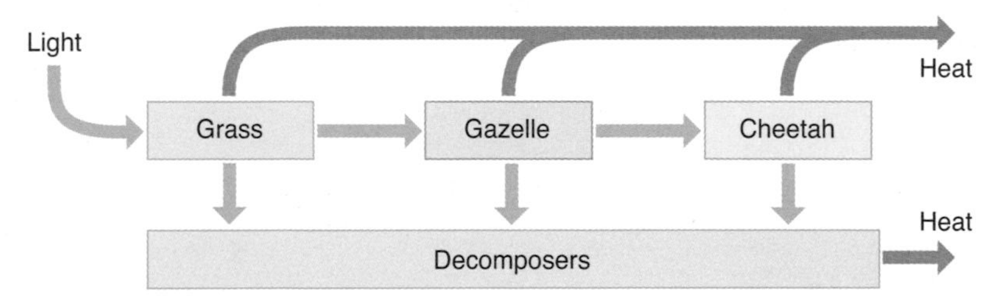

Figure 3.1 Flow of energy through a simple food chain

Of course, food chains rarely exist in isolation. They are linked to other food chains to form a **food web**. Figure 3.2 shows the main energy pathways through the food web of the Arctic tundra.

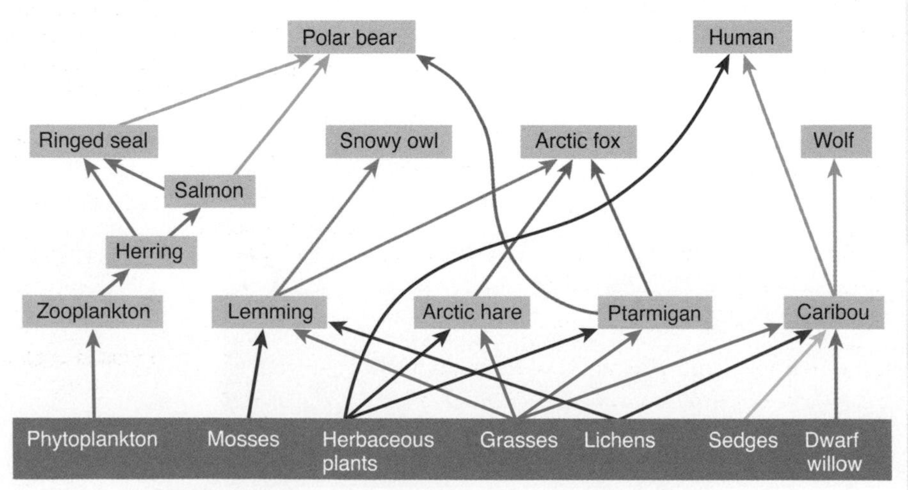

Figure 3.2 Food web of the Arctic tundra

Since the food chains interlink, there are consequences:

- A change in the number of one organism can influence the numbers of another that is not apparently dependent on it. For example, a reduction in the number of caribou because of a disease could mean more lemmings to eat the lichens; it could also lead to fewer mosses but more snowy owls.
- A change in the number of one organism is unlikely to have a major influence on the numbers of other organisms because the effects are dispersed through many food chains.
- The more complex the food web, the more 'links' there are between food chains to absorb the impact of a change in numbers of one organism, and the more stable the food web will be.

How can we represent the amount of energy that passes through an ecosystem?

The amount of energy that enters an ecosystem determines how much energy is passed from one trophic level to the next. The amount of energy passing to a trophic level determines how much biomass can be produced by organisms in that trophic level. The same amount of biomass can result in:

- a large number of small organisms
- a smaller number of larger organisms

We can use these ideas to represent the flow of energy through food chains as **ecological pyramids** — pyramids of numbers, pyramids of biomass and pyramids of energy.

- A pyramid of numbers represents the relative numbers of the organisms in a food chain, at a given moment, irrespective of biomass (size).
- A pyramid of biomass represents the relative biomass of the organisms in a food chain, at a given moment, irrespective of numbers.
- A pyramid of energy represents the amount of energy transferred to each level of a food chain, irrespective of numbers and biomass, in a given period of time.
- An energy-flow diagram represents the amount of energy passing through the various levels of an ecosystem in a given period of time.

Rather than considering these concepts generally, it is helpful to think of specific examples. Consider the food chain:

Grass → Grasshoppers → Frogs → Birds

If we draw a pyramid of numbers and a pyramid of biomass for this food chain, the diagram looks the same (Figure 3.3). This is because numbers and mass are related. The individual organisms increase in mass along the chain, so a decrease in biomass necessarily also means a decrease in numbers.

Figure 3.3 The pyramid of numbers and pyramid of biomass for this food chain look the same

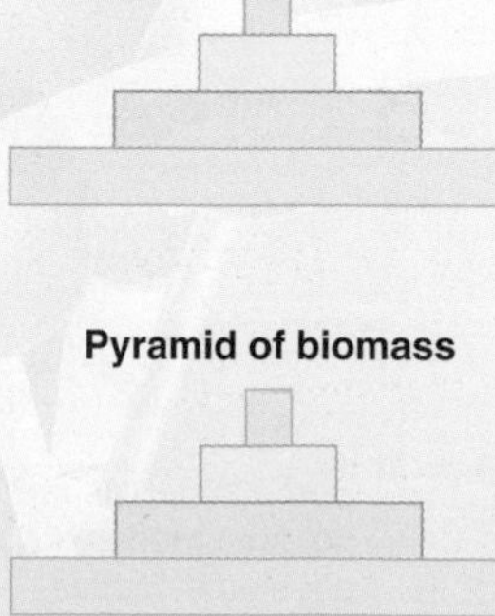

But look what happens when we consider the food chain:

Oak tree ⟶ Aphids ⟶ Ladybirds ⟶ Birds

The pyramids for this food chain are different because of the huge biomass of a single oak tree (Figure 3.4). The *decrease in biomass* to the next stage is accompanied by an *increase in numbers*. Thereafter, numbers and mass are related as in the first food chain.

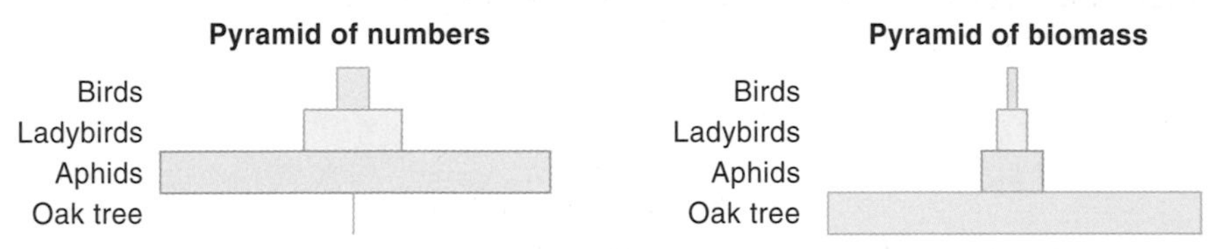

Figure 3.4 The pyramid of numbers and the pyramid of biomass for *this* food chain look very different

Pyramids of numbers can be further complicated if the end consumer is parasitised.

If the birds in the second food chain were parasitised by tiny mites, each bird would have many mites on its body and so the pyramid of numbers would be as shown in Figure 3.5.

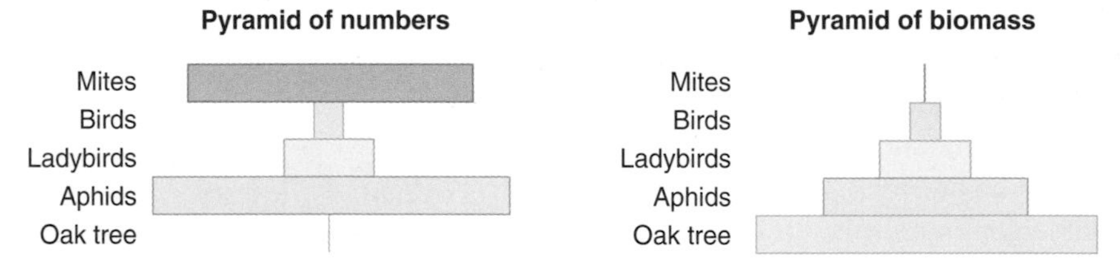

Figure 3.5 Pyramids of numbers and biomass for a food chain in which the tertiary consumer is parasitised

The pyramid of numbers is not very much like a pyramid! However, the biomass pyramid is a true pyramid. Each mite has a much smaller mass than the bird and collectively they still weigh less than the birds they parasitise.

Explaining the shape of pyramids of biomass is relatively straightforward. When a rabbit eats grass, not all of the materials in the grass plant end up as rabbit (Figure 3.6)!

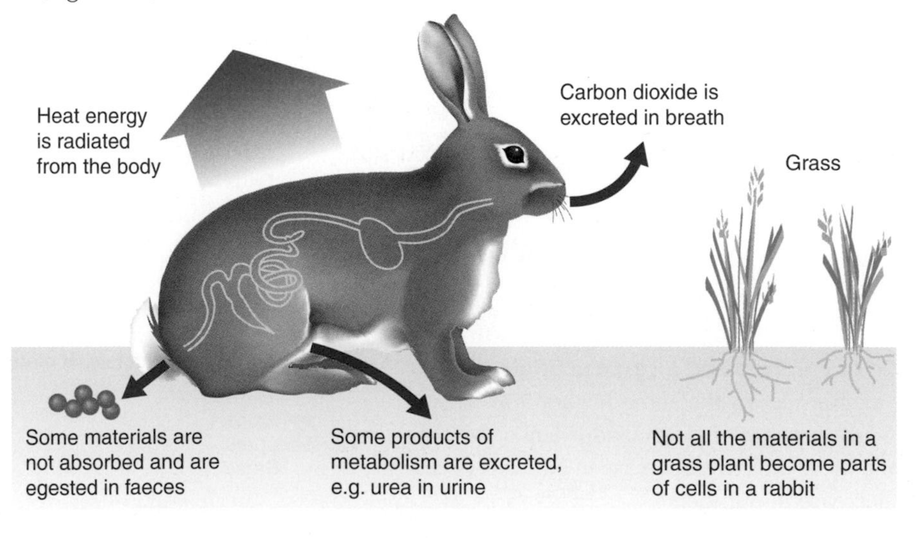

Figure 3.6 Metabolism of a rabbit

There are losses because:

- some parts of the grass are not eaten — the roots for example
- some parts are not digested and so are not absorbed (they pass out in the faeces)
- metabolism of some of the materials absorbed leads to the formation of excretory products, which are released into the environment
- many of the materials are respired to release energy; the carbon dioxide formed is exhaled (also a form of excretion)

Only a small fraction of the materials in the grass becomes incorporated into new cells in the rabbit. Similar losses are repeated at each trophic level in the food chain, so a smaller amount of biomass is available for growth at successive levels. Pyramids of biomass are almost always 'pyramid shaped' to reflect this.

Box 3.1 An unusual pyramid of biomass

Figure 3.7 looks like an impossible pyramid of biomass. How can a small biomass of producers support a larger biomass of primary consumers?

This pyramid of biomass is typical of marine ecosystems in which phytoplankton are the producers. Phytoplankton are unicellular algae. They can photosynthesise and are the basis of many marine ecosystems.

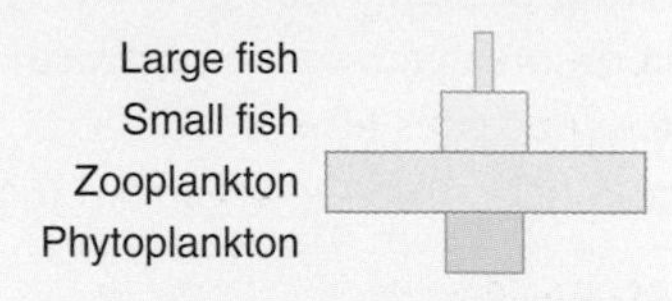

Figure 3.7 Pyramid of biomass for a plankton-based food chain

The answer to the question: 'How can a small biomass of phytoplankton support a larger biomass of zooplankton?' depends on having an understanding what is being measured. Figures for biomass are a 'snapshot in time'. They are an estimate of the total biomass at the time that the sample was taken for analysis. To make sense of the data, the **turnover rate** is needed. This is a measure of how long it takes for the biomass to be produced. For some phytoplankton, this may be less than one day (and can be as little as 8 hours); for zooplankton, it is often nearer to 20 days. If we combine figures for biomass and turnover, we obtain figures that make sense.

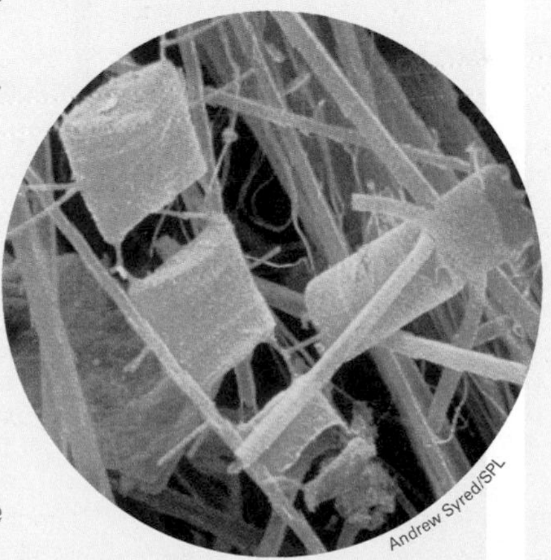

A sample of phytoplankton

Trophic level	Biomass/ $g\,m^{-1}$	Mean turnover rate/days	Daily production/g	Annual production/g
Phytoplankton	4.0	0.5	8.00	2920.0
Zooplankton	21.0	18.0	1.16	423.4

Over the year, there is much more biomass accumulated in the phytoplankton (the producers) than in the zooplankton (the primary consumers).

How productive are ecosystems?

Productivity is a measure of how efficient organisms are at locking up energy in organic molecules in their body cells.

What happens to the energy within a trophic level?

- Photosynthesis stores energy from sunlight in organic chemicals such as glucose and starch.
- Respiration releases energy from organic compounds to generate ATP.
- Almost all other biological processes (e.g. muscle contraction, growth, reproduction, excretion, active transport) use ATP generated in respiration.
- If energy is used to produce new cells (general body cells in growth and sex cells in reproduction), then the energy remains fixed in that organism and can be passed on to:
  - the next trophic level through feeding
  - the decomposers on the death of the organism
- If energy is used for processes other than the production of new cells, then it eventually escapes as heat from the organism (Figure 3.8).

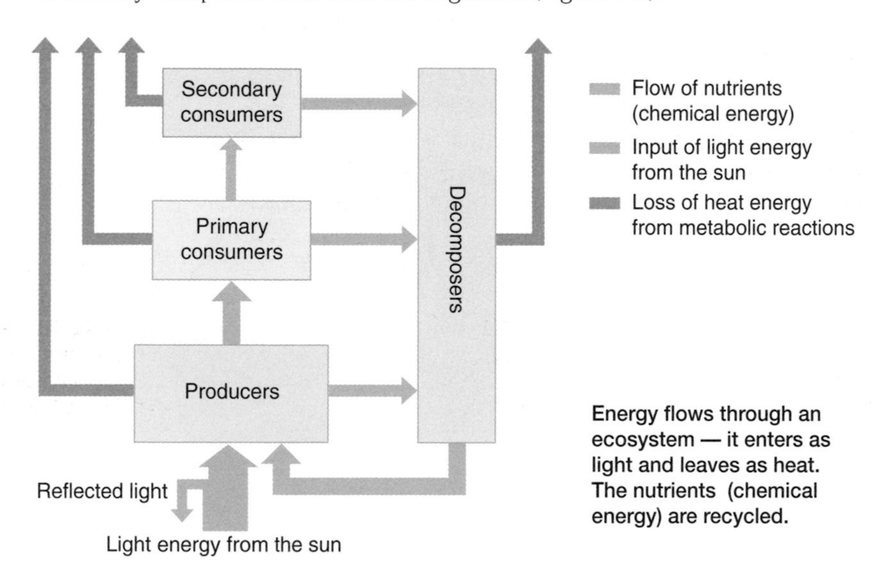

Figure 3.8 Transfer of nutrients and energy in an ecosystem

How energy efficient are food chains?

The efficiency of energy transfer between trophic levels is variable. It depends on a number of factors, which can be broken down into two categories:

- the amount of energy that enters the trophic level
- the amount of energy that is used for metabolic activities that do not result in growth

More energy is retained by an animal and less energy is lost if:

- the animal is an ectotherm and does not use metabolic energy to maintain a high, constant, core body temperature
- the animal is a herbivore and does not have to use much energy simply to obtain its food

An ectotherm is an animal that does not control its body temperature by physiological means; its temperature fluctuates with that of the environment.

The cheetah expends a lot of energy catching its food; the gazelle it feeds on eats stationary plants. This means that more of the energy the cheetah obtains in its food is used in non-growth activities and less is stored as biomass.

- the animal is large — a larger animal has a smaller surface area-to-volume ratio and, therefore, loses less heat per kilogram of body mass than a small animal

The loss of energy between trophic levels explains why food chains often have no more than four trophic levels (producer, primary consumer, secondary consumer and tertiary consumer).

Photosynthesis is energy inefficient. Some areas are better than others, but, often, only 1% of the light energy striking plants is fixed into chemical energy in carbohydrates by photosynthesis.

As a crude rule of thumb, about 10% of the energy fixed into a trophic level is passed to the next. This means that of the light energy that is absorbed and used by a producer in a four-link food chain, 10% passes to the primary consumer, 1% (10% of 10%) passes to the secondary consumer and just 0.1% passes to the tertiary consumer (Figure 3.9). There simply is not enough energy left to support a fifth level.

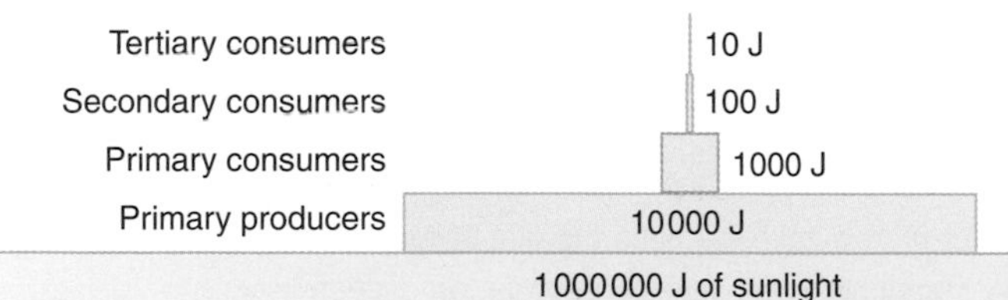

Figure 3.9 Inefficiency of energy transfer between trophic levels limits the number of links in a food chain

How do we measure productivity?

Productivity is defined as the amount of energy input to a trophic level that is converted into biomass:

- in a given period of time (often a year)
- for a given area of the ecosystem (often a square metre)

However, we need to be quite clear about what we mean by 'converted into biomass'.

Plants are the producers and are the first organisms in a food chain. Therefore, we refer to the productivity of plants as **primary productivity**.

Plants absorb light energy and use this to drive the reactions of photosynthesis to produce carbohydrates. They then use some of these carbohydrates in respiration to produce the ATP needed to drive other processes in the plant, such as the synthesis of cellulose and proteins.

So which amount of biomass is the primary productivity? They both are! They are two measures of the primary productivity:

- **Gross primary productivity** is all the biomass produced by the plant per m^2 per year.
- **Net primary productivity** is all the biomass produced by the plant per m^2 per year minus that used in respiration. This is the biomass of the plants that is potentially available to primary consumers.

Using the symbols:

G = gross primary productivity
N = net primary productivity
R = respiration

then

$$N = G - R$$

Box 3.2 Examples of net primary productivity (NPP)

In tropical forests and in marshlands, between 1500 grammes and 3000 grammes of organic material are normally produced per square metre per year. Corresponding figures for other ecosystems are:

- temperate forests, 1100–1500 $g\,m^{-2}\,yr^{-1}$
- dry deserts, 200 $g\,m^{-2}\,yr^{-1}$
- estuaries, coral reefs and sugarcane fields, 3600–9100 $g\,m^{-2}\,yr^{-1}$

Box 3.3 Methods of determining gross and net primary productivity

Gross primary productivity is a measure of the rate of photosynthesis; net primary productivity is a measure of the rate of photosynthesis minus the rate of respiration. One way of determining net primary productivity is to measure oxygen production by photosynthesis and oxygen use in respiration.

This can be done for aquatic producers by using the 'light and dark bottle' technique. A sample of the producers (e.g. unicellular algae) is placed in a transparent bottle (to allow photosynthesis) and a similar sample is placed in a blackened bottle (to prevent photosynthesis).

The concentration of oxygen in the water is measured at the start and at the end of a set time period.

- In the light bottle, the results show the amount of oxygen produced in photosynthesis minus that used in respiration.
- In the dark bottle, the results show the amount of oxygen used in respiration.
- By adding the results for the dark bottle to that for the light bottle, we add back the oxygen used in respiration and obtain the true figure for the amount produced in photosynthesis.

This can then be converted mathematically to show the amount of carbon fixed into biomass during the time of the experiment.

The investigation can also be carried out at different depths to see just how much of the water is productive.

Productivity can also be measured by satellites orbiting Earth (Fgure 3.10). By analysing the wavelengths of light reflected from the Earth, estimates can be made of the concentration of chlorophyll a. This can be related to the primary productivity of the area.

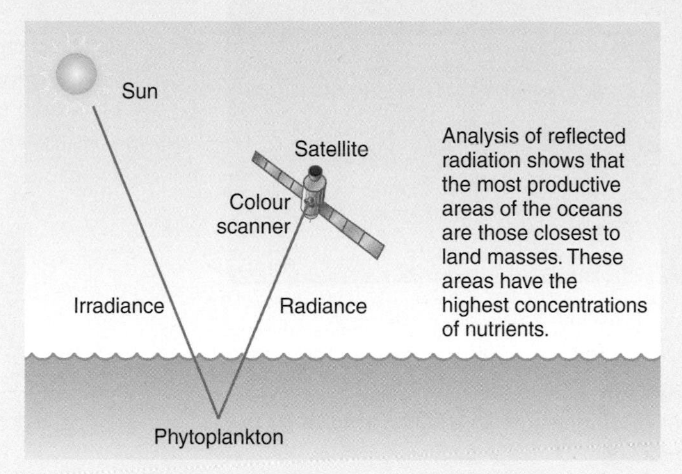

Figure 3.10 Satellites measure reflected radiation, which is then analysed

So far, we have considered only primary productivity. What about secondary productivity?

Secondary productivity is the rate of production of biomass by animals in an ecosystem. As with primary productivity, it is measured in mass units per unit of time (usually a year) per unit of area (usually a square metre). The biomass remaining in the animals is a result of:

- the amount ingested (C)
- the amount used in respiration (R)
- the amount lost in urine (U)
- the amount lost in faeces (F)

So, P (net secondary productivity) $= C - R - U - F$

Both primary productivity and secondary productivity are sometimes expressed as the amount of energy in a trophic level — for example, as kilojoules (or megajoules) per square metre per year.

How do modern farming practices improve productivity?

The aim for the farmer is to get the maximum biomass production for as little input as possible.

In the case of crop plants, the following practices are often adopted:

- monoculture
- the use of organic fertilisers and inorganic fertilisers
- the use of pesticides
- intensive rearing of livestock

Monoculture

In an agricultural context, this means the cultivation of a single crop in a field, usually an extremely large field.

Erica Olsen/FLPA

Harvesting potatoes

The creation of large fields for growing just one crop allows efficient harvesting using specialised machinery.

The main benefits to a farmer of monoculture are:

- reduction of costs by the use of specialised machinery required for arable operations, such as cropping and spraying
- easier control of pests — the number of different pests is often limited by the crop and so specific pesticides can be used
- reduced plant competition for nutrients, space and light — only the crop plant is taking nutrients from the soil and absorbing light energy
- maximised profit from the growing of high gross margin crops

Using fertilisers

In the normal cycle of growth, death and decay of plants, mineral ions removed from the soil for plant growth are returned to the soil when the plant dies. This happens when the plant is decomposed by soil microorganisms.

Harvesting crop plants breaks this cycle and the mineral ions are not returned to the soil, which becomes mineral-poor. If this is not rectified, crops grown in later years will be unable to obtain sufficient nitrates, phosphates and other mineral ions needed for effective growth. The **fertility** of the soil will decrease each year.

Soil fertility is defined as 'that characteristic of soil that allows it to support abundant plant life'.

The use of fertilisers puts back the mineral ions removed by harvesting (Figure 3.11). If an organic fertiliser is used, it must be decomposed in the same way as detritus before the mineral ions are released. It is therefore a **slow-release** fertiliser, releasing only small amounts of ions per kilogram.

Organic fertilisers are materials produced directly from animals, plants and other living organisms. They include materials such as farmyard manure, seaweed, dried blood, sewage sludge and poultry manure.

Inorganic fertilisers do not need to be broken down because they are already in the form of mineral ions. They are **quick-release** fertilisers, yielding large amounts of ions per kilogram.

Because of the different rate of release of the ions, the two types of fertiliser need to be applied at different times:

- To allow time for decay and release of the ions, slow-release organic fertilisers should be applied some time before the main demand from the crop plants.

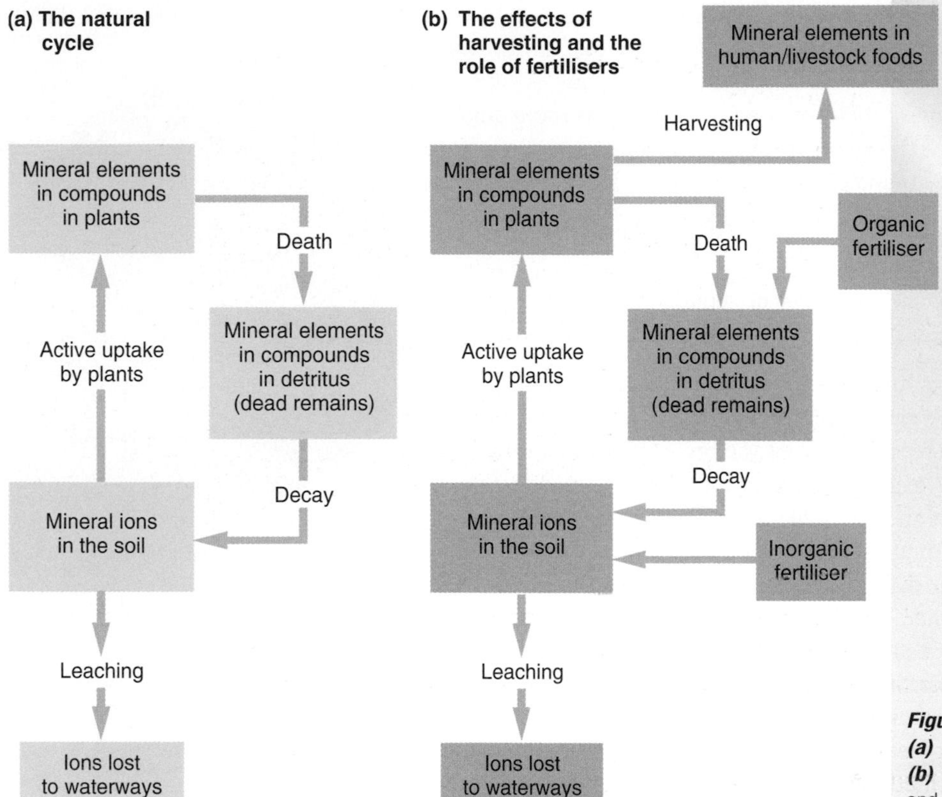

Figure 3.11
(a) The natural cycle
(b) The effects of harvesting and the role of fertilisers

- Quick-release inorganic fertilisers should be applied as the crops approach their peak demand.

Organic fertilisers can improve yields in other ways. The dark colour of organic manure absorbs heat better and helps the soil to warm up faster in spring, aiding germination of seeds. The organic material improves the structure of the soil, giving good drainage and aeration, yet still allowing quite good water retention. This means that there is always adequate oxygen and water in the soil to allow efficient absorption of mineral ions.

Overuse of either type of fertiliser can cause environmental problems, as we shall see in the next chapter.

On the other hand, inorganic fertilisers are much easier to apply (precise amounts can be applied easily to large areas) and to store.

Using pesticides

In agricultural terms, pests are organisms that reduce the productivity of crops. They are not just animals — weeds and some fungi are also pests. Farmers use pesticides to minimise this damage and so maximise productivity.

From the crop growers' point of view, weeds are plants growing in the wrong place. The weeds compete with the crop plants for the available light, water, carbon dioxide and mineral ions. This is interspecific competition because it is competition between members of two different species, i.e. the weed and the crop plant.

Weeds often out-compete crop plants:

- They have higher growth rates than the crops with which they compete.
- They establish their root and shoot systems more quickly.
- They obtain more of the available resources, reducing their availability to the crop plants and so reducing yields.

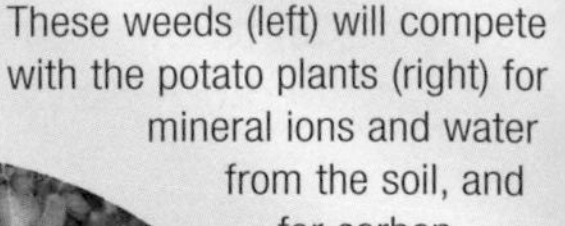
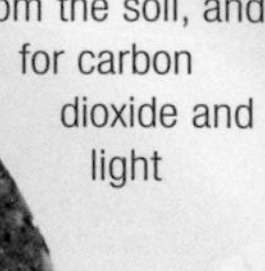

These weeds (left) will compete with the potato plants (right) for mineral ions and water from the soil, and for carbon dioxide and light

The effects of pests can depend on the density of the crop planting. Sometimes there is very little effect at low crop densities, but a much bigger reduction in yield at high crop densities.

Some insects are pests. They can reduce the yield of the crop in a number of ways:

- They can feed directly on the organ of the plant that forms the 'crop' (e.g. the larvae of the carrot fly feeds on the young tap root of the carrot plant and the larvae of the pea moth feed on the peas maturing inside the pea pod).
- They can reduce the yield by feeding on the leaves; this reduces the leaf area and therefore the capacity of the plant for photosynthesis.
- They can feed on and damage the roots, restricting the uptake of mineral ions essential for growth (e.g. the cabbage root fly feeds on cabbage roots, stunting growth).
- They can feed from the phloem and so disrupt the transfer of sugars manufactured in photosynthesis to other organs (aphids reduce crop yields in this way).
- They can spread organisms that cause disease (e.g. the potato aphid can introduce a virus that causes 'leaf-roll'; this reduces the area of leaf exposed to the light and so reduces photosynthesis).

The larvae of the carrot root fly feed on young carrots, leaving tunnels inside

A **pesticide** is a chemical that helps to control the population of a pest. Pesticides can be classified according to the type of organism they control, for example:

- **insecticides** kill insects
- **herbicides** kill plants (they are weedkillers)
- **fungicides** kill fungi
- **molluscicides** kill molluscs (slugs and snails)

Aphids insert their needle-like mouthparts into the phloem of plants and suck out the solution of sugars and amino acids; they can cause considerable damage

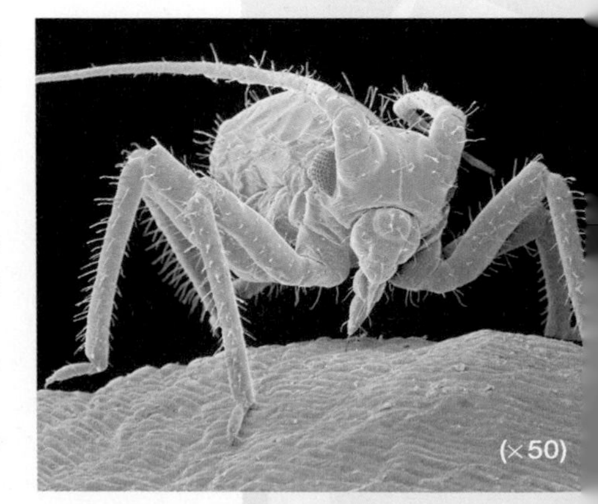

Some pesticides have effects on organisms in the environment other than the pests they are used to control. These effects include the following:

- Insecticides might kill useful insects as well as the targeted harmful insects.
- Pesticides might persist in the environment for many years before they are finally broken down (it can take up to 25 years to break down an application of DDT).
- Herbicides can remain in the soil for long periods, be taken up by crop plants and so enter humans through the food chain.
- Some pesticides (such as DDT) accumulate along food chains (**bioaccumulation** — Figure 3.12).

Sometimes using a chemical pesticide is an ineffective method of pest control because the pests become resistant to it.

Be careful how you describe pesticide resistance in an examination. Some candidates describe it as 'becoming immune' to the pesticide. This is wrong. Immunity results from the exposure of an *individual* to an antigen. This provokes an *immediate* immune reaction and the *individual* becomes immune. Pesticide resistance results from the exposure of a *population* to a pesticide. Natural selection ensures that any resistant types survive to reproduce and pass on their 'resistant' genes. More and more of the *population* become resistant *over a period of time.*

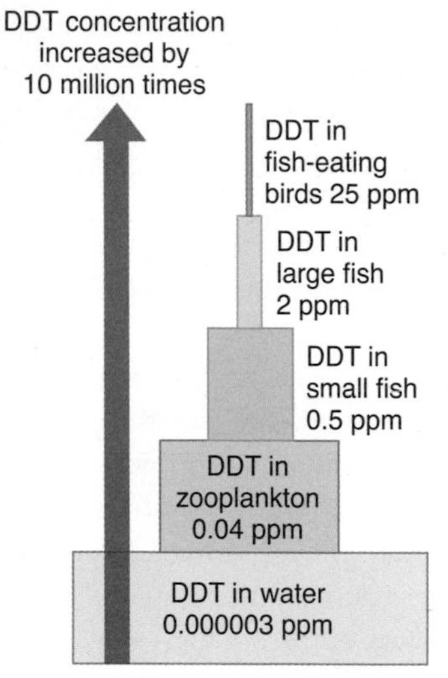

Because DDT is stored in fatty tissue, each organism in the food chain receives the dose stored in all the animals it eats

Figure 3.12 Bioaccumulation

DDT is a pesticide still in use in some countries.

Biological control

A **biological control** method may involve, for example, introducing a natural parasite or predator of the pest into the area with the aim of reducing the pest population. Biological control methods do not aim to eradicate the pest but to reduce pest numbers to a level at which they do not cause major economic damage (Figure 3.13).

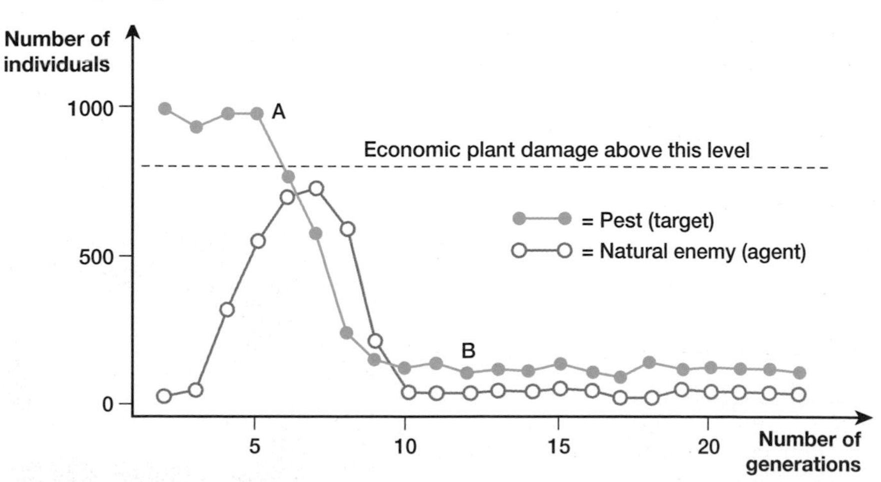

Biological control maintains the numbers of pests at a low level; the pest is not eradicated

Figure 3.13 Biological control

Biological control methods include the following:

- Introducing a **predator** — for example, ladybirds have been introduced to orange groves to control the aphid populations that reduce the yield of oranges.

- Introducing a **herbivore** — for example, a moth native to South America was introduced to Australia to control the 'prickly pear' cactus, which was taking over vast areas of arable land.
- Introducing a **parasite** — for example, larvae of the wasp *Encarsia* parasitise whiteflies that can devastate tomato crops in greenhouses.
- Introducing **sterile males** — this reduces the number of successful matings and so reduces pest numbers.
- Using **pheromones** — these animal sex hormones are used to attract the males or females, which are then destroyed. Reproduction is reduced and numbers decline; male-attracting pheromones are used to control the damson-hop aphid, reducing damage to plum crops.

Claude Nuridsany & Marie Perennou/SPL

Ladybirds are used to control infestations of aphids. Both larvae and adults feed voraciously on aphids.

- Biological control has several advantages over the use of pesticides:
  - Pests do not usually develop resistance to a predator or parasite.
  - Biological control agents are usually much more specific than pesticides; for example, a carefully chosen predator will target only the pest, whereas a pesticide might target all the animals of a particular group (an insecticide might kill many kinds of insect).
  - Once a natural predator or parasite has been introduced, no reintroductions are necessary, whereas pesticides must be re-applied often.

At first glance, it seems that biological control systems cannot fail and should always be used. However, there are a number of potential problems:

- Research is necessary to ensure that the proposed control agent will control only the pest population. It might also target other, similar organisms as well as natural predators of the pest in the area. It must be clear too that any newly introduced control agent will actually survive and reproduce in the new conditions.
- Using biological control to reduce the numbers of one specific pest may allow another pest to fill its ecological niche. This may make matters worse.
- Biological control is not an appropriate method for controlling pests of stored products such as grain. The grain would become contaminated with the dead bodies of pest and control agent alike!

Integrated crop management

Often, neither chemical control nor biological control alone is really effective in controlling pests. Biological control is often enhanced when low levels of pesticides are used at the same time. This is a simple example of an **integrated control system**.

In integrated crop management, most or all of the following would be considered as methods of maximising productivity:

- selecting crops that are adapted to the type of soil and the climate in the area
- selecting crops that have some resistance to known pests in the area
- choosing appropriate methods of pest control
- rotating crops grown in a particular field so that the same pests do not build up in the soil and the same ions are not continually removed by the crop
- using fertilisers (organic, inorganic or a combination) that are appropriate to the conditions, to replace the mineral ions removed by cropping
- appropriate treatment and storage of the final crop (to minimise damage by pests)
- irrigating the soil (where necessary)

By using a combination of these techniques, a farmer can reduce damage by pests and ensure that crops have a continuous supply of mineral ions.

Intensive rearing of livestock

Traditionally, stock animals were allowed to wander where they wished, within the boundaries of fields or pens. More recently, intensive methods of production have become increasingly popular and profitable. These methods of production allow supermarkets to put cheap meat for sale on their shelves.

These intensively reared animals are fed a precisely controlled diet, have their movement restricted and are kept in a warm environment

Scott Sinklier/Agstock USA/SPL

TopFoto

Intensively reared pigs and poultry

The main principles of intensive farming practices include:

- feeding a precisely controlled diet to increase the production of meat with little fat
- using hormone injections or supplements to increase the rate of growth
- restricting the movement of the animals to reduce energy lost this way and increase energy used in growth
- keeping the animals in a warm environment to reduce energy lost as heat to the environment and increase energy used in growth

Box 3.4 Is intensive rearing of livestock ethically acceptable?

Intensive rearing of stock animals allows rapid growth and, by producing many animals in relatively small areas, farmers can have economies of scale and produce meat cheaply.

However, people are becoming increasingly concerned about the way in which meat is produced. Many people believe that we have a 'duty of care' to livestock to:

- rear them in a manner that gives an acceptable level of freedom
- slaughter them as humanely as possible

Jamie Oliver and Hugh Fearnley-Whittingstall have both spoken out about the intensive rearing of poultry. This extract from an article by a member of the animal rights organisation PETA highlights some of the issues.

> '...video footage and photographs taken inside two sheds showed chickens unable to hold up their own body weight, lying in manure and suffering ammonia burns.
>
> Some who were helped to their feet only to topple over, leaned precariously on their beaks before falling to the ground.
>
> Others flapped their wings as they dragged themselves around the floor. Many could be seen with their legs splayed; their young joints unable to take the weight of their huge bodies.
>
> Rotting carcasses of dead birds are trampled and pecked at by other chickens. Small birds, unable to reach food and water troughs, slowly starve to death.
>
> The growers argue that intensive farming methods are the result of intense pressure from processors and consumers to grow cheaper chicken meat. Free-range chicken meat (meat from chickens allowed to roam free and feed at will) is more expensive to produce.
>
> Farmers can legally hold up to 13 chickens per square metre, with some sheds able to hold up to 40 000 birds.
>
> It used to take the industry 20 weeks to grow a chicken but now producers can grow a bird in 6 weeks. Growers concede fast growth meant there was a risk chickens might not be able to hold their own weight...'

◀ The ammonia is formed from the uric acid excreted by birds. The manure is effectively the birds' 'faeces'.

Summary

Energy, food chain, food webs and pyramids

- Ecosystems sustain themselves using energy, which flows through the ecosystem, and nutrients, which are continually recycled.
- Energy cannot be created or destroyed, it can only be converted from one form to another.
- When energy is converted from one form to another, the amount of useful energy always decreases.

- Energy is trapped by producers in photosynthesis. It is transferred to consumers by feeding and assimilation, and to decomposers by decay of dead bodies; energy is lost as heat as a result of respiration.
- Primary consumers feed on producers, secondary consumers feed on primary consumers and decomposers feed from the dead remains of other organisms.
- Food chains rarely exist in isolation; they are linked to other food chains in a food web.
- Food chains can be represented by ecological pyramids, including:
  - a pyramid of numbers, representing the numbers of the organisms in a food chain, at a given moment, irrespective of biomass
  - a pyramid of biomass, representing the biomass of the organisms in a food chain, at a given moment, irrespective of numbers
  - a pyramid of energy, representing the amount of energy transferred to each level of a food chain, irrespective of numbers and biomass, in a given period of time

Productivity

- Productivity is a measure of how efficient organisms are at locking up energy in organic molecules in their body cells.
- Gross primary productivity is the all the biomass produced by a plant per square metre per year.
- Net primary productivity is all the biomass produced (or energy 'trapped') by the plant per square metre per year, minus that used in respiration.
- Secondary productivity is the rate of production of biomass (or gain of energy) by animals in an ecosystem.

net secondary productivity = biomass ingested – (biomass used in respiration) – (biomass lost in faeces) – (biomass lost in urine)

or

net secondary productivity = energy ingested – (energy lost in respiration) – (energy lost in faeces) – (energy lost in urine)

- Productivity in agriculture is increased by:
  - monoculture — this allows cultivation of large areas of just one crop with the use of specialised machinery, specific pesticides and fertilisers
  - the use of organic and inorganic fertilisers to maintain soil fertility and maximise crop yield
  - the use of pesticides, biological control and integrated management systems to minimise the damage to crops caused by pests and so improve yields
  - the intensive rearing of livestock, which allows efficient feeding and minimises energy losses so that more energy is used in the production of meat
- There are serious ethical issues concerning the welfare of animals reared intensively; and many people argue that it is inhumane and that we have a duty of care to our livestock.

Questions

Multiple-choice

1 In the transfer of energy:

A the total amount of energy remains the same and there is still the same amount of useful energy

B the total amount of energy decreases and there is still the same amount of useful energy

C the total amount of energy decreases and the amount of useful energy decreases

D the total amount of energy remains the same and the amount of useful energy decreases

2 To maintain an ecosystem:

A energy and nutrients flow through the system

B energy is recycled and nutrients flow through the system

C nutrient are recycled and energy flows through the system

D energy and nutrients are recycled

3 Most food chains do not have more than five trophic levels because:

A the animals at the end of the food chain are top carnivores

B there is not enough energy in the system to support another trophic level

C the plants contain too much cellulose

D parts of the animals are not eaten

4 Net primary productivity is:

A gross primary productivity minus losses due to respiration

B gross primary productivity minus losses due to evaporation

C gross primary productivity minus losses due to growth

D gross primary productivity minus losses due to excretion

5 A complex food web is more stable than a simple food web because:

A there are many interdependent links

B there are many alternative food sources for organisms

C there are many alternative habitats for organisms

D all of the above

6 The diagrams show two ecological pyramids.

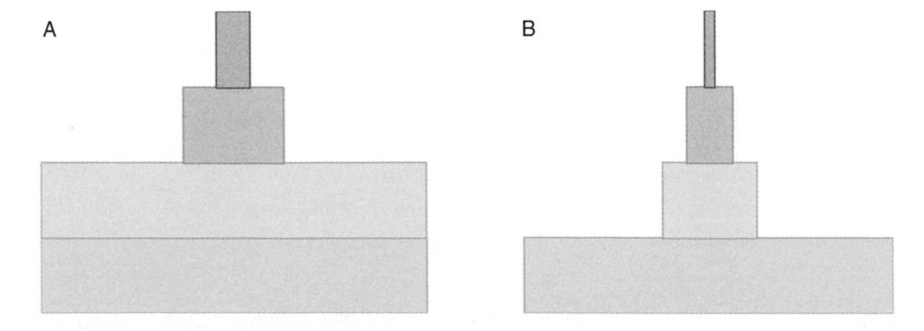

Identify which of the following statements *could* be true:

A pyramid A could be a pyramid of numbers and pyramid B could be a pyramid of biomass.

B pyramid B could be a pyramid of numbers and pyramid A could be a pyramid of biomass.
C they could both be pyramids of biomass
D all of the above

7 Intensive rearing of poultry improves productivity because:
A less energy is lost in movement
B diet is carefully controlled
C less energy is lost in maintaining body temperature
D all of the above

8 In integrated systems of crop management:
A pesticides are never used
B only organic fertilisers are used
C some account is taken of the suitability of crops to the soil type in the area
D crop rotation is rarely practised

9 All the energy entering the last trophic level of a food chain is:
A used in growth
B used in growth or lost in respiration
C used in growth, lost in respiration or lost in excretory products
D used in growth, lost in respiration, lost in excretory products or lost to decomposers

10 Organic fertilisers:
A are derived from dead materials
B are slow-release fertilisers
C improve the structure of the soil
D all of the above

Examination-style

1 Organisms that reduce the yield of a crop plant are called pests. They can be controlled using pesticides, by biological control or by integrated crop management. The diagram shows the effects of repeated pesticide applications on a population of pests.

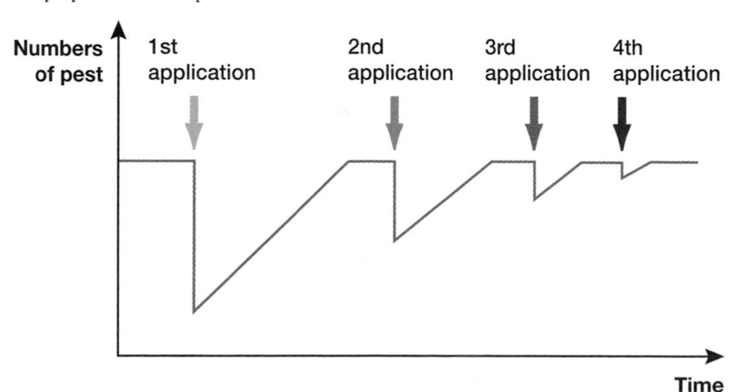

(a) (i) Suggest **two** reasons why the first application of the pesticide does not reduce the pest population to zero. *(2 marks)*

(ii) Explain the reduction in effect of the pesticide at the second, third and fourth applications. *(3 marks)*

(b) Give one benefit of each of the following in an integrated crop management system:

(i) crop rotation (not growing the same crop in the same field in successive years)

(ii) using organic fertilisers (such as farmyard manure) rather than inorganic fertilisers

(iii) planting crops that are tolerant of the local soil pH conditions *(3 marks)*

Total: 8 marks

2 The diagram shows a food web in decaying plant matter. The C:N ratio is the ratio by mass of carbon to nitrogen in the organisms.

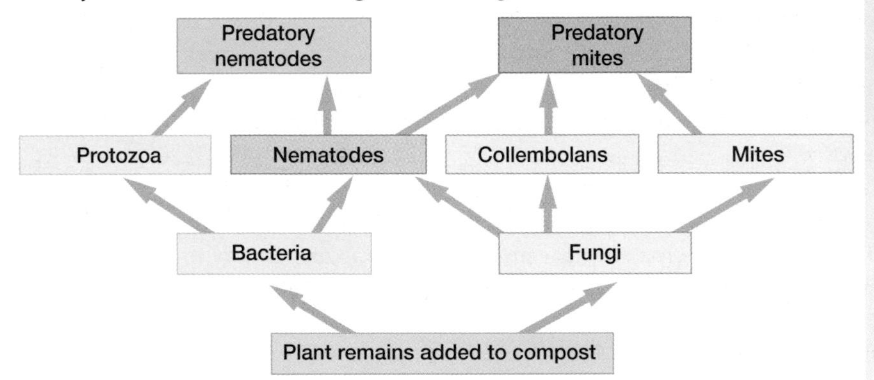

(a) For the food chain:

plant remains → bacteria → nematodes → predatory nematodes

construct:

(i) a pyramid of numbers *(1 mark)*

(ii) a pyramid of biomass *(1 mark)*

(b) Suggest a value for the C:N ratio in the predatory nematodes. Give a *biological* explanation for your answer. *(3 marks)*

(c) Explain why there are no more trophic levels in this food web. *(3 marks)*

Total: 8 marks

3 The table shows what percentage of light falling on plants in an area of coastland marsh is used for various purposes:

Photosynthesis	2.2
Reflection	3.0
Evaporation of water	94.8
Total	**100.0**

It is estimated that about 50% of the energy trapped in photosynthesis is lost in respiration. It is also estimated that the net primary productivity of coastland marsh is about 50 000 kJ m^{-2} y^{-1}.

(a) (i) Explain what is meant by the term net primary productivity. *(2 marks)*

(ii) Calculate the total amount of energy shining on 1 m^2 of this area of marshland per year. Show your working. *(3 marks)*

(b) Explain why some areas can only support ecosystems in which the food chains have few trophic levels. *(3 marks)*

Total: 8 marks

4 The concentration of the insecticide DDT in different organisms in a food chain is shown below.

Algae → Small fish → Large fish → Heron

0.04 ppm 0.8 ppm 1.5 ppm 40 ppm (ppm = parts per million)

(a) (i) What is an insecticide? *(1 mark)*

(ii) Explain why the concentration of DDT increases along the food chain. *(3 marks)*

(iii) Describe and explain one other danger of using insecticide. *(2 marks)*

(b) DDT has been banned in many countries because it is not broken down easily in the soil. The following passage is an extract from a research article into the breakdown of DDT:

DDT labelled with radioactive carbon was added to soil, and the mixture was incubated anaerobically for 2 weeks and 4 weeks. DDT and seven possible decomposition products were separated by chromatography, and the radioactivity of material from individual spots was determined. The DDT was dechlorinated by soil microorganisms to DDD, and only traces of other degradation products were detected. No degradation of DDT was detected in sterile soil.

(i) Suggest why the experiment was carried out with ordinary soil and with sterile soil. *(2 marks)*

(ii) What do the results suggest? *(2 marks)*

(iii) Suggest why DDT was broken down in the investigation but is not naturally broken down easily in the soil. *(2 marks)*

Total 12: marks

5 The diagram shows the flow of energy through an ecosystem at Silver Springs in Florida.

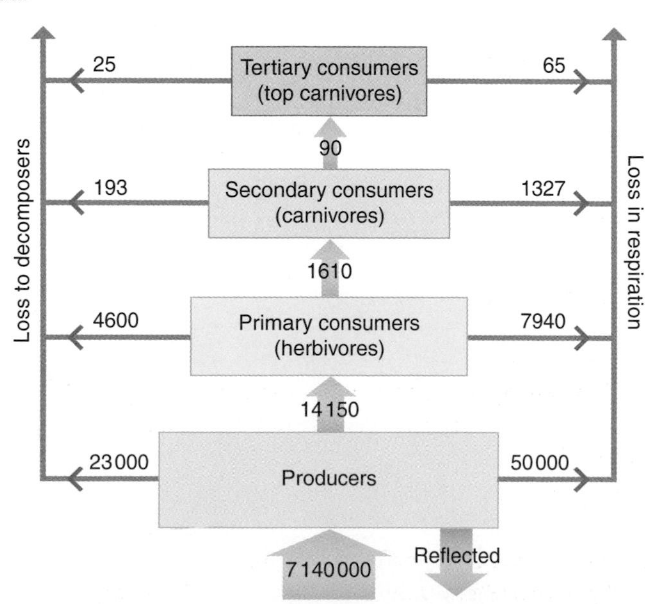

All figures on the diagram are in kilojoules per square metre per year ($kJ\,m^{-2}\,y^{-1}$).

(a) Showing all working, use the data to calculate:

(i) the amount of energy reflected from the plants

(ii) the percentage efficiency of energy transfer between secondary and tertiary consumers *(4 marks)*

(b) (i) The efficiency of the energy transfer between producer and primary consumer is 16.2% while between primary consumer and secondary consumer it is 11.4%. Suggest an explanation for this difference. *(4 marks)*

(ii) Salmon and trout are fish that can be 'farmed'. They are bred in relatively small ponds where they are fed a carefully controlled diet. The energy transfer from their food can be over 30%. Suggest and explain three reasons why their productivity is so high. *(6 marks)*

Total: 14 marks

Chapter 4
How are elements cycled through ecosystems?

This chapter covers:

- the importance of microorganisms in cycling nutrients through ecosystems
- the main stages of the carbon cycle
- the main stages of the nitrogen cycle
- short-term and long-term fluctuations in the concentration of carbon dioxide in the atmosphere
- the roles of carbon dioxide and methane in the greenhouse effect
- the possible causes and effects of global warming
- environmental issues resulting from the overuse of fertilisers
- how ecosystems evolve from simple to more complex in succession

Chapter 3 focused on the flow of energy *through* ecosystems, noting how energy is lost continuously and has to be replaced. Nutrients, however, are *recycled* within ecosystems.

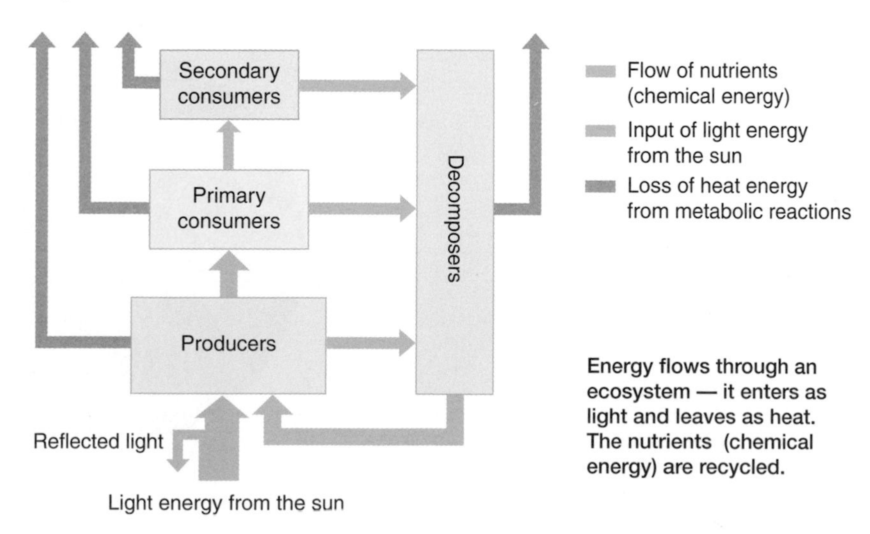

Figure 4.1 Cycling of nutrients

There is a finite amount of each nutrient in an ecosystem and the same atoms are reused repeatedly. This happens at the ecosystem level and also globally. It is possible that some of the carbon atoms in the carbon dioxide from Julius Caesar's last breath as he was murdered might have ended up in the potato you ate the other day and are now part of you.

A similar point is made in the song 'On Ilkley moor baht 'at' in which key lines in successive verses read:

Ilkley is a town in Yorkshire.

'...where have you been since I last saw you?

...you're bound to catch your death of cold

...then we shall have to bury thee

...then worms will come and eat thee up

...then ducks will come and eat up worms

...then we shall come and eat up ducks

...then we shall all have eaten thee — on Ilkley moor baht 'at!'

Most of the words have been translated from the dialect; 'baht 'at' means 'without hat'!

The song is fairly accurate, but it misses out one key biological stage — the role of the **decomposers**. It is these organisms that begin the release of mineral ions from complex organic molecules into the soil in a form that can be used by plants.

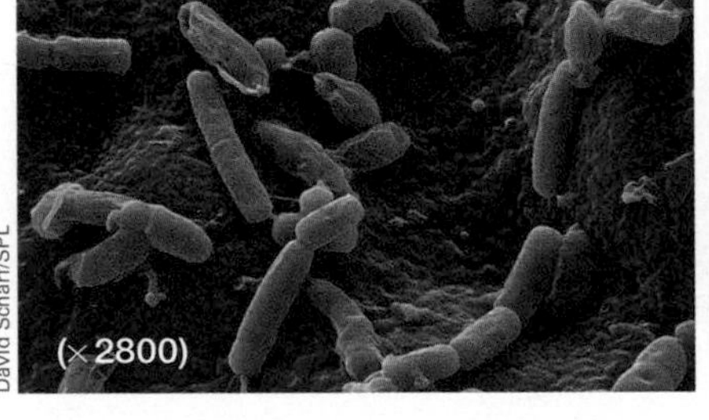

Bacteria (left) and fungi (right). Many bacteria and fungi feed on dead matter, breaking it down to release mineral ions into the soil; they are the decomposers.

Decomposers feed by a method known as **saprobiotic nutrition**. They feed on dead matter — and so do you. You digest the parts of dead animals and plants that you eat and so must the decomposers. To do this, they secrete enzymes onto the dead matter and absorb the products of digestion (just like you). But, unlike you, their digestion does not take place in a gut. It is **extracellular** and takes place in the soil, or wherever the dead matter happens to be (Figure 4.2).

The hyphae of a saprobiotic fungus

Credit: http://en.wikipedia.org/wiki/File:Hyphae.JPG

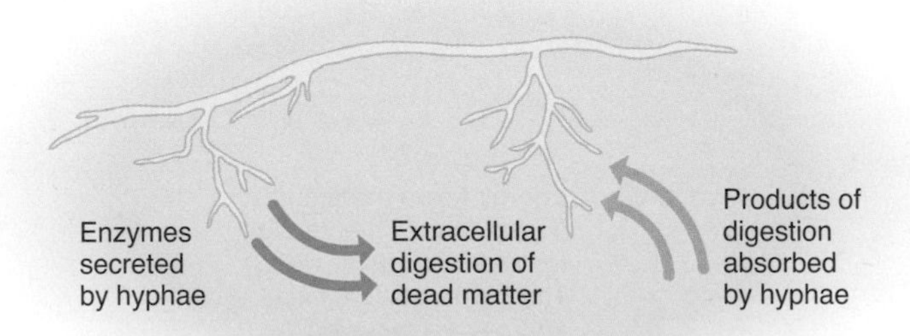

Figure 4.2 Decomposition by a fungus

Decomposers secrete enzymes to digest the organic molecules in the dead matter and then absorb the products of digestion into their cells.

How is carbon recycled?

Carbon is a component of all the major biological molecules: carbohydrates, lipids, proteins, nucleic acids and many others. It is released into the atmosphere as carbon dioxide when any organic molecule is respired or burned. It is removed from the atmosphere in photosynthesis.

The main processes involved in cycling carbon through ecosystems are:

- photosynthesis — the process that fixes carbon atoms from an inorganic source (carbon dioxide) into organic compounds (e.g. glucose)
- feeding and assimilation — feeding passes carbon atoms already in complex molecules to the next trophic level in the food chain where they are assimilated into (become part of) the body of that organism
- respiration — this releases inorganic carbon dioxide from organic compounds
- fossilisation — sometimes living things do not decay fully when they die due to the conditions in the soil, and fossil fuels (e.g. coal, oil and peat) are formed
- combustion — fossil fuels are burned, releasing carbon dioxide into the atmosphere

The carbon cycle is summarised in Figure 4.3.

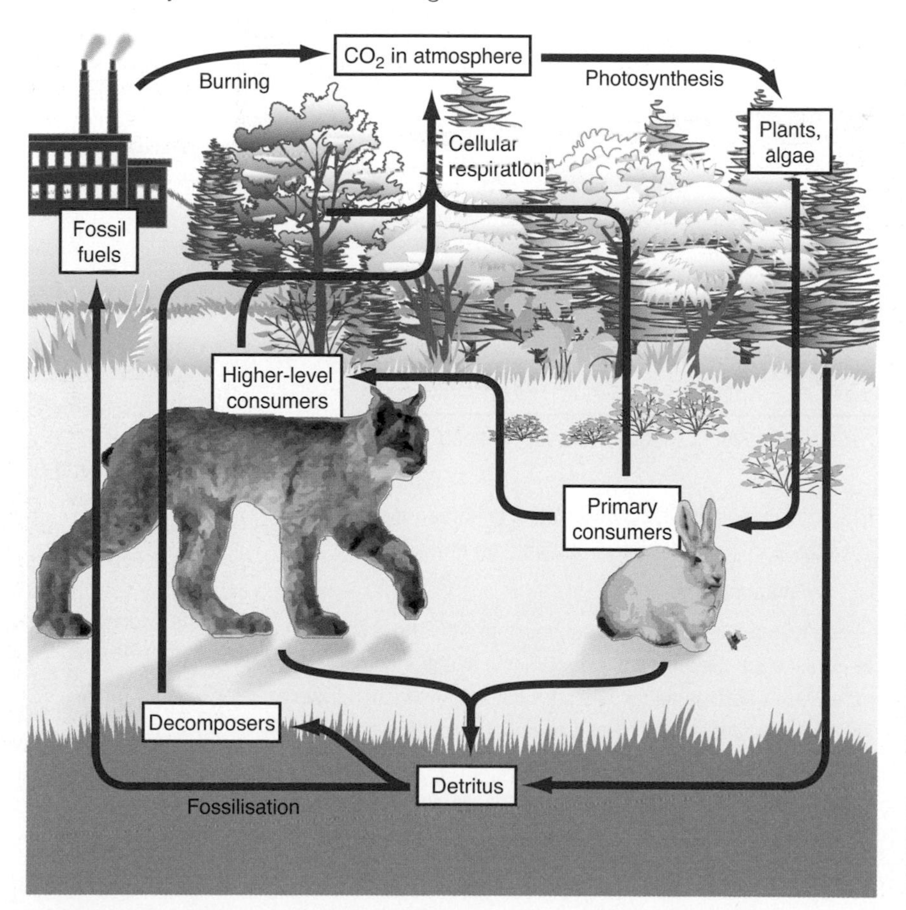

Figure 4.3 The carbon cycle

Be clear in your mind that the only way that carbon present in decomposers in the soil finds its way back into plants is through respiration of the decomposers releasing carbon dioxide into the *atmosphere*, from where plants absorb it and use it in photosynthesis. It is *not* magically absorbed by plant roots!

Box 4.1 Do increasing levels of carbon dioxide in the atmosphere contribute to global warming?

This is a difficult area to be definite about. However, there are some generally undisputed facts.

Fact 1 The level of carbon dioxide in the atmosphere is increasing (Figures 4.4 and 4.5).

The carbon dioxide concentration was fairly stable from 1000 AD to about 1850 AD, when it began to increase dramatically.

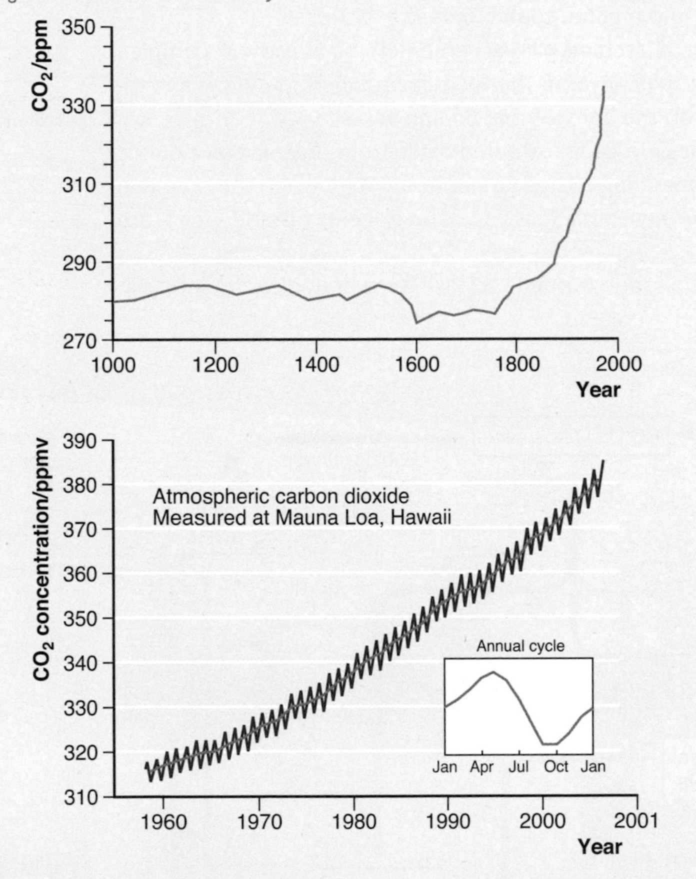

Figure 4.4 Carbon dioxide concentration in the atmosphere over the last 1000 years as measured from ice-core samples

Figure 4.5 Carbon dioxide concentration in the atmosphere over the past 50 years as measured at Mauna Loa using specialised sampling equipment

In winter, the cooler temperatures, shorter day length and loss of leaves by many plants reduce the amount of photosynthesis taking place. Less carbon dioxide is absorbed from the atmosphere. Respiration may also be reduced, but it produces more carbon dioxide than is used in photosynthesis. The level of carbon dioxide rises. In summer, the balance is reversed and the level falls.

This then seems to be true. The increase in the concentration of carbon dioxide seems to be correlated with the increase in industrialisation and burning of fossil fuels.

Fact 2 Carbon dioxide is a **greenhouse gas**.

Carbon dioxide, methane and water vapour are all greenhouse gases. These gases (and some others) combine to produce the **greenhouse effect**. Some of the incoming solar radiation is reflected by the Earth's surface and would pass back into space, if it were not for the greenhouse gases absorbing some of it and re-radiating it towards the Earth's surface (Figure 4.6).

Greenhouse gases prevent excessive loss of reflected heat to space. Without the greenhouse effect, the Earth would be about 33°C cooler. The UK would have a mean summer temperature of –13°C!

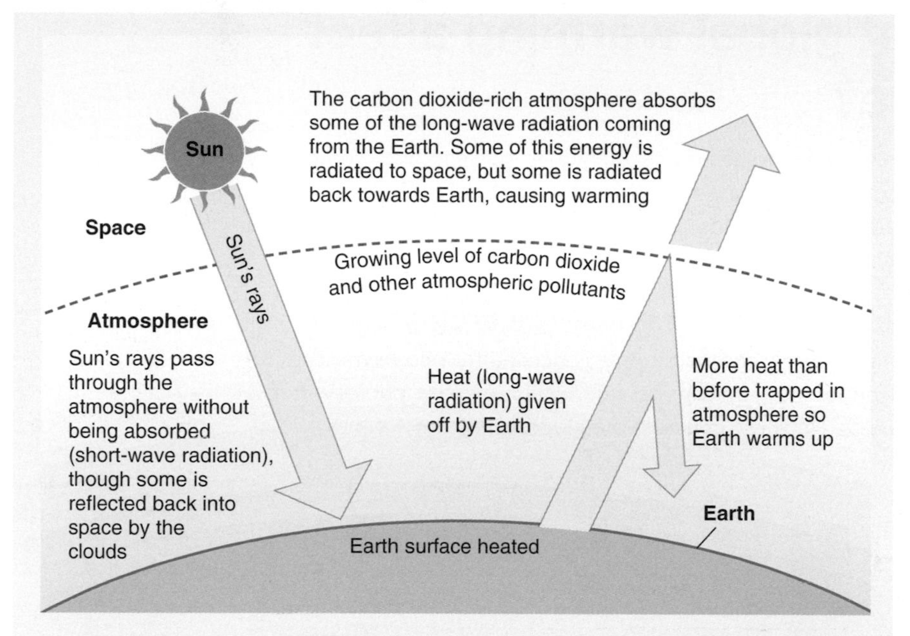

Figure 4.6 The greenhouse effect

There is considerable experimental evidence to show that the greenhouse gases absorb certain wavelengths of radiation and re-radiate them. This seems to be solid ground.

Fact 3 Increased carbon dioxide concentrations are causing **global warming**.

Here, the evidence is not so definite. You should review some of the mass of evidence that is available and at least be aware that the evidence is sometimes unclear.

Many computer simulations predict that by 2080 the temperature in the UK will have risen by between 2°C and 4°C. This is largely on the basis of predictions of the increase in carbon dioxide emissions. This could have an impact on our climate and, as a result, on agriculture. There is no doubt that the temperature of the Earth has increased over the past 30 years, but are we sure that this is due to increased carbon dioxide concentrations? Many scientists believe that this *is* the case and quote evidence like that given above to support their argument.

If the increased concentration of carbon dioxide causes global warming then there should be a direct correlation between the two over a longer period of time. However, although the concentration of carbon dioxide has continued to increase over the past 50 years, the atmospheric temperature has stabilised.

In addition, over the last 500 000 years, there have been periods of:

- high concentrations of carbon dioxide linked with low temperatures
- stable carbon dioxide concentrations linked with temperatures rising and falling
- the current increase in carbon dioxide concentration linked with global warming

So what should we do? Scientists persuaded 37 industrialised nations to commit to reduce the emissions of greenhouse gases in the Kyoto protocol, signed in Japan in 1997. The only major industrialised nation not to commit was the USA.

Is this commitment to reduce greenhouse gas emissions necessary? You should consider what might happen if we do nothing and the climatologists are proved correct. There would be considerable global warming and, by then, it might be too late to do anything.

◀ There is also considerable evidence from both ice-core samples and direct measurements that the concentration of methane in the atmosphere has increased in a similar way to that of carbon dioxide. However, although the global *production* of methane after 2000 continued to increase, the *concentration in the atmosphere* has remained the same. No-one can quite explain this.

◀ There is evidence to suggest that:

- 2000 years ago, it was warm enough in the UK for the Romans to be able to grow grapes near Hadrian's wall in southern Scotland
- during two of the great ice ages, the concentration of carbon dioxide in the atmosphere was much higher than it is today

How might global warming affect agriculture?

As is typical of this whole issue, the evidence is unclear and sometimes conflicting. It is also a complex issue. It is not just a matter of how the increased temperature will affect crop yields, but also of increased carbon dioxide concentration and changed rainfall patterns.

How might global warming influence crop yields?

You might expect that, because carbon dioxide is needed for photosynthesis, an increase in the concentration of carbon dioxide would increase photosynthesis and increase crop yields. However, consider the graphs in Figure 4.7.

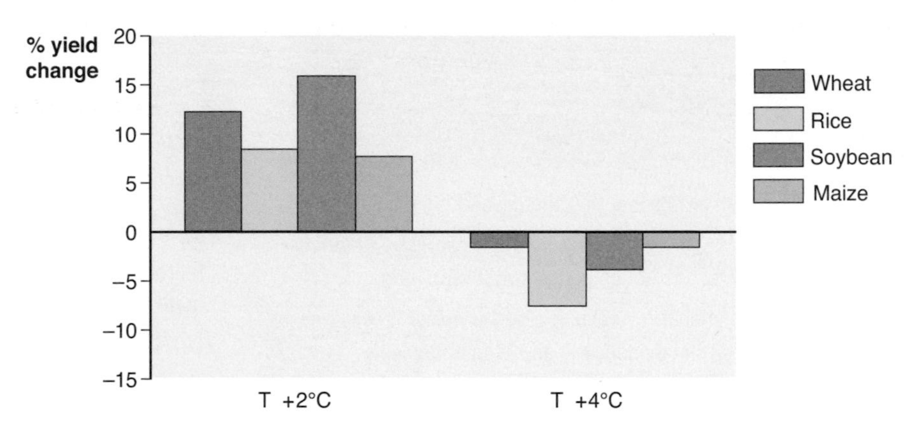

Figure 4.7 The effect of temperature on crop yields with an increased carbon dioxide concentration

Figure 4.8 shows that an increase in temperature of 2°C increases the yield of some major cereal crops, whereas an increase of 4°C decreases the yield. However, it is not that simple. With a temperature increase of 2°C, the yield is increased by 10–15% but the *quality* of the crop is poorer — there is up to 25% less protein in the grain.

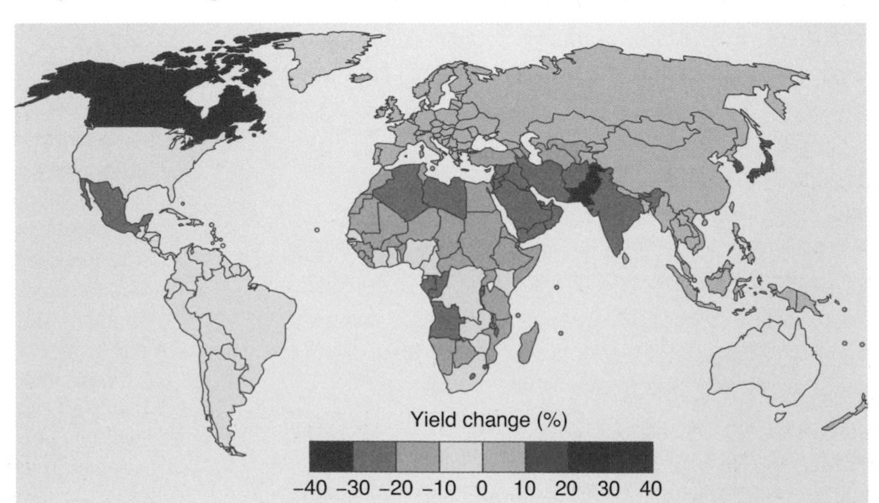

Figure 4.8 A computer prediction of how crop yields will change by 2080

The situation is far from clear, but several research groups have produced predictions similar to that shown in Figure 4.8.

Crop yields in the UK overall are predicted to remain unchanged or to increase slightly. Scotland may experience significant increases in crop yield.

How might global warming influence the distribution and life cycles of insect pests?

As areas become warmer, pests not previously associated with an area may migrate to it. This could include:

- insects that cause damage to crop plants and stock animals
- insects that affect humans more directly

The *Anopheles* mosquito is the vector for the malarial parasite and is currently confined to tropical and sub-tropical areas. Figure 4.9 shows its current distribution and its likely distribution by 2050 as a result of a moderate temperature rise.

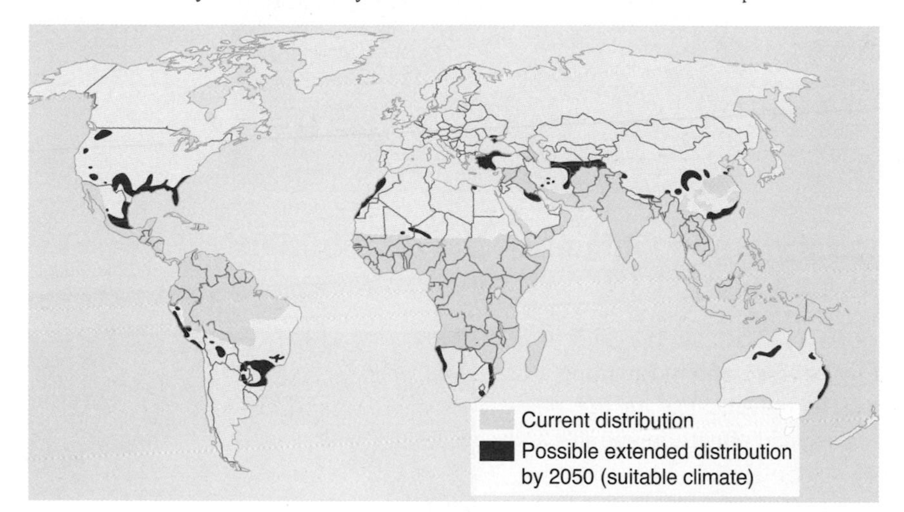

Malaria may become more widespread with global warming as the *Anopheles* mosquito extends its range of habitats

Figure 4.9 Distribution of malaria

An increase in temperature might shorten the life cycles of insects, making more life cycles possible in an extended breeding season. This would mean more damage to crops and stock. Figure 4.10 shows how, in blowflies, increased temperatures shorten the period from an egg hatching to the emergence of an adult from the final larval stage.

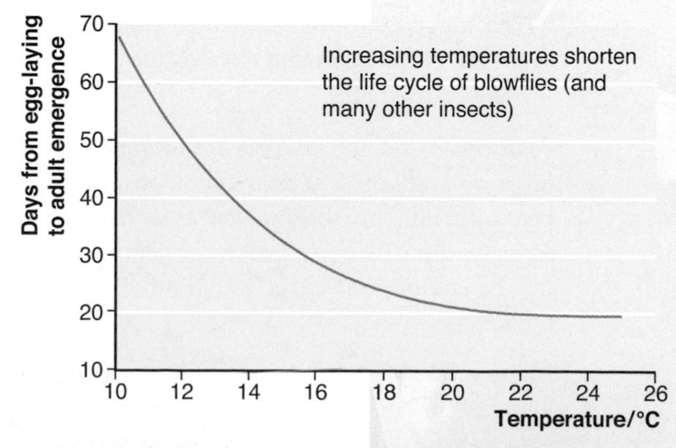

Figure 4.10 Effect of temperature on the life-cycle of the blowfly

In addition, there is fossil evidence from previous warm periods in the Earth's history showing that leaf-eating insects became much more common. There is, therefore, the possibility that this could happen again with leaf-eating insects such as locusts becoming much more common and causing much more damage than they already do.

Box 4.2 Just how much damage do locusts cause?

A single locust can eat its own body mass of 'leaf' each day — this amounts to about 2 g. This does not sound too threatening until you realise that a swarm of locusts may cover 1200 km^2, with each square kilometre containing between 40 million and 80 million locusts. Such a swarm could consume 150 million kg (150 000 tonnes). of 'leaf' per day. We don't need any more locusts!

Adult locusts feeding

TopFoto

How might global warming influence the distribution of wild animals and plants?

Most computer models predict that the effects of global warming will be felt first, and most strongly, at the poles rather than at the equator. As the poles warm, there will be two broad effects on animals and plants:

- The habitat of those adapted to the cold, arctic conditions will be reduced and their numbers will shrink.
- The range of those adapted to warmer conditions will be either extended towards the poles or shifted towards the poles.

The extent of the sea ice at the North Pole shrank considerably between 1979 and 2003

Thomas Nilsen/SPL

The polar bear's sea-ice habitat is shrinking at an alarming rate

Polar bears and spruce trees are examples of organisms adapted to cold conditions. Polar bears depend on the sea ice as a habitat. Already, there is considerable evidence that the extent of sea ice at the North Pole is reducing.

At the moment, the population of polar bears appears to be stable, but it cannot be expected to remain so if the area of sea ice continues to shrink.

Recent measurements of white spruce trees in Alaska show that their growth has been significantly stunted. This is expected to get worse as the climate continues to get warmer.

Studies in the USA and the UK have shown that many migratory birds are appearing earlier in the year than has been normal and they are breeding earlier and staying longer before migrating. Many flowering plants in these countries are blooming earlier and producing seeds earlier than in previous years.

None of this sounds too drastic, but what happens when the distribution of species shifts, rather than just being extended? This is easier for animals than for plants. In the USA, another species of spruce tree is parasitised by budworm caterpillars. At the moment, four different species of bird feed on the caterpillars and keep their population in check. However, there is evidence that the bird population is shifting northwards, away from the region where the spruce trees grow. What will then keep the caterpillars in check?

How is nitrogen recycled?

Nitrogen is found in many biological compounds. It is present in proteins, amino acids, DNA, RNA, adenosine triphosphate (ATP), ADP and some vitamins.

So, without nitrogen organisms could not synthesise:

- genetic material (DNA)
- principal structural materials (proteins)
- the energy transfer molecule (ATP)

Nitrogen is, therefore, a key element in living organisms. Like carbon, nitrogen is continuously recycled. The nitrogen cycle is summarised in Figure 4.11.

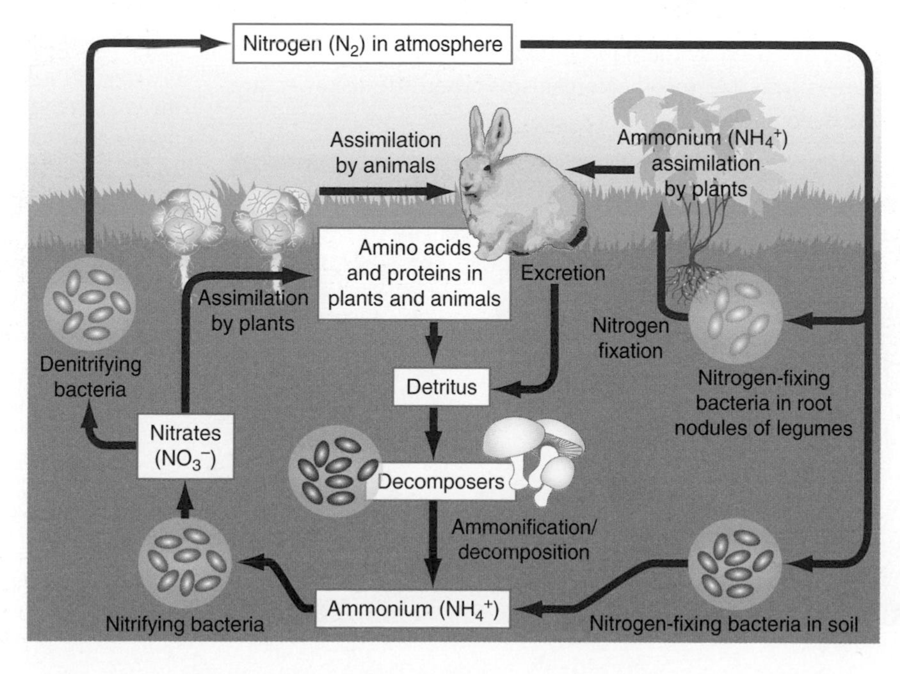

Figure 4.11 The main stages in the nitrogen cycle

The main processes in the cycle are as follows:

- Plants absorb nitrates from the soil.
- The nitrates are then used to form amino acids, which are used to synthesise proteins.
- The plants are eaten by animals, the proteins digested and the amino acids produced are absorbed and assimilated into animal proteins.
- Both plants and animals die, leaving a collection of dead materials (detritus) that contains the nitrogen fixed in organic molecules. In addition, excretory products such as urea also contain nitrogen.
- Decomposers decay the excretory products and detritus, releasing ammonium ions (NH_4^+) into the soil. This process is often referred to as **ammonification.**
- **Nitrifying bacteria** oxidise the ammonium ions to nitrites (NO_2^-) and then to nitrates (NO_3^-), which are taken up by the plants. The conversion of ammonium ions to nitrates is called **nitrification**.

During ammonification, decomposers secrete enzymes that digest proteins into amino acids. These are then absorbed. In reactions in the cells of the decomposers, the amino group of the amino acids is split off from the rest of the molecule to produce ammonium ions.

The conversion of ammonium ions to nitrate ions is oxidation because, in the process, oxygen is gained and hydrogen ions are lost.

These processes keep the nitrogen 'cycling'. However, other processes add to or reduce the total amount of nitrogen that is being cycled through living things.

Denitrifying bacteria reduce nitrate to nitrogen gas, which escapes from the soil. This decreases the total amount of nitrogen available to the plants and, therefore, to all other organisms.

Nitrogen-fixing bacteria 'fix' nitrogen gas into ammonium ions. This happens in two main situations:

- Nitrogen-fixing bacteria living free in the soil reduce nitrogen gas into ammonium ions in the soil. These ions are oxidised into nitrates by nitrifying bacteria, adding to the amount of nitrogen available to the plants and, therefore, to other organisms.
- Nitrogen-fixing bacteria in nodules on the roots of legumes (plants with 'pods', for example peas, beans, lentils and clover) form ammonium ions that are passed to the legumes and used by them to synthesise amino acids. The extra nitrogen only becomes available to other organisms when the legumes are eaten or die and are decomposed.

Legumes are often used in crop rotation systems by organic farmers. They can be grown as a crop (e.g. peas or beans) that is harvested, and then the remains of the plants are ploughed into the soil. Alternatively, a non-crop legume such as clover can be grown and ploughed in at the end of the year. In this way, all of the nitrogen fixed is added to the soil; none is lost in a crop.

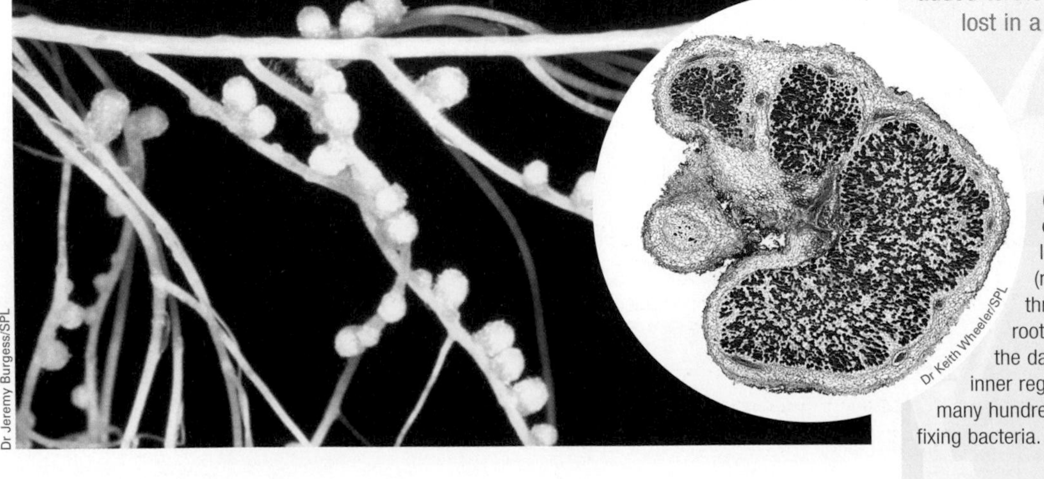

(left) Root nodules on the roots of a legume.
(right) Section through a single root nodule. Each of the darker cells in the inner region contains many hundreds of nitrogen-fixing bacteria.

What can go wrong when using too much nitrogenous fertiliser?

Fertilisers are used to counteract losses of mineral ions. In a natural ecosystem, all the carbon, nitrogen and mineral elements would be recycled. In a farmed ecosystem, some of the minerals are lost with the harvested crops and when stock animals are sent to market.

When organic fertilisers are used, they decompose slowly and release mineral ions over a period of time. There is usually little problem with the use of such fertilisers. However, when inorganic fertilisers are used, the mineral ions are added directly to the soil and are available instantly. Because these ions are soluble (particularly nitrates and phosphates), they can be carried into nearby waterways. This is called **leaching**.

If too many ions are leached, the following can happen:

- Algae in the waterway multiply rapidly because increased synthesis of proteins and nucleic acids is possible.
- The increased algal growth forms a mat over the surface (if the algae are filamentous) or an algal bloom (if the algae are unicellular).
- The algal mat or bloom reduces the transmission of light to lower levels in the waterway.
- Plants growing at these levels cannot photosynthesise and die.
- Algae start to die as the mineral ions are used up.
- Microorganisms decompose the dead plants and algae and reproduce rapidly as the amounts of dead algae and plants increase.
- The microorganisms use up oxygen, in aerobic respiration, in increasing amounts as their numbers increase.
- The level of oxygen in the water falls dramatically and many animals die.

Robert Brook/SPL

The increased concentration of nitrate ions has caused the filamentous algae to reproduce rapidly and form a floating mat

This whole process is known as **eutrophication** (Figure 4.12).

Eutrophication is *more* likely to occur in hot weather because:

- the mineral ions become more concentrated as a result of increased evaporation of water
- metabolic processes are speeded up because of increased enzyme activity

Eutrophication is *less* likely to occur in moving water than still water because:

- the mineral ions are diluted rapidly
- the water is being re-oxygenated continuously

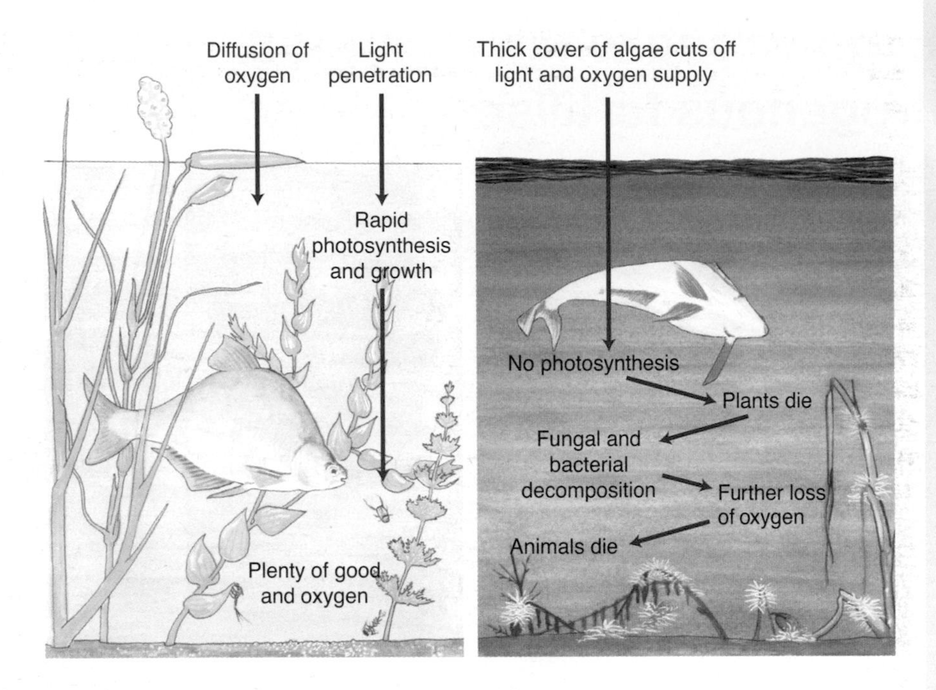

Figure 4.12 Effects of eutrophication

Although organic fertilisers are less likely to cause eutrophication, **organic pollution** of waterways can still occur. If sewage or any other organic matter enters a waterway, the following can happen:

- Microorganisms start to decompose the organic matter and reproduce rapidly.
- The microorganisms use up oxygen, in aerobic respiration, in increasing amounts as their numbers increase
- The level of oxygen in the water falls and many animals die.

As with eutrophication, the consequences of organic pollution are more serious in hot weather and in still water.

Why do ecosystems change over time?

The ecosystems that exist today did not always exist. They have developed from other previous systems by **succession**. Bare rock does not remain bare for long. Soon, lichens can be seen growing on the surface of the rock. These extremely resilient organisms are able to colonise harsh environments and reproduce there. They are **pioneer species**.

Bare rock is a harsh environment that few organisms can inhabit. Lichens colonise the rock, dissolving some of the mineral ions and crumbling the surface of the rock into a primitive soil.

Through the natural recycling processes already discussed, the presence of the lichens changes the conditions, making them less harsh. The living lichens grow into the rock causing it to crumble. When the lichens die, decomposers act on their remains and release mineral elements into the crumbled rock. The mixture of dead remains, crumbled rock and mineral ions forms a primitive soil in which mosses can grow. Spores of mosses that land there can now 'germinate' and the mosses grow, outcompeting the lichens in the changed environment.

As lichens die, they are decomposed and, together with the crumbled rock particles, form a soil that can support mosses that now colonise the area

This is the essence of succession. Organisms colonise an area, changing the **abiotic** (physical) conditions; the changed abiotic conditions allow other species to colonise the area. The new species compete with the ones there previously and become dominant. The new species in turn change the abiotic conditions, allowing other species to enter and so the process continues. The various stages in a succession are called **seres**.

As successive producers colonise an area, they create more and different niches for other organisms to occupy. As a consequence, succession usually involves an increase in the complexity of food webs. The final, most complex state of a succession is the **climax community** (Figure 4.13).

In the example given, you can see that as different types of vegetation enter the area, they affect the amount and depth of soil. This, in turn, allows other types of plant to enter. The increasing complexity of the plant community creates more and more ecological niches and so more animals will also enter the area. The species diversity will rise through the succession, until the climax is reached. The climax is the most complex community that can exist under the prevailing environmental conditions.

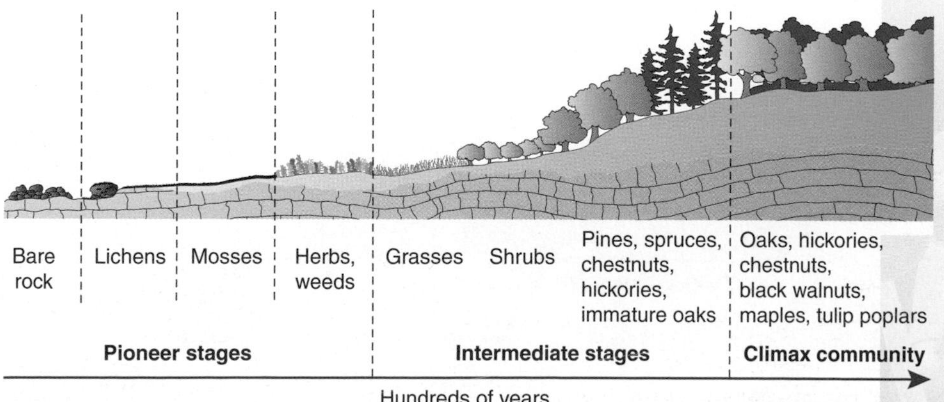

Figure 4.13 Succession

A woodland climax can arise through a different succession. The water in a lake or pond can be replaced by sediments that allow land plants to grow and give rise to a succession that results in woodland (Figure 4.14).

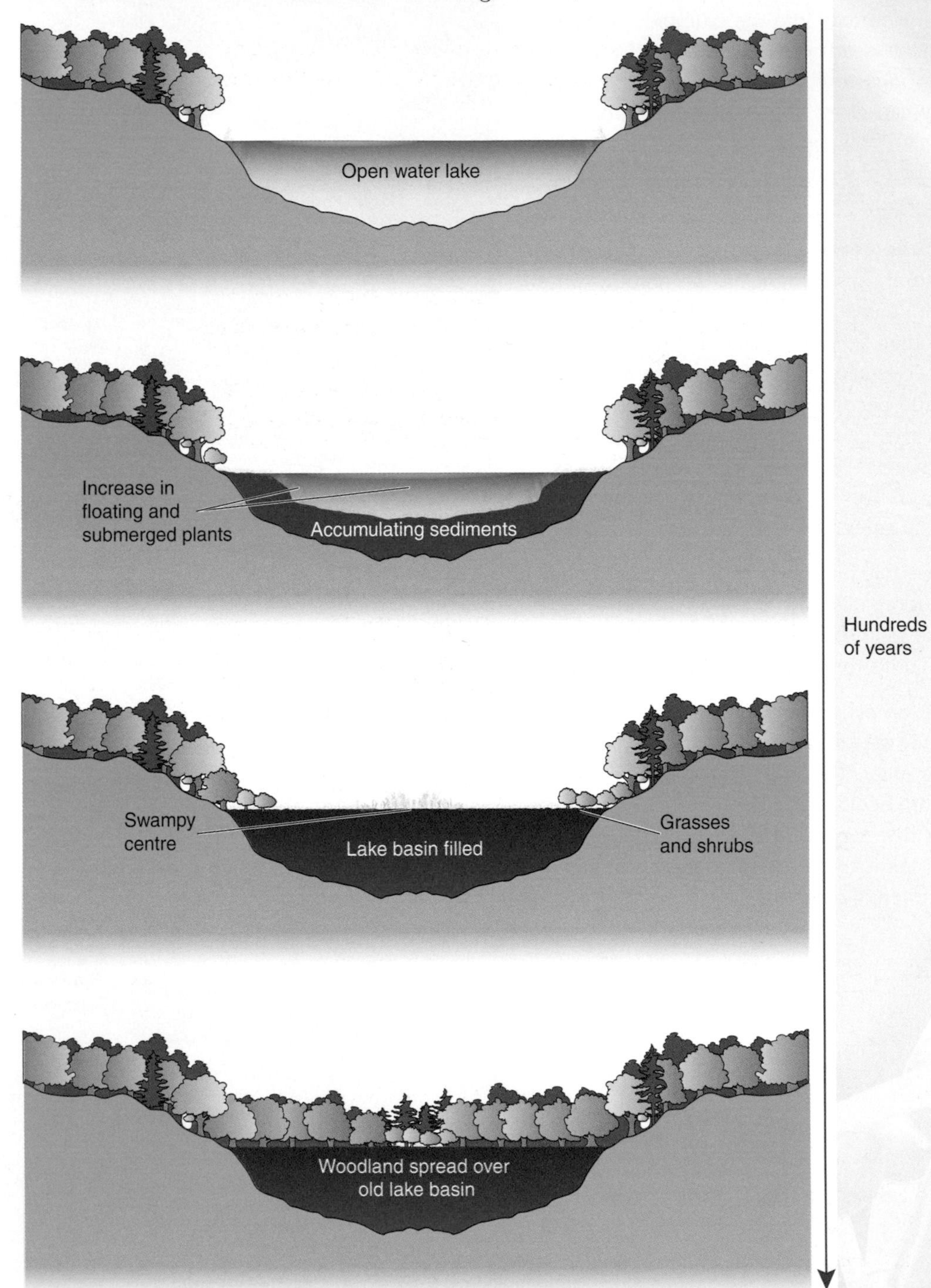

Figure 4.14 A woodland climax can arise from bare water

Why do different areas have different climax communities?

Forest climax communities in Europe do not become as complex as tropical rain forest because of the climate. Our mixed forests are said to be a **climatic climax** community.

Grassland in much of Europe would revert to woodland or forest if sheep and cattle did not graze it. They nip off the growing points at the tips of young tree shoots, preventing them from growing. Grasses grow from ground level, not from the tips of shoots, and so can re-grow. These grasslands are a **grazing climax**.

Where a succession starts from bare, previously uncolonised ground, or from a newly formed pond with no life, the succession is a **primary succession**. Sometimes, communities are destroyed by fire or by a farmer ploughing a field. When a new succession begins in such an area it is a **secondary succession** (Figure 4.15). This may result in the same climax as was originally present or in a different one.

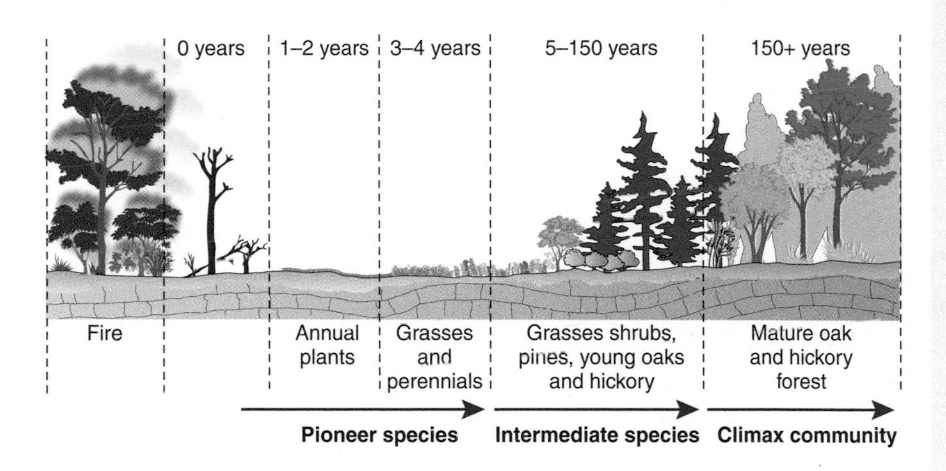

Figure 4.15 A fire destroys the original climax community and allows a secondary succession

How does succession affect conservation?

Succession and conservation often pull in different directions. Succession leads to change in an ecosystem, whereas in conservation we may want to maintain the ecosystem in its present condition. This means that we must interfere and manage the process of succession in such a way that we maintain the ecosystem.

An example of an ecosystem that would develop into another if it were not maintained is heather moorland, which is often maintained as a habitat for grouse. It also provides a habitat for other animals that would not exist elsewhere.

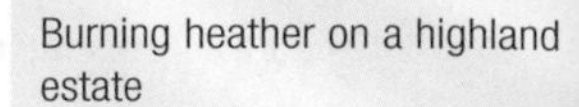
Burning heather on a highland estate

Heather plants are small woody shrubs. When there is a dense cover of heather, it provides an ideal habitat for grouse. As the heather grows taller, the cover becomes less dense and the habitat is less suitable for grouse (Figure 4.16).

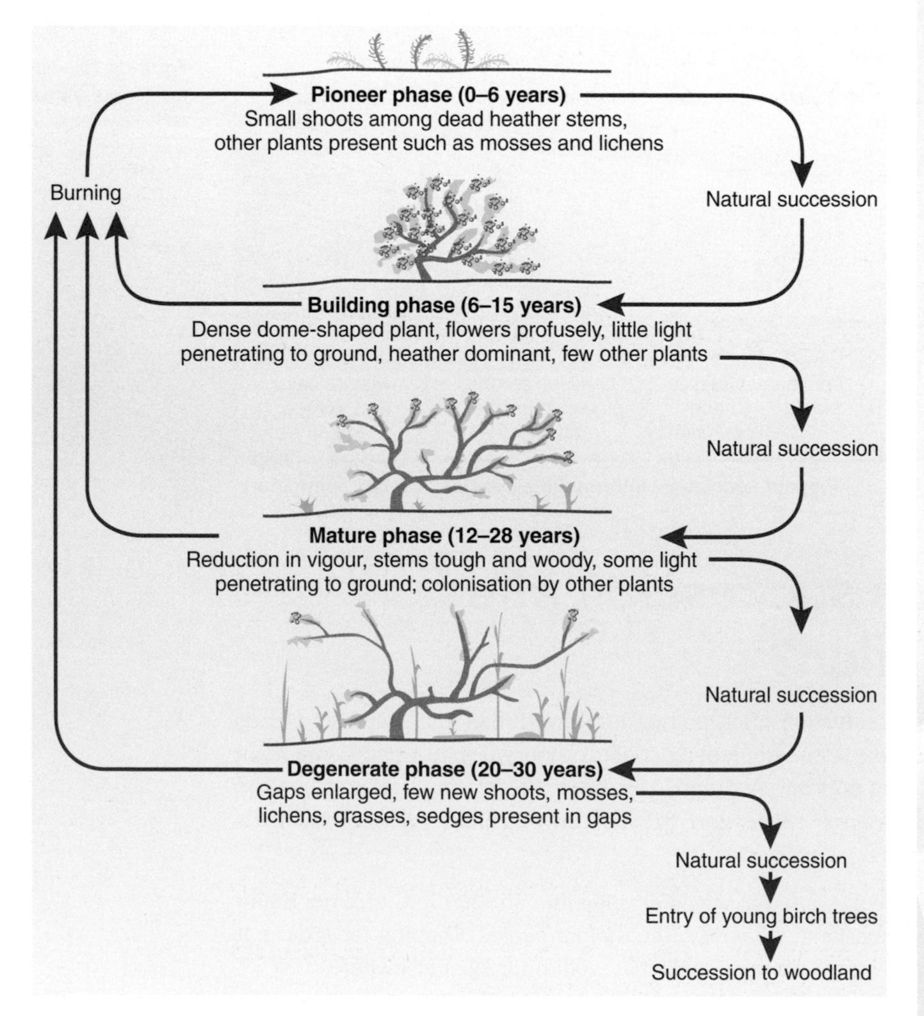

Figure 4.16 The appearance of heather at different phases of growth

As gaps appear during the senescent (degenerate) phase, other plant species, for example grasses, can enter. If nothing is done, the whole area will become colonised by grasses and then by small trees, such as birch.

To maintain the ecosystem as heather moorland, sections are burned in a controlled way once every 10–15 years. The burning has several effects:

- It kills developing trees.
- The heat encourages germination of heather seeds in the soil.
- It acidifies the soil, making it more suitable for heather and less suitable for other plant species.

As a result, new heather plants grow vigorously and once more form a dense network of heather plants. The ecosystem is maintained.

Summary

Decomposition

- Decomposers include organisms such as bacteria and fungi. They feed saprobiotically and break down the dead remains of organisms.
- This decomposition allows the release of mineral ions from organic molecules in the dead matter.

The carbon cycle

- In the carbon cycle:
  - carbon dioxide is removed from the air and fixed into organic molecules in photosynthesis
  - carbon dioxide is replaced in the air when any organism respires or when fossil fuels are burned
  - organic molecules are passed from animals to plants by feeding and assimilation
  - organic molecules are passed from animals and plants to decomposers by saprobiotic feeding on the dead remains of organisms

The greenhouse effect and global warming

- Carbon dioxide and methane are greenhouses gases; they absorb long-wave radiation passing out of the atmosphere and re-radiate it back towards the Earth.
- Because of the greenhouse effect, the Earth is warmer by 33°C than it otherwise would be and life is possible.
- There is considerable evidence that the concentration of carbon dioxide in the atmosphere has increased in the past 150 years.
- The evidence that increased carbon dioxide concentrations are responsible for global warming is conflicting.
- The influence of increased carbon dioxide concentrations on crop yields is unclear. Based on experimental predictions it appears that:
  - at a small increase in temperature (2°C), the increased carbon dioxide concentration could result in an increase in crop yields

- at a higher temperature increase (4°C), the crop yields are reduced
  - different crops respond differently
- Global warming may affect insect pests by:
  - extending their breeding ranges
  - shortening their life cycles so that more cycles are possible in the breeding season
- Global warming may affect wild animals and plants by:
  - changing their habitats
  - extending their breeding ranges
  - shifting their breeding ranges
  - extending their breeding seasons
  - shifting their breeding seasons

The importance of nitrogen

- Nitrogen is found in key biological molecules, such as proteins, DNA and ATP.
- In the nitrogen cycle:
  - nitrates are absorbed from the soil by plants and used to make proteins
  - animals eat the plants and use the plant proteins to make animal proteins
  - when the animals and plants die, the proteins pass into the detritus; in addition, excretory products such as urea pass into the detritus
  - decomposers release ammonium ions from nitrogen-containing organic molecules
  - nitrifying bacteria oxidise the ammonium ions to nitrates
  - denitrifying bacteria reduce nitrates to nitrogen gas
  - nitrogen-fixing bacteria (free in the soil or in root nodules of legumes) reduce nitrogen gas to ammonium ions

Eutrophication

- If inorganic fertilisers are overused, nitrate ions may leach into waterways and cause eutrophication:
  - the increased concentration of nitrates causes increased growth of algae
  - the increased growth of algae blocks out light so that submerged plants cannot photosynthesise and die
  - the algae die as the ions are used up
  - the dead plants and algae are decomposed by bacteria that respire aerobically, using up increased amounts of oxygen from the water
  - the concentration of oxygen in the water falls and many organisms die

Succession

- Complex ecosystems develop from simpler ones in the process of succession:
  - a hostile area is colonised by pioneer species
  - the pioneer species change the abiotic conditions and allow other species to colonise the area
  - the new species also change the abiotic conditions allowing another stage of colonisation
  - the different stages are called seres
  - as more plant species colonise, they create more niches for other organisms

- the most complex ecosystem to develop is the climax community
- A secondary succession can occur when an existing ecosystem is destroyed.
- Conservation of ecosystems sometimes requires successions to be managed.

Questions

Multiple-choice

1 The microorganisms that break down dead matter are:
 A herbivores
 B detritivores
 C decomposers
 D nitrogen-fixing bacteria

2 Greenhouse gases:
 A include carbon dioxide and methane
 B re-radiate long-wave radiation back to the Earth
 C help to make life possible on Earth
 D all of the above

3 Carbon dioxide is returned to the atmosphere:
 A when animals and plants respire
 B when animals respire and when plants photosynthesise
 C both A and B
 D neither A nor B

4 Decomposers feed by:
 A parasitic nutrition
 B saprobiotic nutrition
 C intracellular digestion
 D autotrophic nutrition

5 Which of the following does not occur during the rapid eutrophication of waterways?
 A nitrate ions are leached into waterways
 B there could be an algal bloom
 C algae use up the oxygen in the waterway
 D plants die as a result of lack of light

6 Which of the following is/are true?
 A in the past 150 years, the concentration of carbon dioxide in the atmosphere has increased year on year
 B in any one year, the concentration of carbon dioxide in the atmosphere is higher in winter than in summer
 C both A and B
 D neither A nor B

7 In the nitrogen cycle, nitrogen gas is returned to the atmosphere by:
 A nitrogen-fixing bacteria
 B ammonifying bacteria/decomposers
 C nitrifying bacteria
 D denitrifying bacteria

8 Global warming could affect the yield of crops by:

A increasing the yield of some crops

B decreasing the nutritional quality of some crops

C decreasing the yield of some crops

D all of the above

9 In an ecological succession:

A pioneer species are the first to colonise an area

B the most complex community to develop is the climax community

C each stage in the succession alters the abiotic conditions so that other species can enter

D all of the above

10 Eutrophication occurs faster in:

A still cold water

B still warm water

C moving warm water

D moving cold water

Examination-style

1 The diagram represents the circulation of nitrogen in an aquarium. High concentrations of nitrates are toxic to many animals.

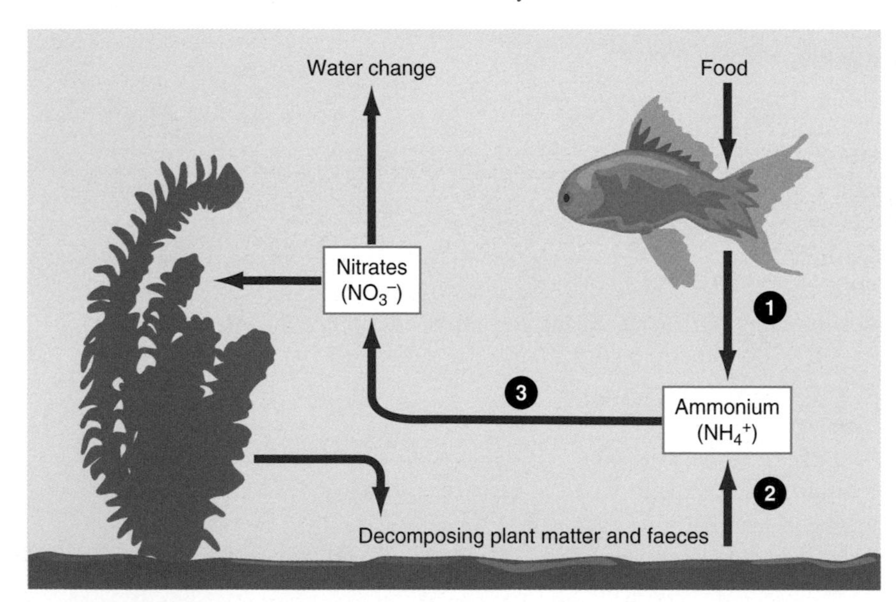

(a) Name the process labelled **1**. *(1 mark)*

(b) Describe how ammonium ions are formed in the process labelled **2**. *(3 marks)*

(c) Name the process labelled **3** and explain why it can be described as oxidation. *(2 marks)*

(d) Explain why the water in the aquarium should be changed regularly. *(2 marks)*

Total: 8 marks

2 The diagram shows the main stages of the carbon cycle.

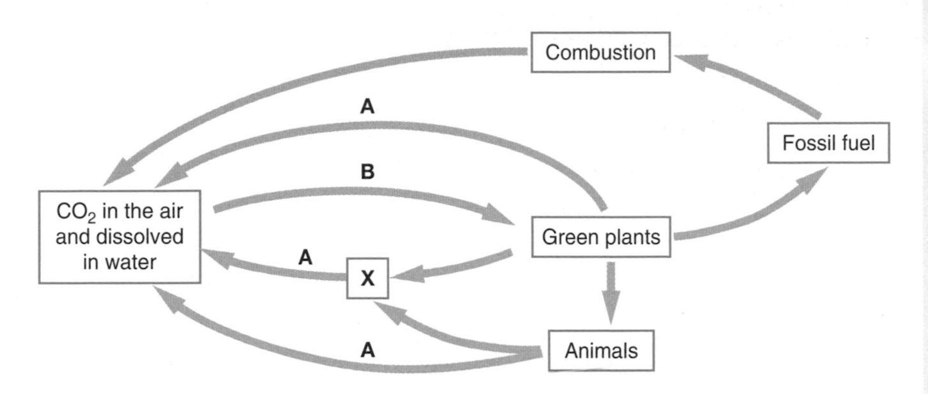

(a) (i) Name the processes labelled **A** and **B**. *(2 marks)*

(ii) Name the organisms represented by the box labelled **X**. *(1 mark)*

(b) Explain how a carbon atom in glucose in the organisms in **X** could eventually become part of a glucose molecule in an animal. *(5 marks)*

Total: 8 marks

3 The diagram shows the energy exchanges between the sun, Earth and space in a situation where the average temperature of the Earth is stable at 14°C. (Figures for energy exchange are watts per square metre of the Earth's surface.)

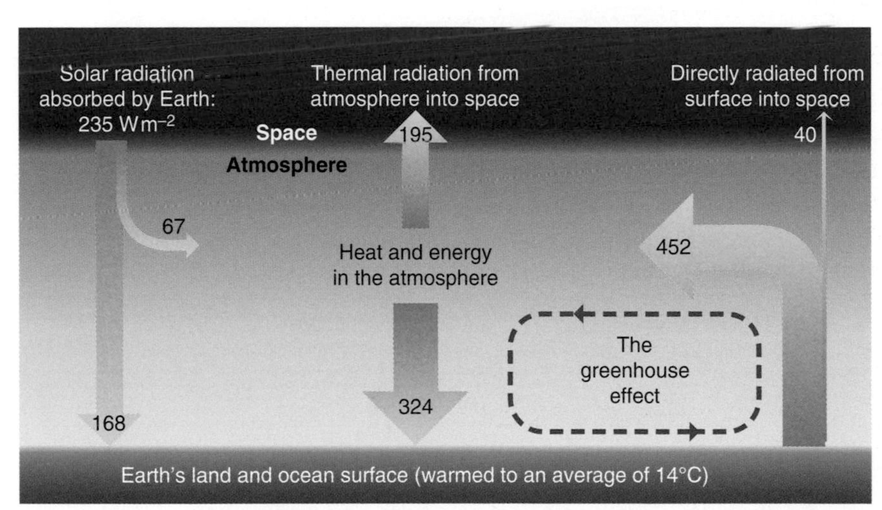

(a) Name *two* greenhouse gases. *(2 marks)*

(b) Use figures from the diagram to explain why the surface of the Earth stays at a constant 14°C. *(4 marks)*

(c) Use figures from the diagram to explain what would happen to the temperature of the Earth if there were no greenhouse effect. *(4 marks)*

Total: 10 marks

4 Over a period of 4 years, a biologist investigated the plants growing in an area. The results are shown below. In the diagram, the height of the bars represents the mean height of the plants and the width represents the mean spread of plant stems.

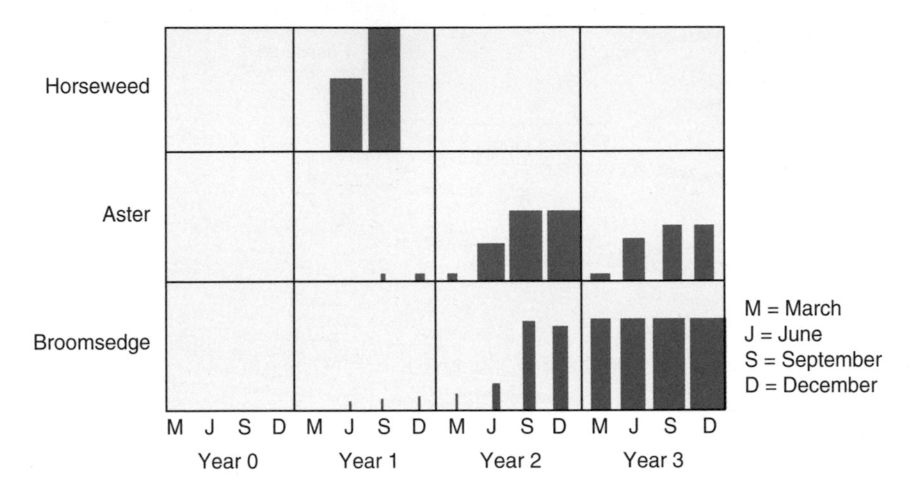

The biologist concluded that there was evidence of succession taking place.

(a) Describe three pieces of evidence that support the biologists conclusion that a succession is taking place in this area. *(6 marks)*

(b) Explain why ecosystems change through the process of succession. *(3 marks)*

Total: 9 marks

5 Many scientists believe that the Earth is entering a period of global warming.

(a) Describe *three* possible effects of global warming on living organisms. *(3 marks)*

(b) Some scientists believe that the increase in the temperature of the Earth is due to increasing concentrations of carbon dioxide in the atmosphere. The graph shows the changes in Earth's temperature and concentration of carbon dioxide over the past 500 000 years.

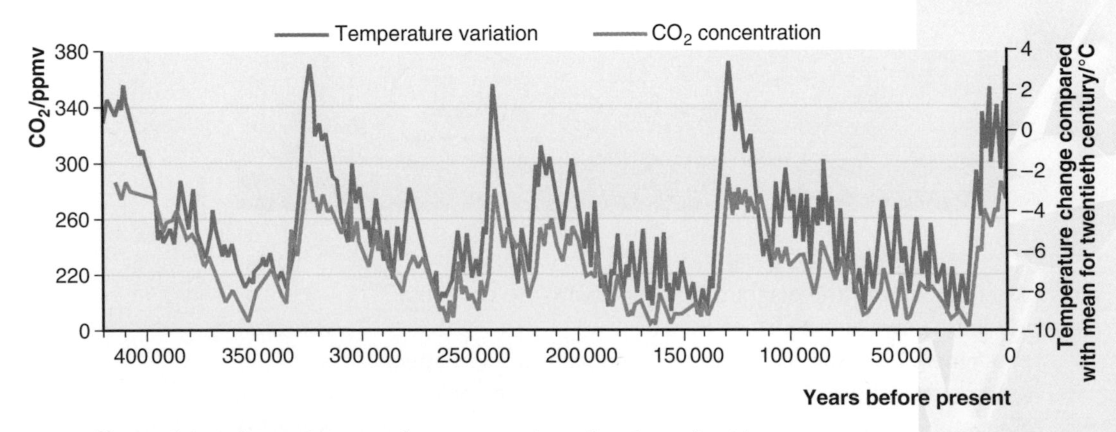

(i) Explain why an increased concentration of carbon dioxide in the atmosphere could be responsible for global warming. *(3 marks)*

(ii) Give one way in which the data support the idea of carbon dioxide causing global warming and one way in which the data do not support this idea. *(2 marks)*

(c) Other scientists suggest that the Earth's temperature is more likely to be affected by changes in how much energy is radiated towards the Earth by the sun. The graph shows changes in solar energy output and the Earth's temperature from 1750 to 2000.

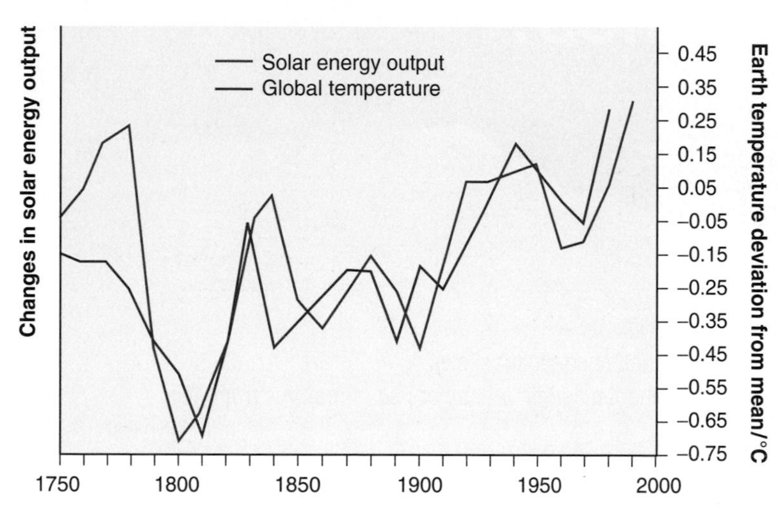

(i) Describe the trend shown by the graph. *(2 marks)*

(ii) Scientists who look critically at this graph have said that it does not show that solar activity causes an increase in global temperature. Explain why. *(2 marks)*

(d) Neither set of data quoted in (b) and (c) can offer firm proof of a cause of global warming. Explain why. *(3 marks)*

Total: 15 marks

Chapter 5

How do organisms pass on their genes?

This chapter covers:

- the distinction between genes and alleles
- the nature of dominant, recessive and codominant alleles
- the probability that offspring will inherit a certain feature or pair of features from their parents
- how the existence of multiple alleles of a gene affects patterns of inheritance
- how gender is determined
- the pattern of inheritance of characteristics determined by genes on the sex chromosomes

Genes are found on the chromosomes in all the cells of our bodies, apart from mature red blood cells. They are passed on from one generation to the next in the sex cells. There are 'rules' that govern the way that genes are inherited. These result in predictable outcomes, if the genes present in the parents are known. This chapter looks at some of the more common patterns of inheritance.

◀ Mature red blood cells do not have a nucleus and so do not have any chromosomes or genes.

What are genes like?

In the AS course, we defined a gene as follows:

A gene is a sequence of nucleotides at a fixed position on a strand of DNA that codes for a sequence of amino acids that forms the primary structure of a protein.

Here is a shorter, working definition:

A gene is a section of a chromosome that determines a particular feature.

Diploid human cells contain 46 chromosomes.

◀ Haploid sex cells contain 23 chromosomes.

Of the 46 chromosomes, 44 are called **autosomes** and have no influence on gender. These autosomes exist as 22 pairs; one chromosome from each pair is derived from the father (**paternal**), the other is derived from the mother (**maternal**). The other two chromosomes are **sex chromosomes** and, among other things, determine gender. Males have two dissimilar chromosomes — one X chromosome and one Y chromosome. Females have two X chromosomes.

◀ Diploid cells have two copies of each chromosome. These pairs of chromosomes are called **homologous pairs**.

Because chromosomes exist in homologous pairs, and genes are carried on the chromosomes, we must have two copies of each gene. Sometimes, genes have two (or sometimes more) different 'versions'. These different versions of the same gene are called **alleles**. Homologous chromosomes have alleles of the same gene in the same sequence along the chromosome. The place where a gene is found is the **locus** of the gene on the chromosome.

(×2500)

L Willatt, East Anglian Regional Genetics Service/SPL

The 46 chromosomes of a cell from a man; there are 22 pairs of chromosomes and two, the X and the Y chromosomes, which do not make a true pair

There are different alleles for quite a number of human genes, producing different outcomes. Some examples are shown in the photographs below.

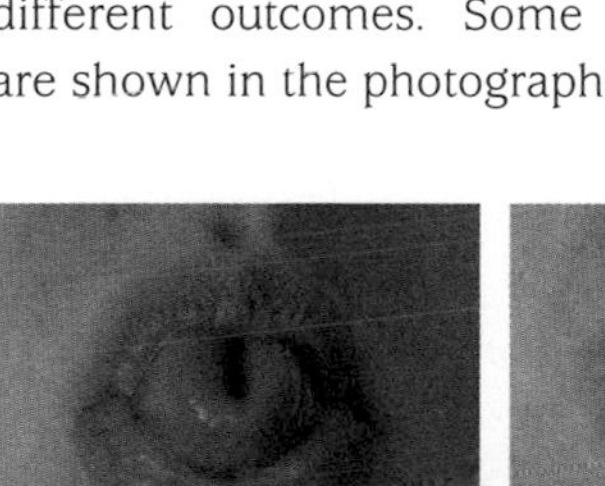

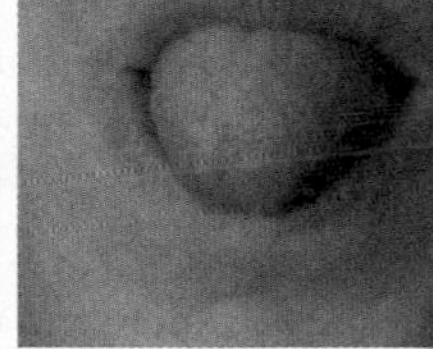

Can you roll your tongue?

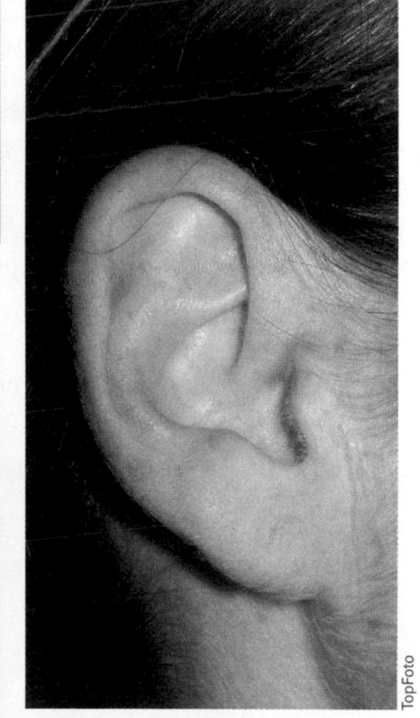

TopFoto

Do you have attached earlobes?

TopFoto

Do you have a widow's peak?

People have different 'versions' of the same feature because they have different alleles of the gene that controls the feature.

How can parents showing one version of a feature have children with the other version?

This all hinges on the idea that where there are two alleles of a gene, one of the alleles is usually **dominant** and the other is **recessive**.

What do 'dominant' and 'recessive' mean?

The gene for earlobes has two alleles expressed as free earlobes and attached earlobes. Everyone has two alleles for this feature in all their cells (except the sex cells), because they have two copies of the chromosome that carries it. A person could have:

- two alleles for attached earlobe per cell
- two alleles for free earlobe per cell
- one allele for attached earlobes and one allele for free earlobes per cell

These conditions are called **homozygous** because the alleles are the same.

This condition is called **heterozygous** because the alleles are different.

What would be the results of these situations? The first two are fairly obvious — two alleles for attached earlobes results in attached earlobes and two alleles for free earlobes results in free earlobes.

What happens when there is one allele for attached earlobes and one allele for free earlobes? Individuals that are heterozygous for this gene *always* have free earlobes. It seems the 'attached' allele just does not work when the 'free' allele is also present. The allele for free earlobes is dominant over the allele for attached earlobes. So we can now say what we mean by dominant and recessive:

A dominant allele expresses itself in the heterozygous condition
A recessive allele only expresses itself in the homozygous condition.

How are genes passed on?

The only physical link between the generations is the sex cells. Genes are passed on in male and female sex cells.

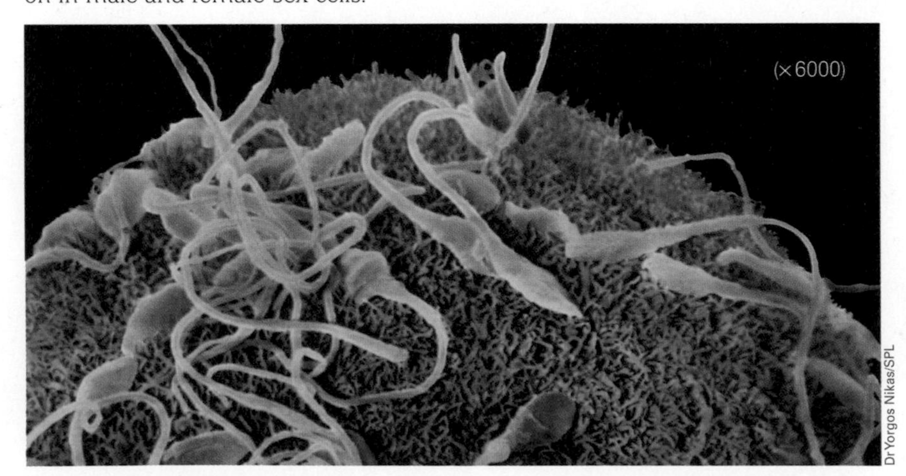

Half the genes of a new individual come from the father's sperm; half come from the mother's egg cell

Sex cells are **haploid** cells — they contain only one copy of each chromosome. Therefore, they contain only one copy of each gene. When the two sex cells fuse at fertilisation, a **diploid** cell is formed with two copies of each chromosome and two copies of each gene.

The ferti

These genes are copied into all the cells of the offspring as the zygote divides by mitosis, over and over again.

We can represent this process diagrammatically using the symbols **E** and **e** for the different alleles:

E represents the dominant allele for free earlobes

e represents the recessive allele for attached earlobes

Suppose a man homozygous for free earlobes and a woman homozygous for attached earlobes have a child.

	Father	Mother	
Parents	**EE**	**ee**	Both have two identical alleles
Sex cells	E	e	The sex cells have only one allele
Child	**Ee**		The zygote, and all the cells of the child have two different alleles

The alleles an individual carries (in this case, **EE**, **ee** or **Ee**) is the **genotype**. The result of those genotypes (free or attached earlobes) is the **phenotype**.

What would be the outcome if both parents had had free earlobes with the genotype **Ee**?

	Father	Mother	
Parents	**Ee**	**Ee**	Both are heterozygous (have different alleles)
Sex cells	**E and e**	**E and e**	Parents produce two types of sex cell in equal numbers, each having one allele

		Male sex cells	
		E (1/2)	e (1/2)
Female sex cells	E (1/2)	**EE** (1/4)	**Ee** (1/4)
	e (1/2)	**Ee** (1/4)	**ee** (1/4)

A sperm carrying either allele could fertilise an egg carrying either allele because fertilisation is a random process. We can work out the possible genotypes in the children by drawing a **Punnett square**.

Possible genotypes of the children	**EE**	**Ee**	**Ee**	**ee**
Possible phenotypes of the children	Free	Free	Free	Attached

If the parents are both heterozygotes, we would expect three out of four children to have free earlobes and one out of four to have attached earlobes — a ratio of 3:1.

Put another way, there is a 75% or 3/4 probability that any particular child will have free earlobes and a 25% or 1/4 probability that a child will have attached earlobes.

This pattern occurs in all organisms when two heterozygotes cross-breed (Figure 5.1).

Box 5.1 Ratios and the real world!

Genetic diagrams show the probability that a certain genotype or phenotype will be produced. For instance, in a cross between two heterozygotes, it is predicted that one quarter, 25%, will be homozygous recessive and show the feature determined by the recessive allele. A moment's thought should make you realise that this prediction may not be realised in any particular situation.

You might predict that with any toss of a coin there is a 50% chance of the coin landing head-side up. If you toss a coin ten times you would, therefore, predict five heads and five tails. However, if you actually did this, you might get, for example, seven heads and three tails or four heads and six tails. However, if ten people tossed a coin ten times, the outcome would probably come close to 50 heads and 50 tails; if 100 people tossed a coin ten times, the outcome would probably be even closer to 500 heads and 500 tails.

Predicted ratios from genetic diagrams are only likely to be realised with large numbers of offspring. When the numbers are small, the laws of chance can have a disproportionate effect.

In fruit flies, the allele for grey body colour is dominant to that for black body colour. When two heterozygous grey flies are crossbred, a ratio of 3 grey-bodied flies to 1 black-bodied fly is found in their offspring.

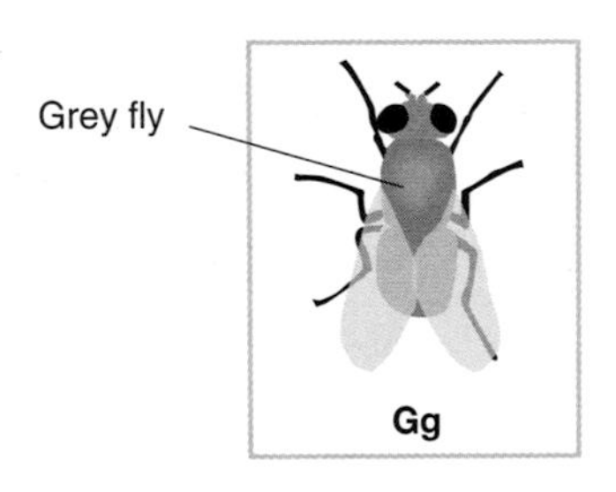

Figure 5.1 (a) Heterozygous fruit flies

Grey fly

Gg

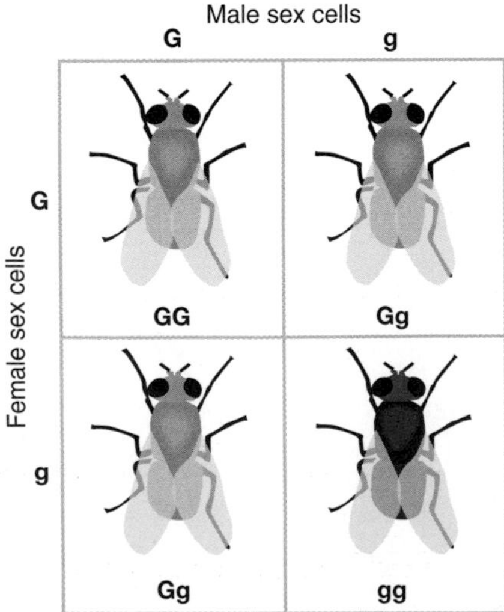

In pea plants, the allele for purple flower colour is dominant to that for white flower colour. When two heterozygous purple-flowered plants are crossbred, a ratio of 3 purple-flowered plants to 1 white-flowered plant is found in their offspring.

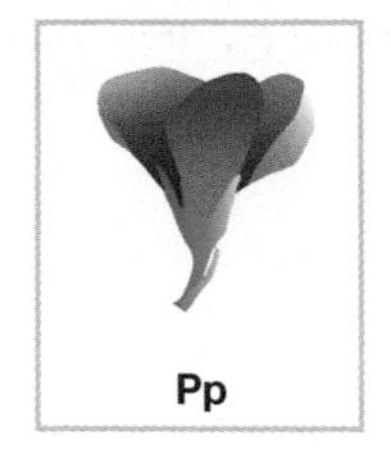

Male sex cells

Pp

Female sex cells	P	p
P	PP	Pp
p	Pp	pp

Figure 5.1 (b) Flower colour in pea plants

So now we have the answer to the question: 'How can parents showing one version of a feature have children with the other version?' If both the parents show the feature determined by the dominant allele, but are heterozygous, they can produce children who show the recessive feature.

What can we deduce from family trees?

Can we find out which allele is dominant?

The answer is yes — usually. To do this, we have to apply the principle we have just learned — that two individuals who are heterozygous show the dominant feature but can have offspring who show the recessive feature. It can never happen the other way round. Individuals who show the recessive feature have only recessive alleles and so that is all they can pass on. So two parents showing the recessive feature can only have children who show the same recessive feature.

So, whenever there is a situation in which two parents show the same feature but have children with a different feature, the feature shown by the parents must be controlled by the dominant allele.

Figure 5.2 shows a family tree (we call them **pedigrees**) for the ability to taste phenylthiocarbamide (PTC).

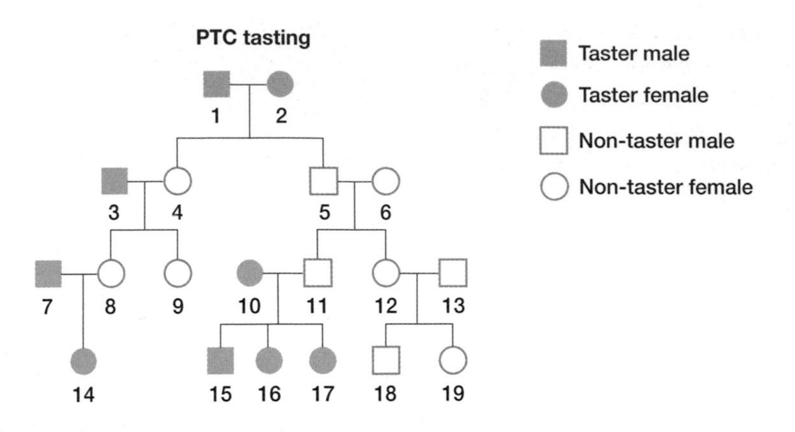

Figure 5.2 PTC tasting

PTC is a bitter-tasting chemical, yet some people cannot taste it. The ability is determined by a single gene with two alleles — one for tasting and one for non-tasting.

In Figure 5.2, individuals 1 and 2 can taste, but their children cannot. If tasting were recessive, then:

- individuals 1 and 2 would be homozygous (they could not carry *any* dominant alleles)
- they would pass on *only* recessive alleles to *all* their children
- the children would have the same genotype and phenotype as their parents

As this is not the case, tasting cannot be recessive. It *must* be determined by a dominant allele and individuals 1 and 2 must be heterozygous. They each pass on their recessive allele to both their children. This is not very likely, but quite possible.

Can we tell if an individual is homozygous or heterozygous for the dominant allele?

Again, the answer is yes — usually. There are two possibilities. We will be able to work it out if:

- we know the parents' genotypes
- we know the genotypes of their offspring

However, we may not *know* the genotypes — we may have to work those out. Let us look at a pedigree that shows the inheritance of albinism in a family (Figure 5.3).

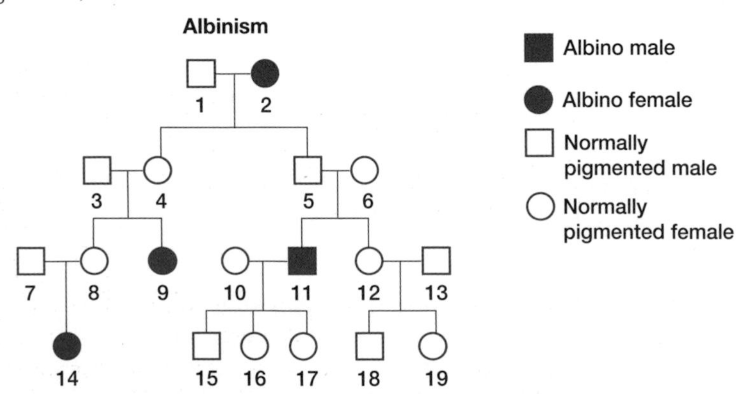

Figure 5.3 Albinism

First, we can work out that the allele that causes albinism is recessive. We do this in the usual way. In Figure 5.3, individuals 3 and 4 are unaffected, but have a child (individual 9) who is an albino. The alleles causing this child's albinism came from the parents but were not expressed in the parents. Therefore, the allele for albinism must be recessive. The same situation is found with individuals 5 and 6 and their son, individual 11.

A human albino

Let us represent the alleles as:

- **A** — for normal pigmentation
- **a** — for albinism

What is the genotype of individual 8? She is unaffected so could be **AA** (homozygous) or **Aa** (heterozygous). However, she has a child (individual 14) with albinism — genotype **aa**. One of these alleles must have come from individual 8, so she must be heterozygous.

What about individual 15? She also is unaffected and so could be **AA** (homozygous) or **Aa** (heterozygous). But her father (individual 11) is affected and so must be **aa**. He must have passed an **a** allele onto individual 15. She must be heterozygous.

However, with individuals 12, 13, 18 and 19 we cannot be sure. It is likely that they are homozygous, but the recessive allele may be present in any or all of them. We can prove that an individual is heterozygous; we cannot be 100% sure from evidence like this that an individual is homozygous for the dominant allele.

Remember, if you are asked how you would discover whether an organism is heterozygous or homozygous, you must breed it with one showing the recessive feature (carry out a test cross).

Box 5.2 The test cross

The test cross (sometimes called the back cross) is a practical way to find out whether an individual is homozygous or heterozygous for the dominant allele.

Pea plants with purple flowers can be homozygous for the dominant purple allele, or heterozygous. To find out which is which, a breeding experiment must be carried out that will allow the genotype of the original purple-flowered plant to be deduced from the results. Possible outcomes can be predicted if the purple-flowered plant is crossed with a plant whose genotype is definitely known. This can only be a white-flowered plant, which *must* have two recessive alleles.

- First, use a genetic diagram to make the predictions (Figure 5.4). Represent the alleles as:
  - **P** — for purple flowers
  - **p** — for white flowers
- Carry out the cross, collect the seeds and germinate them.
- If *any* white-flowered plants are produced, the original purple-flowered plant must have been heterozygous.

Any white-flowered plants have inherited *two* recessive white alleles and one of these *must* have come from the purple-flowered parent. If all the seeds produce purple-flowered plants, it is likely that the original plant was homozygous.

<table>
<tr><th></th><th>If the plant is homozygous:</th><th>If the plant is heterozygous:</th></tr>
<tr><td>Genotypes of parents</td><td>PP × pp</td><td>Pp × pp</td></tr>
<tr><td>Genotypes of gametes</td><td>P p</td><td>P p p</td></tr>
<tr><td>Genotypes of offspring</td><td>Pp</td><td><table><tr><td></td><td>P</td><td>p</td></tr><tr><td>p</td><td>Pp</td><td>pp</td></tr></table></td></tr>
<tr><td>Phenotypes of offspring</td><td>All have purple flowers</td><td>Half purple flowers, half white flowers</td></tr>
</table>

Figure 5.4 A cross with the homozygous plant should produce all purple-flowered offspring; a cross with the heterozygous plant should produce a ratio of 1 purple-flowered plant:1 white-flowered plant.

Are alleles always dominant or recessive?

The short answer to this is 'no'. Sometimes alleles are **codominant**. This is a fairly straightforward idea; the two alleles of a gene are both equally dominant and so, in the heterozygote, both 'still work'— they are both expressed.

An example of this is flower colour in snapdragons, which is controlled by a single gene with two alleles:

- $\mathbf{C}^R$ — determines red-coloured petals
- $\mathbf{C}^W$ — determines white-coloured petals

The possible genotypes are:

- $\mathbf{C}^R\mathbf{C}^R$ — plants with red flowers
- $\mathbf{C}^W\mathbf{C}^W$ — plants with white flowers
- $\mathbf{C}^R\mathbf{C}^W$ — plants with pink flowers; both alleles express themselves; some red and some white pigment is produced, resulting in pink flowers

If a red-flowered plant and a white-flowered plant are cross-bred, all the offspring have pink flowers (Figure 5.5).

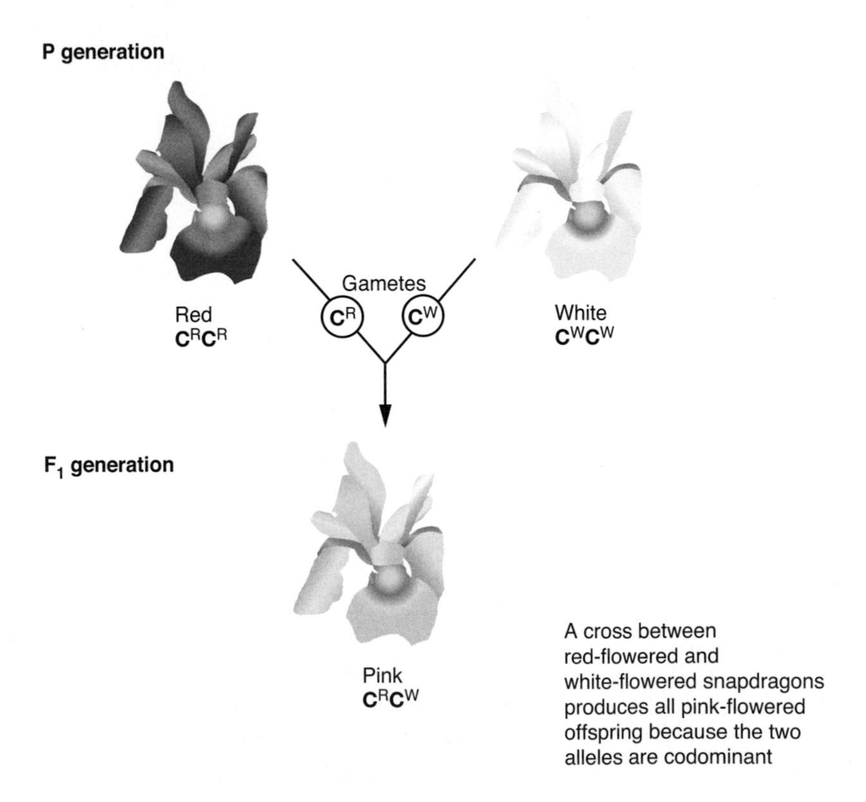

Figure 5.5 A cross between red and white snapdragons

If two pink-flowered snapdragons are crossbred (or if one is self-fertilised) the resulting offspring would be a mixture of red, pink and white-flowered plants in the ratio 1:2:1. Figure 5.6 shows why.

Genotypes of parents	C^RC^W		C^RC^W
Genotypes of gametes	C^R C^W		C^R C^W
Genotypes of offspring		C^R	C^W
	C^R	C^RC^R	C^RC^W
	C^W	C^RC^W	C^WC^W

When two pink snapdragons are crossed, there are three phenotypes in the offspring, in the ratio of 1 red to 2 pink to 1 white

Figure 5.6 A cross between two pink snapdragons

Box 5.3 Roan cattle: another example of codominance

In cattle, a single gene controls the colour of hair:

- C^R — determines red hairs
- C^W— determines white hairs

As the alleles are codominant, there are three possible phenotypes (Figure 5.7).

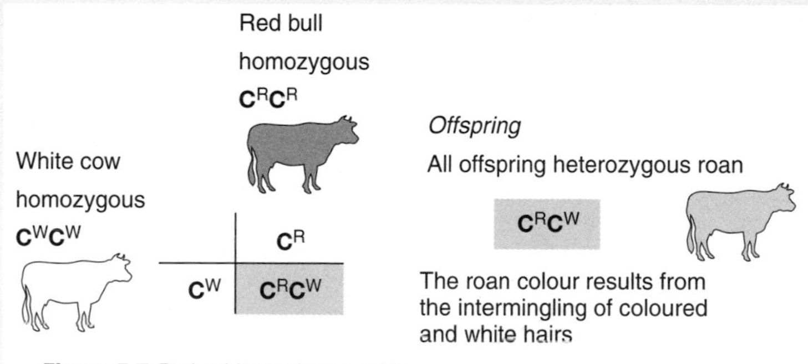

Figure 5.7 Red, white and roan cattle

Are there always two alleles of a gene?

Again, the short answer is 'no'. Some genes have more than two alleles, in which case the pattern of inheritance is a little more complex. This situation is called **multiple-allele inheritance**. However, the basic rules are the same — alleles are dominant or recessive or codominant.

Although a particular gene has more than two alleles, an individual will still only have a maximum of two of the alleles. This is because the different alleles are found at the same locus (position) on homologous chromosomes. Because there are two copies of each chromosome, a person can have only two alleles of the gene.

An example of multiple-allele inheritance is the inheritance of the ABO blood groups. This is an interesting example, because it also involves codominance.

In the ABO blood grouping system, there are four blood groups, determined by the presence or absence of two antigens (A and B) on the surface of the red blood cells (Figure 5.8).

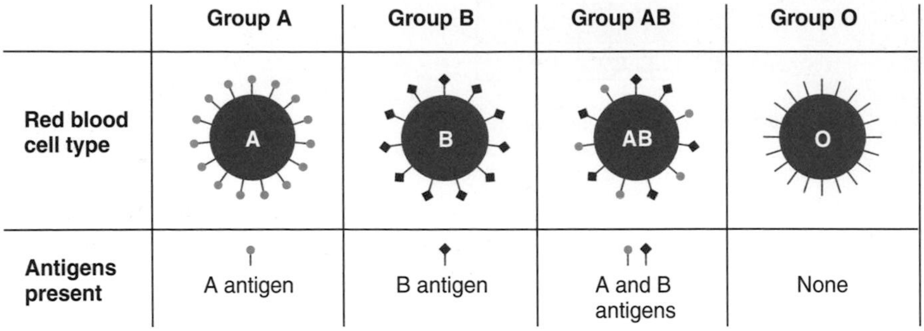

	Group A	Group B	Group AB	Group O
Red blood cell type	A	B	AB	O
Antigens present	A antigen	B antigen	A and B antigens	None

Figure 5.8 The A, B and O blood groups. The antigens present on the red blood cells determine the blood group.

There are three alleles involved in the inheritance of these blood groups:

- $\mathbf{I^A}$ which determines the production of the A antigen.
- $\mathbf{I^B}$ which determines the production of the B antigen.
- $\mathbf{I^O}$ which determines that neither antigen is produced.

Alleles $\mathbf{I^A}$ and $\mathbf{I^B}$ are codominant; $\mathbf{I^O}$ is recessive to both. The possible genotypes and phenotypes (blood groups) are shown below.

Genotype	Blood group
I^AI^A, I^AI^O	A
I^BI^B, I^BI^O	B
I^AI^B	AB
I^OI^O	O

It is possible for two parents, with blood groups A and B, to have four children, each with a different blood group!

Genotypes of parents: I^AI^O I^BI^O

Genotypes of gametes: I^A I^O I^B I^O

Genotypes of offspring:

	I^B	I^O
I^A	I^AI^B	I^AI^O
I^O	I^BI^O	I^OI^O

Phenotypes of offspring: 1 AB : 1 A : 1 B : 1 O

Figure 5.9 Inheritance of blood groups. Parents with blood groups A and B could have four children, each with a different blood group.

In an examination, you may be given a pedigree of a feature determined by multiple alleles. In such a question, you will be told that the feature is an example of multiple-allele inheritance. The same principles hold true as for determining genotypes in other pedigrees — start from a known genotype and work forwards to children and back to parents.

Box 5.4 Solving problems involving multiple-allele inheritance

In the inheritance of blood groups, there are two blood groups that can have only one genotype. Blood group AB has genotype I^AI^B and blood group O has genotype I^OI^O.

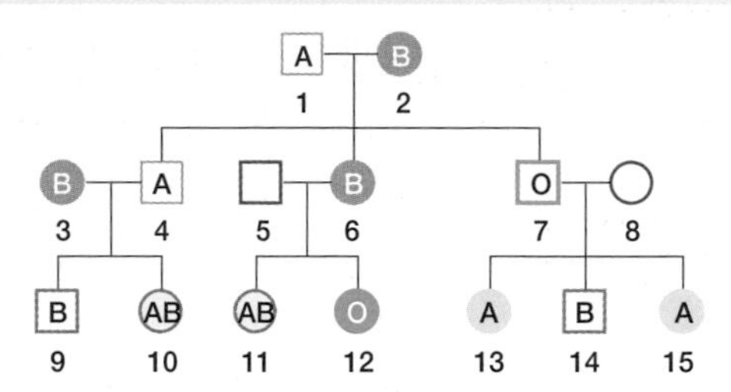

Figure 5.10 A pedigree of inheritance of blood groups in a family

What are the blood groups of individuals 5 and 8?

In Figure 5.10 individual 5 must pass on an I^O allele to one daughter (individual 12) who must have two. He must also pass on an I^A allele to his other daughter (individual 11) who cannot have inherited this from her mother who is blood group B. Individual 5 therefore has the genotype I^AI^O and is blood group A.

Individual 8 must be blood group AB. Individual 7 is blood group O and so can only pass on I^O alleles. The I^A and I^B alleles evident in children 13 and 14 must therefore have come from individual 8.

What are the genotypes of individuals 1 and 2?

Individual 1 is blood group A, so the genotype could be I^AI^A or I^AI^O.

Individual 2 is blood group B, so the genotype could be I^BI^B or I^BI^O.

However, one of their sons, individual 7, is blood group O and so must have the genotype I^OI^O. One of these alleles came from each parent, so both parents must be heterozygous.

What is sex-linked inheritance?

Some features are inherited more often by one sex than the other; these features are said to be sex-linked.

Gender, in humans, is determined by the X and Y chromosomes (the sex chromosomes).

The genotype of males is **XY** and that of females is **XX**. In every conception, there is a 50% probability that the child will be a girl (or a boy) — see Figure 5.11.

In every case, there is a 50% probability that a child will be a girl (or boy)

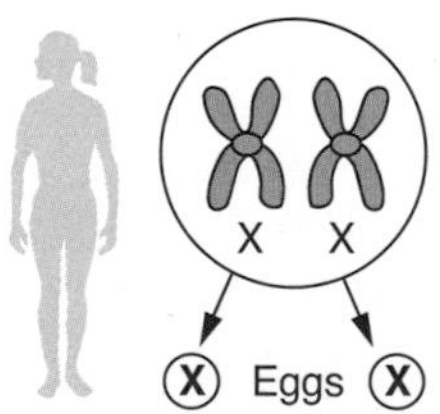

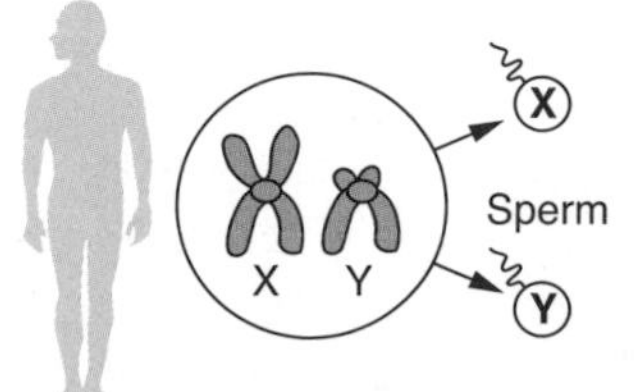

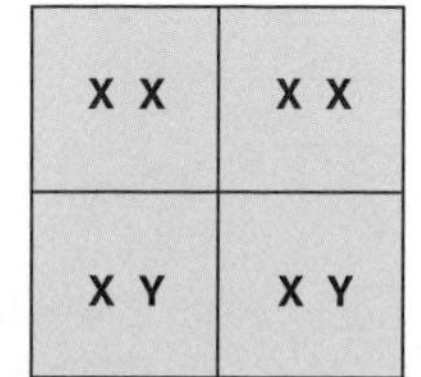

50% female

50% male

X and Y chromosomes — note how dissimilar they are

(×11500)

Biophoto Associates/SPL

Figure 5.11 Gender determination

The difference in size of the sex chromosomes means that there are genes on the X chromosome that are not present on the Y chromosome. Because of this, a recessive allele on the single X chromosome in a male is expressed — there is no equivalent dominant allele on the Y chromosome.

In females, a recessive allele must be present on *both* X chromosomes for it to be expressed.

Males can only inherit such alleles from their mothers, since fathers pass on the Y chromosome to their sons.

Examples of features determined by recessive alleles carried on the X chromosome include red–green colour blindness and haemophilia.

Sex-linked features determined by recessive alleles on the X chromosome share the following characteristics:

- They are much more common among males (because females must inherit two chromosomes carrying the recessive allele, whereas males must inherit only one).
- Affected males inherit the allele from their mothers.
- Affected females inherit one allele from each parent (so the father will be affected).
- Females who are heterozygous for the condition are carriers.
- The condition may 'skip' a generation and then appear *in the males only* (Figure 5.12).

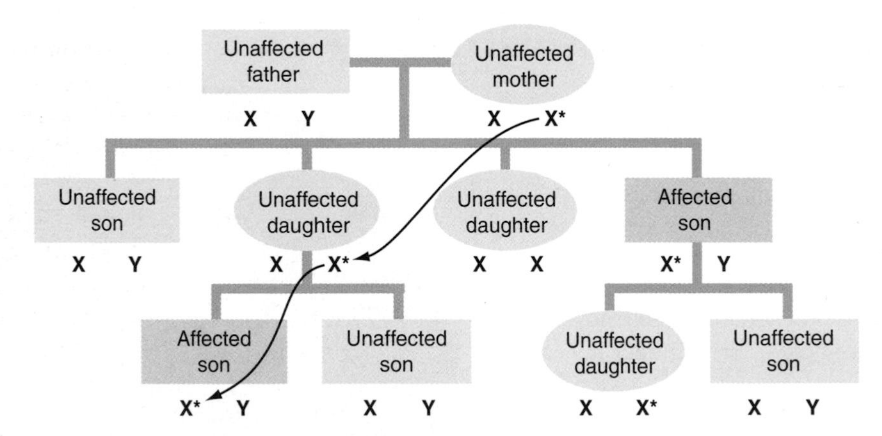

Figure 5.12 The main features of sex-linked inheritance

Genotypes of sex-linked features include the appropriate sex chromosome as well as the allele. For example, in red–green colour blindness, **B** represents the allele for normal vision and **b** represents the allele for red–green colour blindness. The possible genotypes and phenotypes are:

- $\mathbf{X^BY}$ — normal male
- $\mathbf{X^bY}$ — affected male
- $\mathbf{X^BX^B}$ — normal female
- $\mathbf{X^BX^b}$ — carrier female (not colour blind)
- $\mathbf{X^bX^b}$ — affected female

To determine the genotypes of individuals in a pedigree of a sex-linked feature begin with an individual who has a genotype of which you can be certain. This can only be either an affected male (e.g. genotype X^bY — with the affected X chromosome inherited from his mother) or an affected female (e.g. genotype X^bX^b — each parent has passed on one affected X chromosome).

You can now work back and work forward from this known starting point. For example, what are the genotypes of individuals 9 and 10 in Figure 5.13?

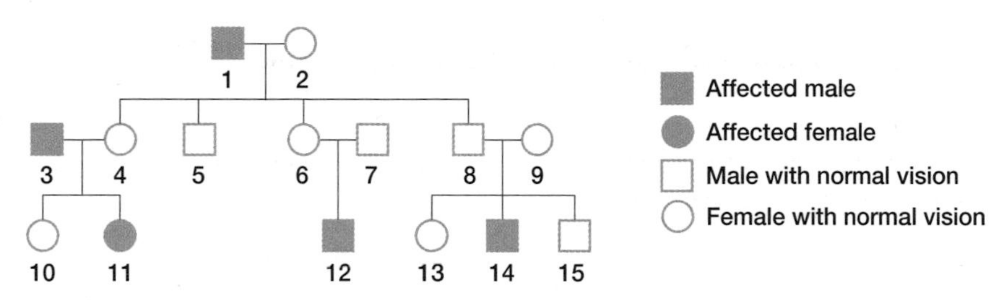

Figure 5.13 Inheritance of red-green colour blindness in one family

If you were not told that this was a sex-linked feature, there are several hints:

- It skips a generation.
- It is more common in the males.
- The only affected female has an affected father.

Individual 9 is the mother of individual 14 — an affected male (X^bY). The X^b chromosome can have come only from the mother (individual 9) who is unaffected and so must have the genotype X^BX^b.

Individual 10 is the daughter of individual 3 and is unaffected. She could be X^BX^B (a normal female) or X^BX^b (a carrier female). She inherits an X chromosome from each parent and so must inherit X^b from individual 3 (affected male). She must have genotype X^BX^b — a carrier female.

When you are solving pedigrees such as this, write on the pedigree diagram the genotypes of which you can be certain at the outset. This will help you to see where the various affected chromosomes have come from.

Summary

- Genes are sections of DNA in a chromosome that determine a particular feature.
- Alleles are different 'versions' of a gene (for example, pea plants have purple and white alleles of the gene for flower colour).
- Homologous chromosomes carry alleles of the same genes at the same loci.

Dominant and recessive alleles

- Dominant alleles are expressed in the homozygote and in the heterozygote.
- Recessive alleles are only expressed in the homozygote.
- When one allele is completely dominant, heterozygotes show the feature determined by the dominant allele.
- Two heterozygotes can produce offspring showing the recessive feature as well as offspring showing the dominant feature.
- Two individuals showing the recessive feature must be homozygous and can pass on only recessive alleles; all their offspring will show the recessive features.
- To find out whether an organism showing the dominant feature is homozygous or heterozygous we carry out a test cross. This involves breeding the organism with one showing the recessive feature. Then:

- if the original organism was homozygous, all the offspring will show the dominant feature
- if the original organism was heterozygous, some of the offspring will show the dominant feature and some will show the recessive feature (a 1:1 ratio is predicted)

Codominance

- Some alleles are codominant; both alleles are expressed in the heterozygote.

Multiple-allele inheritance

- Some genes have more than two alleles; this is multiple-allele inheritance. In multiple-allele inheritance, although there are more than two alleles in the population as a whole, any individual still has only two alleles of that particular gene.
- The inheritance of the ABO blood groups is an example of multiple-allele inheritance, with the $\mathbf{I}^A$ and $\mathbf{I}^B$ alleles also being codominant.

Gender determination

- Gender, in humans, is determined by the X and Y chromosomes; males have the genotype **XY** and females are **XX**.

Sex-linked conditions

- Some conditions are determined by recessive alleles carried on the X chromosome; these called sex-linked conditions.
- More males than females are affected by sex-linked conditions because males need only inherit one X chromosome with the recessive allele while females must inherit two such chromosomes.

Questions

Multiple-choice

1 A gene can be defined as:
   A a section of DNA coding for a particular protein
   B a section of a chromosome that determines a particular feature
   C both A and B
   D neither A nor B

2 A dominant allele is:
   A always inherited more frequently than a recessive allele
   B only expressed in the homozygote
   C expressed in the homozygote and the heterozygote
   D only expressed in the heterozygote

3 Homologous chromosomes:
   A exist in pairs
   B are the same size and shape
   C carry alleles of the same genes at the same loci
   D all of the above

4 In a cross between two heterozygotes in which allele **A** is completely dominant over allele **a**, the ratio of offspring showing the feature determined by allele **A** to that determined by allele **a** will be:

A 1:1
B 1:2:1
C 4:0
D 3:1

5 Multiple-allele inheritance is best defined as a condition in which:

A more than one gene controls a feature
B more than one allele controls a feature
C more than two alleles control a feature
D more than two genes control a feature

6 In a cross between two heterozygotes in which allele **A** is codominant with allele **B**, the ratio of different phenotypes in the offspring will be:

A 1:1
B 1:2:1
C 1:3
D 3:1

7 An individual showing a feature determined by a recessive allele that is not sex-linked:

A must be homozygous for the allele
B has inherited one of the alleles from each parent
C will pass on one recessive allele to all his/her offspring
D all of the above

8 A recessive allele **a** is carried on the X chromosome. In a particular family, neither parent shows the condition determined by allele **a**, but one of the sons does. This is possible because:

A both parents carry one copy of the recessive allele
B both parents carry two copies of the recessive allele
C the father carries one copy of the recessive allele, but the mother does not carry any copies
D the mother carries one copy of the recessive allele, but the father does not carry any copies

9 In snapdragons, the allele for red flower colour, $\mathbf{C}^R$, is codominant with the allele for white flower colour, $\mathbf{C}^W$. Heterozygotes are pink. If a pink-flowered snapdragon is crossbred with a white-flowered snapdragon, the ratio of different coloured flowers in the offspring will be:

A 1 red flowered:1 white flowered
B 3 red flowered:1 white flowered
C 1 pink flowered:1 white flowered
D 3 pink flowered:1 white flowered

10 Two parents are blood group A and blood group B. Which of the following is *not* possible?

A none of their parents was blood group B
B all of their parents were blood group B
C all of their children could be blood group B
D none of their children could be blood group B

Examination-style

1 The pedigree shows the inheritance of ABO blood groups in a family. The inheritance of ABO blood groups is an example of multiple-allele inheritance.

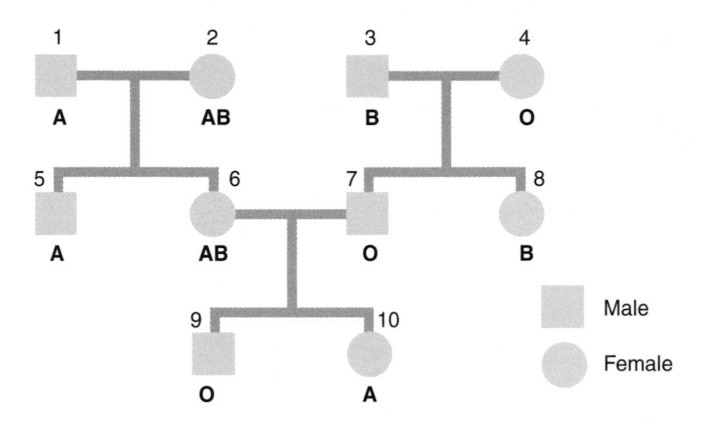

(a) What is meant by multiple-allele inheritance? *(2 marks)*

(b) What is the genotype of individual 3? Explain your answer. *(3 marks)*

(c) If individual 8 were to have children with someone of blood group AB, what is the probability that their first child would be blood group AB? Explain your answer. *(4 marks)*

Total: 9 marks

2 The graph shows the variation in height of pea plants. Height in pea plants is controlled by a single gene with two alleles. The tall allele is completely dominant over the dwarf allele.

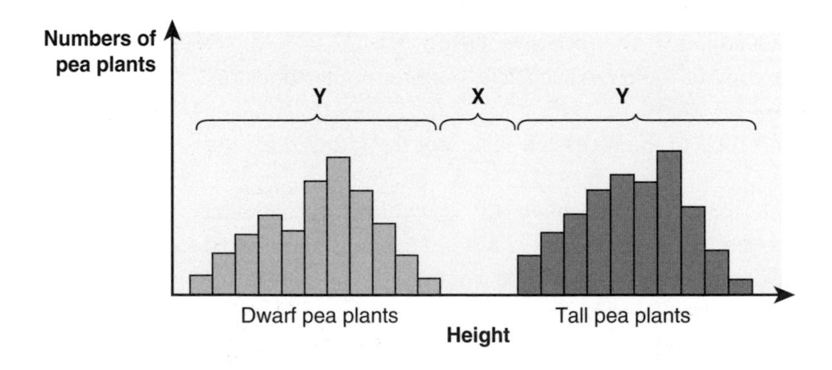

(a) Suggest a reason for the difference in height between the two groups (**X**). Explain your answer. *(2 marks)*

(b) Suggest a reason for the variation in height within each group (**Y**). Explain your answer. *(2 marks)*

(c) Describe and explain the results you would expect if two tall pea plants, both heterozygous for the gene for height, were crossbred. *(3 marks)*

Total: 7 marks

3 Flower colour in pea plants is controlled by a single gene with two alleles. The allele for purple flowers is completely dominant over the allele for white flowers.

(a) Explain what is meant by the following terms:

(i) gene

(ii) allele

(iii) dominant *(3 marks)*

(b) How could you find the genotype of a purple-flowered plant of unknown parents? Explain your answer. *(4 marks)*

Total: 7 marks

4 Andalusian fowl can have plumage with have three distinct colours:

- black
- white
- blue

In breeding experiments, the following results were obtained:

Parents	black × white
Offspring	all blue

Parents:	blue × blue
Offspring:	1 black:2 blue:1 white

(a) Suggest an explanation for these results. Use evidence from the crosses to support your explanation. *(4 marks)*

(b) If a blue fowl were bred with a white fowl, what offspring would you expect? Explain your answer. *(4 marks)*

Total: 8 marks

5 The pedigree shows the inheritance of red-green colour blindness in a family. Red-green colour blindness is a sex-linked condition, determined by a recessive allele on the X chromosome. Carriers are heterozygous for the condition but are unaffected.

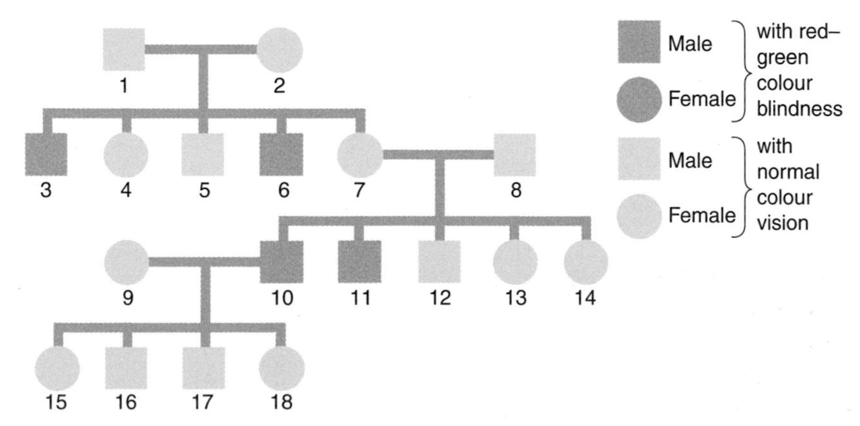

(a) What evidence in the pedigree suggests that red-green colour blindness is:

(i) sex linked? *(1 mark)*

(ii) recessive? *(1 mark)*

(b) Give the number of **one** individual who must be a carrier.
Explain your answer. *(3 marks)*

(c) If individual 17 were to have children with a female with red-green colour blindness, what would be the genotypes of:

(i) their sons

(ii) their daughters

Explain your answers. *(4 marks)*

Total: 9 marks

Chapter 6

How do the frequencies of genes change when the environment changes?

This chapter covers:

- the concept of a gene pool within a population and within a species
- how biologists estimate the frequencies of alleles in a population
- how natural selection can lead to a change in allele frequency
- different types of selection
- how natural selection may result in speciation

In his book *The Selfish Gene*, Richard Dawkins observes that most people think of reproduction as a method by which genes are passed on from generation to generation, allowing the organism to perpetuate itself through time.

He suggests that we should shift our thinking and think of our bodies as the result of the way in which the genes are organised to perpetuate *themselves* through time. A change in the environment can make it more or less likely that a particular gene will be passed on because of the way in which the gene affects the body. If the gene codes for a feature that is not adapted to the change in the environment, the body the gene has created for itself may not survive to pass on that gene.

Box 6.1

The selfish gene/body can be compared to the chicken and egg paradox (Figure 6.1). Is the egg the chicken's way of making more chickens or is the chicken the way that the genes in the egg make more genes?

Figure 6.1 Which came first?

What is a gene pool?

The **gene pool** of a group of organisms is the sum of all the alleles of all the genes present in that group. We usually talk about the gene pool of a population or the gene pool of a species. Figure 6.2 shows how the gene pools of populations and the gene pool of a species relate to each other.

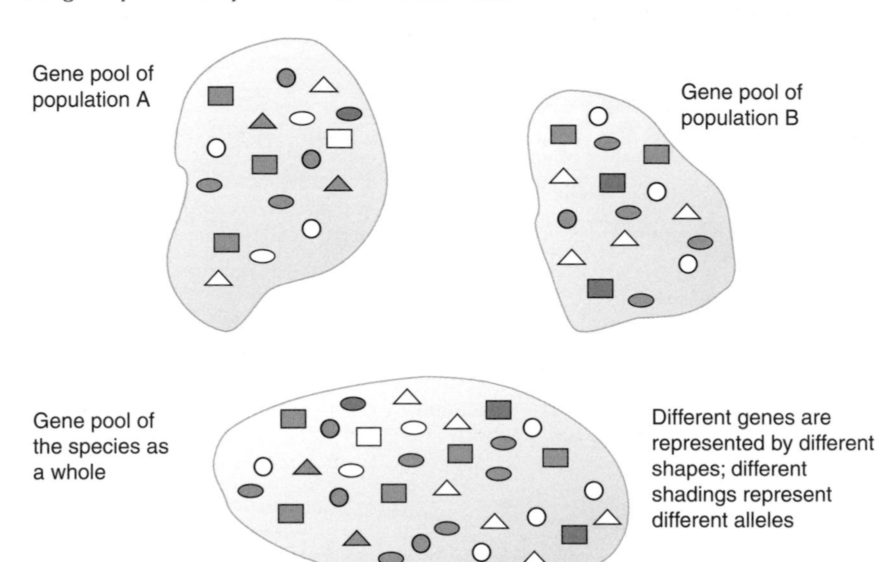

Figure 6.2 Gene pools of two populations of a species and the gene pool of the entire species

A population is all the organisms of a species living in an area at a particular time.

The gene pool of a species defines what a species is like. The different combinations of the various alleles in the gene pool produce the range of phenotypes for that species. However, the gene pool is not static. Mutations are always occurring and introducing new genes into the gene pool. Disadvantageous genes are lost from the gene pool by natural selection. So the gene pool of a species is in a constant state of flux.

How can we estimate the frequency of an allele in a population?

What if we know the numbers of the different genotypes in the population?

The frequency of a particular allele is the proportion it represents of all the alleles of that gene, expressed as a decimal (e.g. 0.2, 0.74). It is always less than 1.

Consider a population of diploid individuals and consider just two alleles of the same gene:

- allele **A** (the dominant allele)
- allele **a** (the recessive allele)

Be clear that we are estimating the frequency of **alleles**. The frequency of a gene in a population or species is always 100%. All members of a species have the same genes; the individuals are different because they have different combinations of alleles of the common genes.

Suppose also that:

- there are 1000 individuals in the population
- 360 are homozygous dominant (**AA**)
- 480 are heterozygous (**Aa**)
- 160 are homozygous recessive (**aa**)

What is the frequency of allele **A**? There are:

- 1000 individuals, each with two alleles of this gene, making a total of 2000 alleles
- 360 people carrying two **A** alleles, making 720 **A** alleles
- 480 people (the heterozygotes) carrying one **A** allele, making a further 480 **A** alleles

This gives 1200 **A** alleles, out of a total of 2000 alleles for this gene.

So, the frequency of allele **A** = $\frac{1200}{2000} = 0.6$

What is the frequency of allele **a**? There are:

- 1000 individuals, each with two alleles of this gene, making a total of 2000 alleles
- 160 people carrying two **a** alleles, making 320 **a** alleles
- 480 people (the heterozygotes) carrying one **a** allele, making a further 480 **a** alleles

This gives 800 **a** alleles, out of a total of 2000 alleles for this gene.

So the frequency of allele **a** = $\frac{800}{2000} = 0.4$

We could work out of the frequency of allele **a** more easily. Common sense tells us that the frequencies of the two alleles of the gene added together must equal 1.0. There aren't any other alleles, so these two must account for all the alleles.

So if the frequency of allele **A** = 0.6 and the two frequencies together = 1.0, the frequency of allele **a** must equal 1 – 0.6 = 0.4.

Biologists generalise this into an equation that can be applied to populations whether or not the numbers of all the genotypes are known.

In this equation:

p = the frequency of the dominant allele

q = the frequency of the recessive allele

$p + q = 1$

This is the first Hardy–Weinberg equation, derived by two biologists, Hardy and Weinberg.

What if we do not know the numbers of the different genotypes in the population?

If we do not know the numbers of the different genotypes, then we have to find a way of estimating them. To do so, we use the second Hardy–Weinberg equation, together with the one we have just established.

Genotype frequencies in a population are calculated as follows:

- If the frequency of an allele in the population is known, we assume that it will be the same in males and females.
- The frequency of the dominant allele is p (in both males and females) and the frequency of the recessive allele is q (in both males and females).
- There are no other alleles of this particular gene, so: $p + q = 1$.
- The frequency of the dominant homozygote is p (male) multiplied by p (female), i.e. p^2.
- The frequency of the recessive homozygote is q (male) multiplied by q (female), i.e. q^2.
- The frequency of the heterozygote is [p (male) $\times$ q (female)] + [p (female) $\times$ q (male)], i.e. $pq + pq = 2pq$.
- This accounts for all the possible genotypes, so:

$$p^2 + 2pq + q^2 = 1$$

This is the second Hardy–Weinberg equation.

Using this equation, if you know any of the values p, q, p^2 or q^2, you can calculate all the others.

Let us consider the original population again and see if we obtain the same answer using this method. The gene has two alleles **A** and **a**, with **A** being dominant over **a**. There are three genotypes in the population:

- **AA**
- **Aa**
- **aa**

But this time we *don't* know how many of each genotype there are, so we cannot calculate allele frequencies in the same way as previously.

However, we *do* know how many of the **aa** genotype there are, because these individuals show the recessive feature.

Using the original figures (page 119), there were 160 **aa** genotypes, giving an **aa** genotype frequency of:

$$\frac{160}{1000} = 0.16$$

This time we divide by 1000, not 2000. This is because, although there are 2000 alleles in the population, we are now dealing with the genotypes of organisms, of which there are 1000.

In the second Hardy–Weinberg equation, the frequency of the recessive homozygotes is q^2. So,

$$q^2 = 0.16$$

We can now calculate q by taking the square root of 0.16 (q^2).

$$q = \sqrt{0.16} = 0.4$$

Since $p + q = 1$

$$p = 1 - 0.4 = 0.6$$

These are the figures obtained previously.

We can also calculate the frequency of the heterozygotes ($2pq$):

$$2pq = 2 \times 0.6 \times 0.4 = 0.48$$

and the homozygous dominant genotypes (p^2):

$$p^2 = 0.6 \times 0.6 = 0.36$$

Now that we have the frequencies of the genotypes, we can work out actual numbers and see if they tally.

There are 1000 individuals in the population of which:

- **AA** has a frequency of $0.36 = 0.36 \times 1000 = 360$ individuals
- **Aa** has a frequency of $0.48 = 0.48 \times 1000 = 480$ individuals
- **aa** has a frequency of $0.16 = 0.16 \times 1000 = 160$ individuals

So it works! But, a word of caution: the second Hardy–Weinberg equation is only strictly valid if a number of conditions are met. These include:

- a large population
- individual organisms are diploid
- the individuals reproduce sexually and mating is random
- mutation does not occur
- there is no natural selection
- there is no migration

In practice, only the first two of these conditions can be met, as there is usually some sexual selection in mating and mutations *do* occur, as do natural selection and migration. Even so, the Hardy–Weinberg equation allows useful estimates of allele frequencies to be made.

The types of question you might expect in an examination, and the solutions you should provide, are illustrated in the worked examples below.

Worked example 1

A dominant allele, **A**, has a frequency of 0.6 in a population. Calculate the frequency of the heterozygotes in the population.

Answer

$p = 0.6$
but $p + q = 1$, so $q = (1 - 0.6) = 0.4$
The frequency of the heterozygote is $2pq$, i.e. $2 \times 0.6 \times 0.4 = 0.48$

Worked example 2

In a population, 75% of people can taste PTC. PTC tasting is determined by a dominant allele. Calculate the frequency of the dominant allele.

Answer

The PTC tasters include the dominant homozygotes and the heterozygotes, so we cannot deduce anything directly from this. But the recessive homozygotes make up the rest of the population (100 – 75 = 25%).The frequency of the recessive homozygotes is 25% or 0.25.
$q^2 = 0.25$
$q = \sqrt{0.25} = 0.5$
but $p + q = 1$, so p (frequency of dominant allele) $= (1.0 - 0.5) = 0.5$

If you have to calculate the numbers of individuals with a certain genotype, calculate the genotype frequencies first, as in the worked examples, then multiply the frequency by the number of *individuals* in the population. This will give you the number of that genotype in the population. For example, if the frequency of the recessive genotype is 0.6 and the population is 500, the number showing the recessive genotype is $0.6 \times 500 = 300$.

How can natural selection influence the frequencies of alleles in a population?

Box 6.2 The origin of the theory of natural selection

HOW SCIENCE WORKS

We owe much of our current thinking on natural selection to the ideas of Charles Darwin, who put forward the idea to the Royal Society in 1858. His paper suggested that those organisms best adapted to their environment would have an advantage and be able to reproduce in greater numbers than others and, therefore, pass on the advantageous adaptations. He knew nothing of genetics, so he was unable to suggest how this might take place.

However, Darwin wasn't the first with such ideas. For many years in Europe, the Christian belief had been that the Earth and all species had been created about 6000 years ago. In the mid-1700s, George Buffon challenged this idea, suggesting that:

- the Earth was much older
- organisms changed over time in response to environmental pressures and random events

These ideas are similar to what we believe today. However, Buffon had no evidence to back up his ideas and, as a result, could not convince people.

At the start of the nineteenth century, Jean Lamarck put forward a theory which included the idea that those best adapted to their environment would survive to reproduce in greater numbers. However, he suggested that characteristics acquired during a lifetime could be passed on. He suggested, for example, that modern giraffes had acquired their long necks because ancestral giraffes, with shorter necks, had repeatedly stretched them to reach food and this acquired characteristic of longer necks was then passed on to the next generation. Repeated over many generations, this resulted in the long-necked giraffes of modern times. This idea is not generally accepted today, although some environmental effects on DNA can be inherited by future generations.

In 1858, Darwin published his paper on natural selection. He had developed the idea some 20 years earlier, but was afraid of the ridicule he might receive. In 1858, another biologist, Alfred Russel Wallace had come to similar conclusions and they both presented papers at the same meeting of the Linnean Society on natural selection. These papers would change thinking on the origin of species forever.

Some of Darwin's evidence came from a visit to the Galapagos Islands. This is a small group of islands in the Pacific Ocean about 600 miles off the coast of Ecuador in South America.

Darwin visited five of the Galapagos Islands. He studied the different finches that he found on the islands. Darwin noted that there were many similarities between them, as well as the obvious differences. He concluded that the simplest explanation was that an 'ancestral finch' had colonised the islands from the mainland and, in the absence of predators, had been able to adapt to the different conditions on the islands and, eventually, evolve into different species (Figures 6.3).

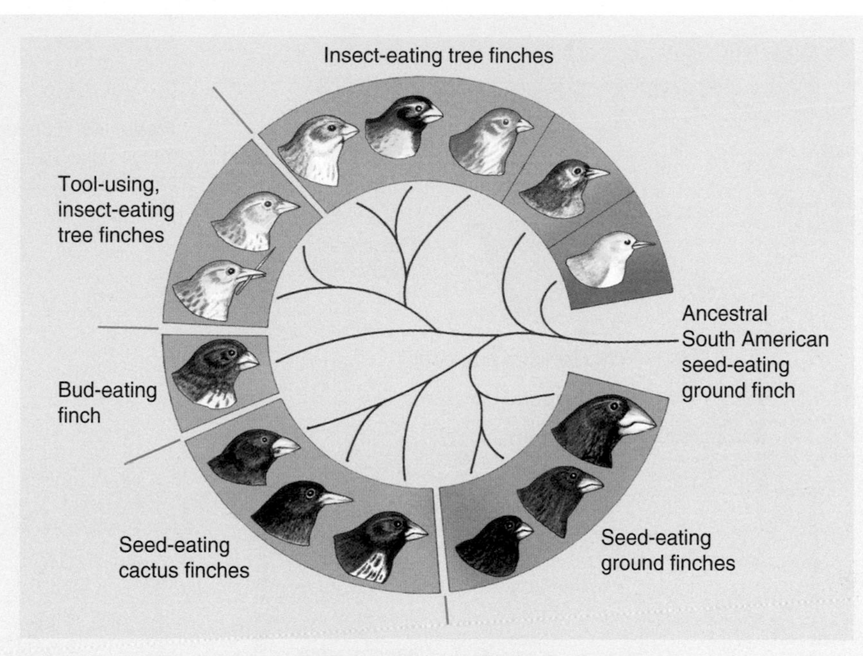

Figure 6.3 The Galapagos finches. Darwin believed that one type of ancestral finch colonised the Galapagos Islands from mainland Ecuador. The different conditions on the islands led to this ancestor evolving into the different species of finch that Darwin found.

At the time, Darwin called this 'descent with modification' and believed it to be key evidence in support of his theory of natural selection. Based on his work in the Galapagos and elsewhere, Darwin summarised his observation as two main ideas:

- All species tend to produce more offspring than can possibly survive.
- There is variation among the offspring.

From these observations, he deduced that:

- there will be a 'struggle for existence' between members of a species, because they over-reproduce, and resources are limited
- because of variation, some members of a species will be better adapted than others to their environment

Combining these two deductions, Darwin proposed that:

> 'those members of a species which are best adapted to their environment will survive and reproduce in greater numbers than others less well adapted'

This is his theory of **natural selection**. It is summarised in Figure 6.4.

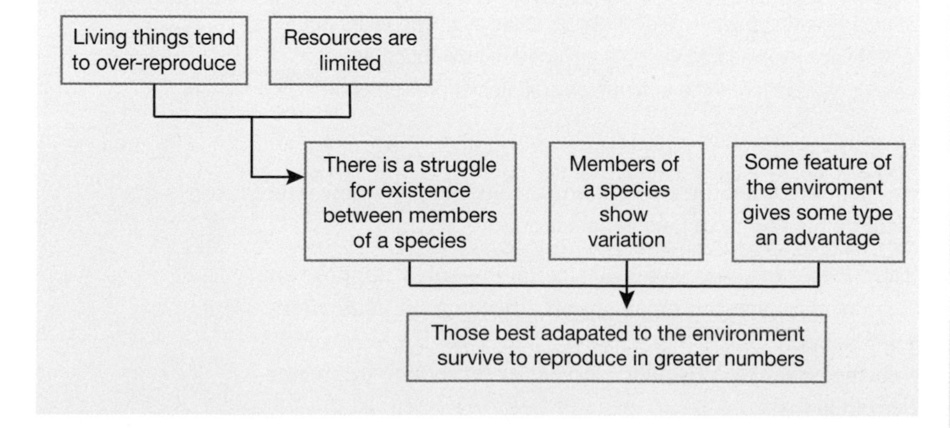

Figure 6.4 A flow chart summarising the logic of Darwin's ideas

Is there any other evidence to back up Darwin's ideas? Recently, biologists have analysed the DNA from the finches and produced a phylogenetic tree (Figure 6.5).

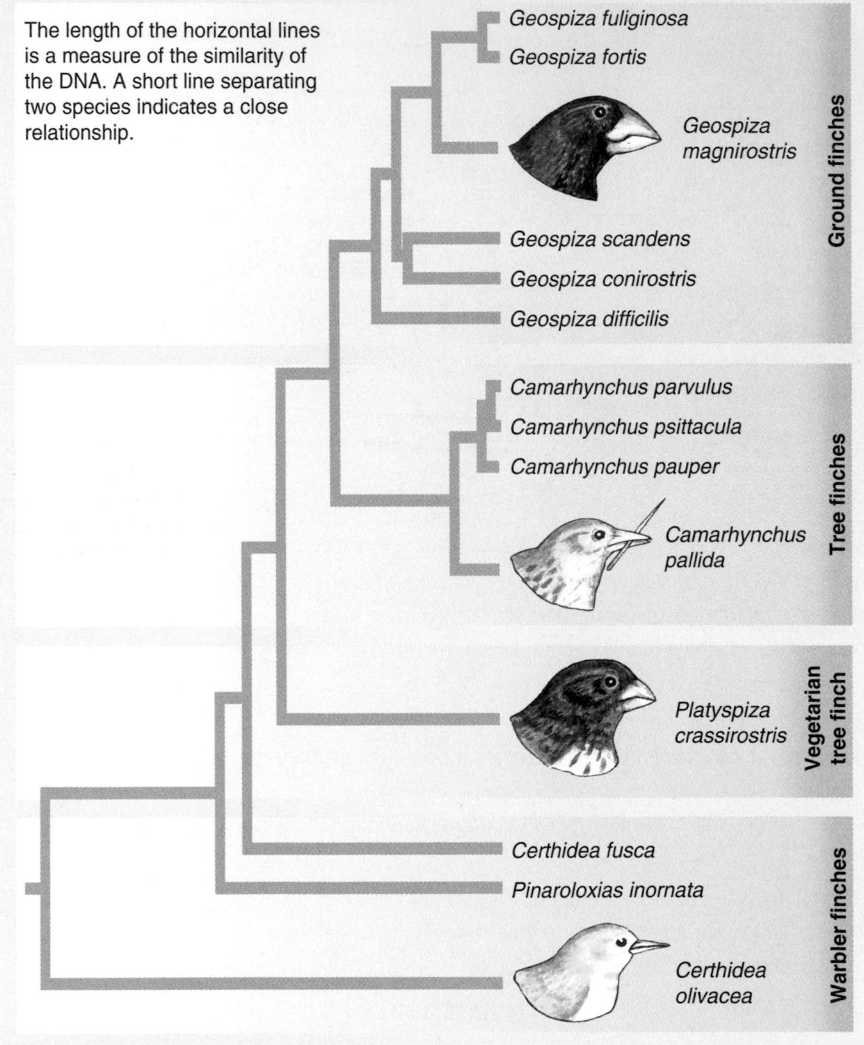

Figure 6.5 A phylogenetic tree of Darwin's finches based on analysis of their DNA

This modern evidence seems to support Darwin's deductions, made over 150 years ago. It suggests that an ancestral finch evolved into the different species now found in the Galapagos Islands as a result of adapting to the different environmental pressures on the islands.

When Charles Darwin first developed the idea of natural selection, he knew little of genetics. We can modify his theory to take gene action into account.

Genes or, more precisely, alleles of genes, determine certain features. Suppose an allele determines a feature that gives an organism an advantage in its environment. The following will happen:

- The individuals with the advantageous allele survive to reproduce in greater numbers than other individuals.

- They pass on the advantageous allele in greater numbers than other types pass on their alleles for the same gene.
- The frequency of the advantageous allele in the population will, therefore, be higher in the next generation.
- As this process repeats over many generations, the frequency of the advantageous allele increases.

Mutations are important in introducing variation into populations. Any mutation could produce an allele that:
- confers a selective advantage — the frequency of the allele will increase over time
- is neutral in its overall effect — the frequency may increase slowly, remain stable or decrease (the change in frequency will depend on which other genes/alleles are associated with the mutation)
- is disadvantageous — the frequency of the allele will be low and the allele could disappear from the population (Figure 6.6)

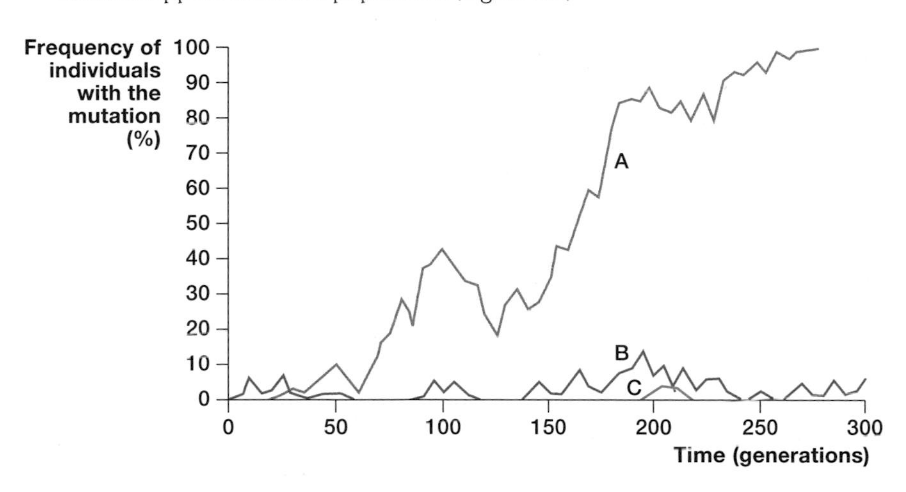

Figure 6.6 The change in frequency of advantageous (A), neutral (B) and harmful (C) mutations in a population over 300 generations

Box 6.3 Antibiotic resistance: a modern example of selection in action

Mutations in bacteria can make them resistant to an antibiotic. This can be shown by culturing bacteria on nutrient agar on which discs containing various antibiotics are placed (Figure 6.7). The antibiotic diffuses from the disc into the agar. If the bacteria are sensitive to the antibiotic, they cannot grow where it is present. If they are resistant to the antibiotic, it has no effect on their growth.

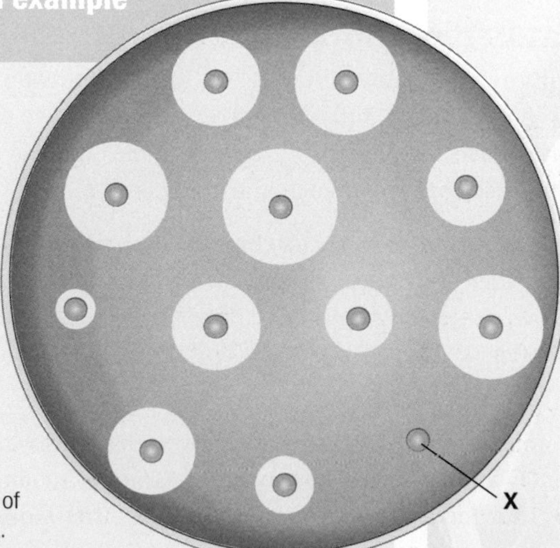

Figure 6.7 Bacteria growing on nutrient agar with discs that contain antibiotics. The bacteria are sensitive to most of the antibiotics, but are resistant to the antibiotic labelled X.

What happens next depends on whether or not the bacterial population is exposed to the antibiotic. This is summarised in Figure 6.8.

Figure 6.8

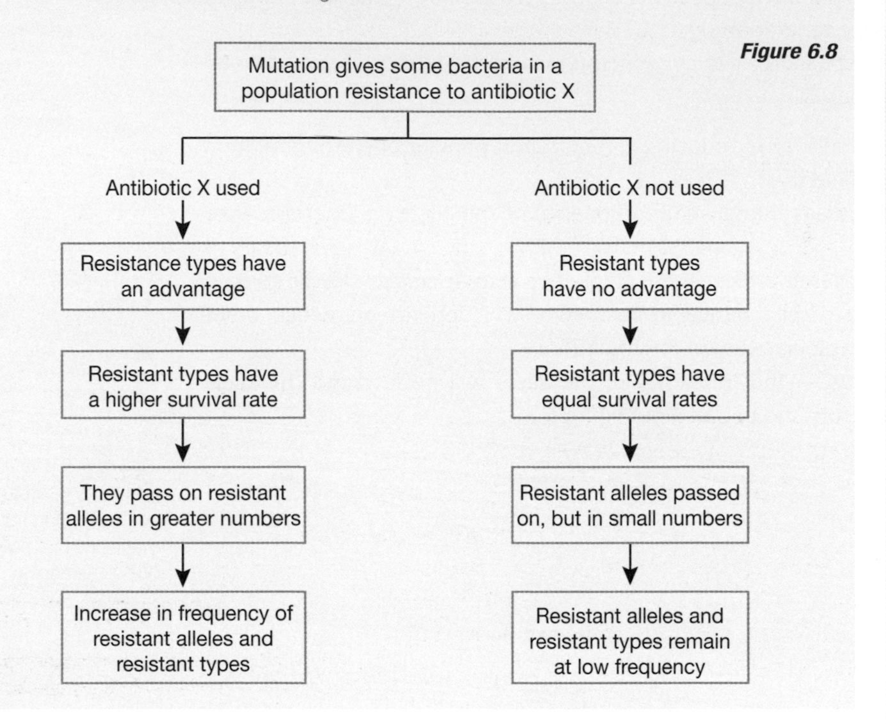

Are there different types of selection?

There are several types of selection. These include **directional selection** and **stabilising selection**.

What is directional selection?

A feature may show a range of values. Individuals at one extreme could have a disadvantage, whereas those at the other extreme have an advantage. For example, thicker fur (longer hair) in mice is an advantage in a cold climate.

If the environment were to change so that it became significantly colder, or a group of mice were to establish a population in a new, colder environment, there would be a selection pressure in favour of the mice with long fur and against those with short fur.

Over time, selection operates against the disadvantaged extreme and in favour of the other extreme. The mean and range of values shift towards the favoured extreme. The frequency of the alleles causing long fur will increase (Figure 6.9).

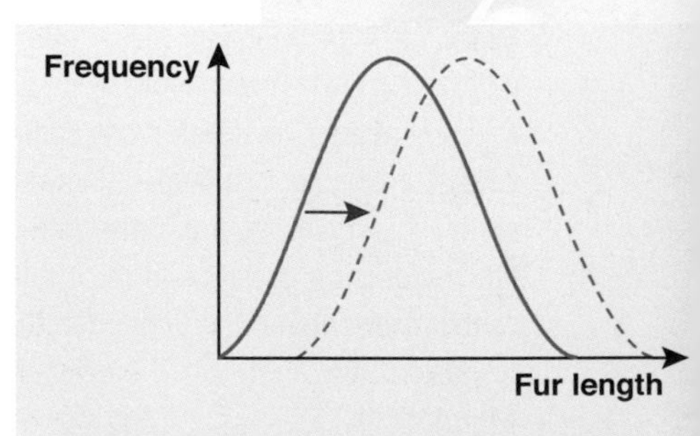

Environmental changes favour the selection of longer fur, causing the normal distribution to shift

Figure 6.9 Directional selection

Looking at the graph carefully, we can see that the distribution has shifted. As we might expect, there are now no mice with the very shortest fur — they could not survive in the new environment. However, there are mice with longer fur lengths than any of those in the original distribution. Where have they come from? They must be the result of either:

- mutations

or

- new combinations of alleles

In the original, warmer environment, such mutations/combinations of alleles would have been disadvantageous. Any mice with fur this long in the original, warmer environment would have been unable to cool themselves effectively and would have died.

What is stabilising selection?

In a stable environment, selection often operates *against both extremes* of a range. It operates to reduce the variability in the population and to make the population more uniform.

Birth mass in humans is an example. Babies who are very heavy or very light show a higher neonatal mortality rate (die more frequently at, or just after, birth) than those of medium mass. Over time, selection operates to reduce the numbers of heavy and light babies born (Figure 6.10).

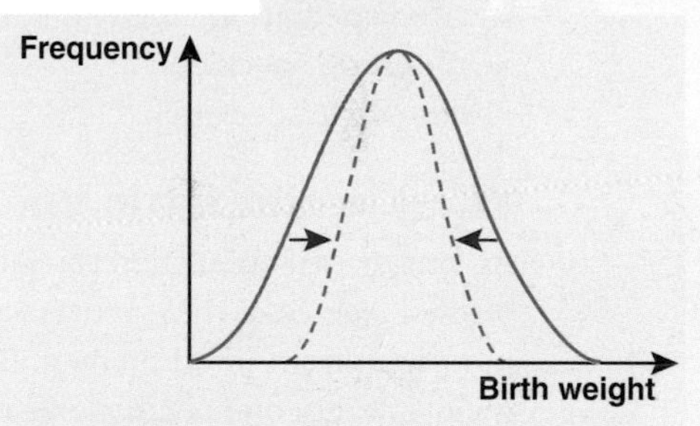

In a stable environment, selection operates to reduce the numbers of heavy and light babies born

Figure 6.10 Stabilising selection

Stabilising selection is illustrated in Table 6.1, using the examples of the evolution of resistance to pesticides or to antibiotics and the selective advantage of sickle-cell anaemia heterozygotes.

Table 6.1

	Pest resistance to pesticide (or bacterial resistance to antibiotics)	**Sickle-cell anaemia (determined by a recessive allele)**
Selection pressure	Repeated use of pesticide/antibiotic	Incidence of malarial parasite
Variation in the population	Chance mutations in some individuals confer resistance	Homozygous dominants (no sickle-cells and no resistance), heterozygotes (partial sickling but resistance to malaria) and homozygous recessives (fatal condition)
Which are at an advantage?	Resistant forms	Heterozygotes have resistance to malaria and only partial sickling of red blood cells
Consequences for phenotype	Resistant forms survive to reproduce in greater numbers — with time, more of the population are resistant	Incidence of sickle-cell anaemia remains higher than in areas with no malaria
Consequences for allele frequencies	Alleles conferring resistance are passed on in increasing numbers with each generation — frequency increases	Allele for sickle-cell anaemia remains high where malarial parasite is common, because heterozygotes have resistance

How can natural selection lead to the formation of new species?

Natural selection provides a mechanism by which new populations of a species can arise. At what point can these populations be considered to be distinct species?

A species is a group of similar, interbreeding organisms that produce fertile offspring.

If two populations become so different that they cannot interbreed to produce fertile offspring, then we can consider them to be different species. This process is called **speciation**. The two main ways that speciation occurs are:
- **allopatric** speciation
- **sympatric** speciation

What is allopatric speciation?

As long as two populations are able to interbreed, they are unlikely to evolve into distinct species. They must somehow go through a period when they are prevented from interbreeding. During this period, mutations that arise and are maintained in one population cannot be passed to the other.

As a result of this, and different selection pressures in the different environments, genetic differences between the populations increase. Eventually, the two populations are unable to interbreed and have become distinct species.

In allopatric speciation, the species become isolated by some physical feature — a river might change course or a mountain range rise to separate the populations. Interbreeding is impossible and speciation may result.

An example of allopatric speciation occurred in the shrimp populations of the Caribbean Sea and the North Pacific Ocean. These two bodies of water were once joined. About 3 million years ago, the isthmus of Panama was formed and separated the two, creating two populations of shrimps, one on each side of the isthmus.

◀ An isthmus is a narrow strip of land connecting two larger land masses, in this case North America and South America.

The shrimps from either side of the isthmus look remarkably similar. However, not only can they not interbreed, they are extremely aggressive towards each other. Two new species have evolved from one original species as a result of geographical isolation and allopatric speciation.

What is sympatric speciation?

Speciation need not involve physical separation. Two diverging populations may inhabit the same area, but be prevented from breeding in a number of ways, including:
- seasonal isolation — members of the two populations reproduce at different times of the year

- temporal isolation — members of the two populations reproduce at different times of the day
- behavioural isolation — members of the two populations have different courtship patterns

Speciation following any of these methods of isolation is referred to as **sympatric speciation**.

An example of sympatric speciation is found in palm trees growing on Lord Howe Island off the east coast of Australia. The soil in some parts of the island is volcanic and in other parts it is calcareous. Palms growing on the different soils developed different breeding seasons (as a result of nutrient availability). They were, therefore, reproductively isolated and developed into two different species (Figure 6.11).

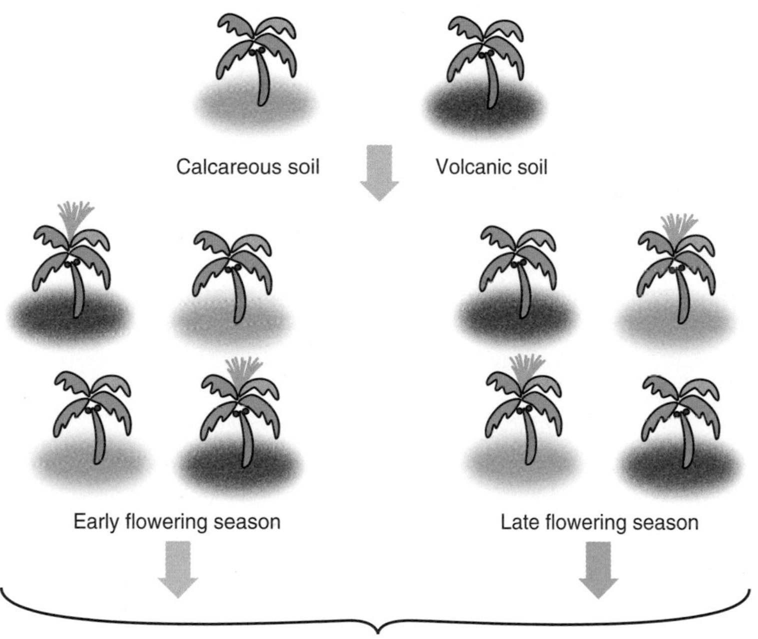

Figure 6.11 Lord Howe Island palms. Palms grew on two different types of soil and developed different breeding seasons. The resulting reproductive isolation led to the evolution of two new species.

Summary

The gene pool

- The gene pool is the sum of all the alleles of all the genes in a population or species.
- The gene pool changes constantly as a result of mutations introducing new genes into the population and disadvantageous alleles being lost through natural selection.

Allele frequency

- Allele frequency is the fraction of all the alleles of a gene that are of one type, expressed as a decimal.
- The Hardy–Weinberg equations are used to estimate the frequency of alleles in a large population in which, ideally, the following conditions are met:
  - individual organisms are diploid
  - individuals reproduce sexually and mating is random
  - mutation does not occur
  - there is no natural selection
  - there is no migration
- The Hardy–Weinberg equations are:

$$p + q = 1$$
$$p^2 + 2pq + q^2 = 1$$

where p = the frequency of the dominant allele
q = the frequency of the recessive allele
p^2 = the frequency of the dominant homozygotes
q^2 = the frequency of the recessive homozygotes
$2pq$ = the frequency of the heterozygotes

Selection

- In natural selection:
  - individuals with an advantageous allele survive to reproduce in greater numbers than other types
  - they pass on their advantageous allele in greater numbers than other types
  - the frequency of the advantageous allele in the population increases in the next generation
  - the process repeats over many generations, with the frequency of the advantageous allele increasing in each generation
- In directional selection, one extreme of a range for a given characteristic has a survival advantage compared with the other extreme; the range of values for the population shifts towards the extreme with the selective advantage.
- In stabilising selection, the two extremes are at a selective disadvantage compared with those showing the mean values for a particular feature.

Speciation

- If two populations of the same species are isolated for sufficient time, they may become so different genetically that they evolve into separate species.
- Speciation involving geographical separation is called allopatric speciation.
- Speciation in which the separation occurs within one area and is a result of different breeding strategies is called sympatric speciation.

Questions

Multiple-choice

1 The gene pool is:

A all the genes in an individual
B all the alleles in an individual

C all the alleles in a population
D all the genes in a population

2 In the Hardy–Weinberg equations:
A $p + q = 1$ and $p^2 + pq + q^2 = 1$
B $p^2 + q^2 = 1$ and $p^2 + 2pq + q^2 = 1$
C $p + q = 1$ and $p^2 + 2pq + q^2 = 1$
D $p^2 + q^2 = 1$ and $p^2 + pq + q^2 = 1$

3 In natural selection, those most adapted to an environment survive to reproduce because:
A resources are limited
B resources are limited and there is a natural variation between members of a population
C resources are limited, there is a natural variation between members of a population and living things tend to over-reproduce
D none of the above

4 Conditions for the Hardy–Weinberg equations to be valid include:
A a large population, diploid individuals and random mating
B a large population, haploid individuals and random mating
C a small population, diploid individuals and random mating
D a large population, diploid individuals and non-random mating

5 Allopatric speciation involves:
A a period when individuals of two populations are prevented from interbreeding
B geographical isolation
C a period of increasing genetic diversity of two populations
D all of the above

6 In directional selection, the selection pressure operates:
A in favour of those individuals showing the mean values for a feature
B in favour of those individuals at one extreme of the range of values for a feature
C in favour of those individuals showing both extremes of the range of values for a feature
D none of the above

7 New alleles arising from mutations in a population will:
A increase in frequency if they are beneficial in their effect and decrease in frequency if they are neutral in their effect
B increase in frequency if they are neutral in their effect and decrease in frequency if they are harmful in their effect
C increase in frequency if they are beneficial in their effect and increase in frequency if they are neutral in their effect
D increase in frequency if they are beneficial in their effect and decrease in frequency if they are harmful in their effect

8 Sympatric speciation involves:
A a period when individuals of two populations are prevented from interbreeding
B geographical isolation
C a period of decreasing genetic diversity of two populations
D all of the above

9 In stabilising selection, the selection pressure operates:

A in favour of those individuals that show the mean values for a feature

B in favour of those individuals at one extreme of the range of values for a feature

C in favour of those individuals showing both extremes of the range of values for a feature

D none of the above

10 Bacterial populations can develop resistance to antibiotics. In this case, the selection pressure is the result of:

A random mutations in the bacterial population

B the use of the antibiotic on the bacterial population

C both A and B

D neither A nor B

Examination-style

1 Gene mutations can be advantageous, harmful or neutral. The graph shows the change in frequency of the individuals possessing some mutations over a number of generations.

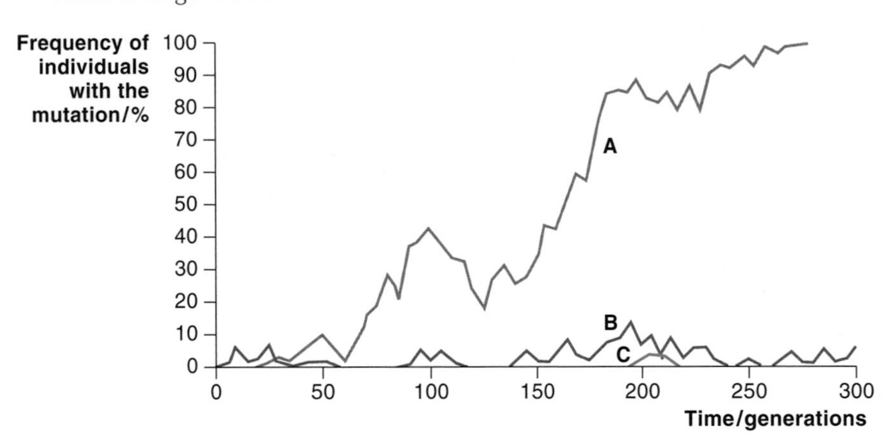

(a) Give *two* ways in which point mutations of DNA can occur. *(1 mark)*

(b) Which of the lines **A**, **B** and **C** represent a harmful, neutral and beneficial mutation? Explain your answer. *(3 marks)*

(c) Explain why it is impossible to use the Hardy–Weinberg equation to calculate genotype frequencies using data from the graphs. *(2 marks)*

Total: 6 marks

2 King cheetahs have a different pattern of spots from ordinary cheetahs. The king cheetah coat pattern is the result of a mutation; the resulting allele is recessive to that for normal coat pattern. A population of 100 cheetahs contained nine king cheetahs.

(a) At first, it was thought that the two might be different species. How could it have been proved that they were members of the same species? *(2 marks)*

(b) Use the Hardy–Weinberg equation to calculate:

(i) the *frequency* of the dominant allele (2 marks)

(ii) the *number* of heterozygotes in the population (2 marks)

(c) Give *two* reasons why the use of the Hardy–Weinberg equation might not be valid on this occasion. (2 marks)

Total 8 marks

3 Biologists investigated stinging nettles in two areas of a large National Park in Japan. There had been a large population of deer in one area for more than 1200 years. The other area had had deer in it only rarely.

Nettles from the area that contained the deer were found to have, on average, 100 times more stinging hairs than the nettles in the other area. When seeds from these plants were grown in the laboratory, they developed into plants that also had high numbers of stinging hairs.

(a) Explain the evidence that suggests that:

(i) the numbers of stinging hairs per leaf is controlled genetically (2 marks)

(ii) the difference in the number of stinging hairs on the leaves of the nettles in the two populations is a result of natural selection (3 marks)

(b) Despite their differences, the two populations have not evolved into different species. Suggest why. (2 marks)

Total: 7 marks

4 (a) In each of the following examples of natural selection, identify the selection pressure and the variant within the population that is best adapted:

(i) wildebeest hunted by lions

(ii) bacteria in a hospital where penicillin is widely used

(iii) nettle plants with different-sized leaves in a shaded woodland area (3 marks)

(b) Allopatric speciation and sympatric speciation are two processes by which new species can evolve. Explain:

(i) one similarity between the two processes (2 marks)

(ii) one difference between the two processes (2 marks)

Total: 7 marks

5 The graph below shows the distribution of root length in a population of a species of grass. The population inhabits an area in which the soil water is held mainly in the top 20 cm.

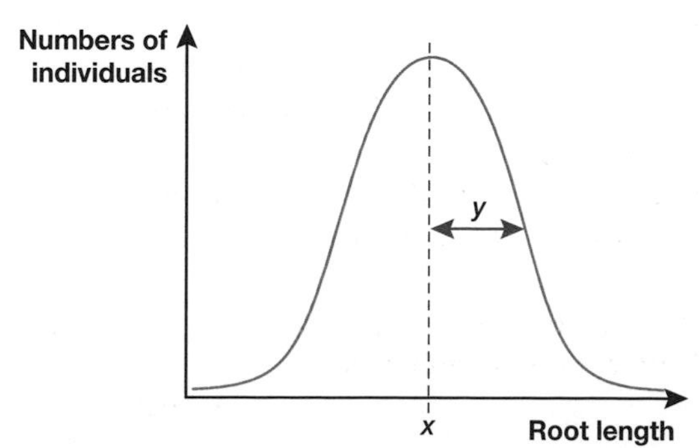

(a) (i) What does the term 'population' mean? *(1 mark)*

(ii) Name the features of the distribution labelled x and y. *(2 marks)*

(b) (i) Sketch, on the graph, the distribution of root lengths you would expect if some of these plants now colonised a different area in which the soil water was held mainly below 20 cm. *(1 mark)*

(ii) Name, with a reason, the type of selection operating in this example. *(1 mark)*

(iii) Describe the evolutionary mechanisms that would lead to this change in the distribution of root lengths. *(4 marks)*

(c) In time, these populations may evolve into different species.

(i) Would this be an example of sympatric or allopatric speciation? Explain your answer. *(1 mark)*

(ii) Describe and explain the conditions essential for speciation to occur. *(3 marks)*

(d) The two new species of grass would be members of the same genus and the same kingdom.

(i) Name three other taxonomic groups to which they would both belong. *(1 mark)*

(ii) Name the kingdom to which both species belong. *(1 mark)*

Total: 15 marks

Chapter 7

How do organisms detect changes in their environment?

This chapter covers:
- the components needed to produce a response to a stimulus and the survival value of being able to produce such responses
- the way in which plants respond to unidirectional stimuli and the benefits of such responses
- the way in which simple animals respond to a range of stimuli and the benefits of such responses
- receptors that respond to internal stimuli in humans, such as chemoreceptors and baroreceptors
- receptors that respond to external stimuli, such as the Pacinian corpuscles in the skin (respond to physical pressure) and the rods and cones in the eye (respond to light)

A stimulus is a change in the external or internal environment of an organism.

In responding to any **stimulus**, there are a number of processes that are common to all, no matter how simple or complex. There must be:
- a **receptor** of some kind to detect the stimulus
- an **effector** of some kind to produce the response
- some kind of **linking system** or **coordinating system** that is influenced by the receptor and can influence the effector

We can represent this diagrammatically, as shown in Figure 7.1.

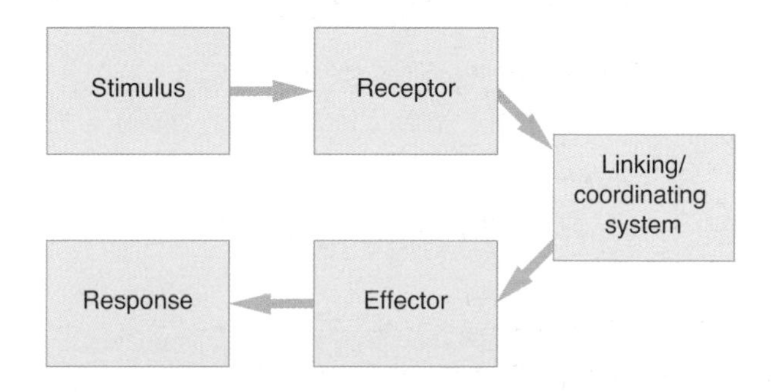

Figure 7.1 Generalised flow chart of behaviour — the processes involved in responding to a stimulus

How do plants respond to unidirectional stimuli?

Plant stems growing towards the area of greatest light intensity

You have probably noticed that, if a plant is placed on a windowsill (where the intensity of the light is greater on the window side than on the other side), it grow towards the window, i.e. towards the greater light intensity. The response of a plant to light is called **phototropism**; plant shoots are **positively phototropic** because they grow towards light. The response is even more marked in young seedlings.

The benefit in plant stems growing towards the greatest intensity of light is that plant stems automatically also direct their leaves in this direction. This means that the pigments in the leaf cells can absorb the maximum amount of light for photosynthesis.

Everyone also knows that plant stems grow upwards and roots grow more or less downwards. The unidirectional stimulus producing this response is gravity. The response by plants to gravity is called **gravitropism**. Plant roots are **positively gravitropic**, because they grow towards gravity; plant shoots are **negatively gravitropic**, because they grow away from gravity.

The benefit of this is that the roots automatically grow towards an environment in which they can anchor the plant, and absorb water and mineral ions.

How do simple animals respond to stimuli?

Some plant responses serve to maintain the plant in a favourable environment; some responses of simple animals do the same. Two types of response to stimuli in simple organisms are:

- **taxis** (plural **taxes**), in which the animal moves along a gradient of intensity of a stimulus, either towards the greatest intensity of the stimulus (a positive taxis) or away from the greatest intensity (a negative taxis); there is a directional response to a directional stimulus
- **kinesis** (plural **kineses**), in which a change in the intensity of the stimulus brings about a change in the rate of movement, *not* a change in the direction of movement

How do woodlice respond to a change in the intensity of light?

Woodlice are small land-dwelling crustaceans. There are several different species, but all are quite similar.

Woodlice are small crustaceans that show a kinetic response to light

Dr Jeremy Burgess/SPL

Because of their flattened shape and small size, woodlice have a relatively large surface area-to-volume ratio. This means that they tend to lose water quickly through their body surface. This is made more significant because there is no waxy cuticle covering their bodies to limit water loss.

They are found typically under logs, stones, bark and amongst leaf litter. These areas are all humid, which limits the rate of water loss from the woodlice. They are all also dark.

When brought into the light, woodlice start to move around much more quickly. This increased rate of movement is a response to the increased intensity of light — it is a kinesis.

The increased movement makes it more likely that the woodlice will, by chance, move into dark, humid conditions once more. When they do, their rate of movement decreases again, making it more likely that they will remain in these more favourable conditions.

Box 7.1 Investigating kinesis in woodlice

The responses of woodlice can be investigated using a **choice chamber** (Figure 7.2).

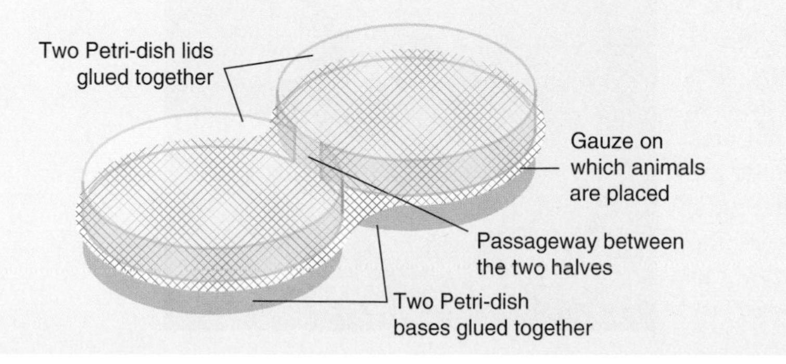

Figure 7.2 A simple choice chamber

By covering one area of the choice chamber to make it dark and leaving the other area uncovered, the choice chamber can be used to test the environmental preferences of woodlice. For example:

- Twenty woodlice are placed at random in the choice chamber.
- One area of the choice chamber is covered.
- The numbers of woodlice in the two areas are recorded every minute for 10 minutes.

Alternatively, the areas of the choice chamber could be made to have either a humid or a dry atmosphere. This can be done by placing wet filter paper under the gauze in one half and anhydrous calcium chloride under the gauze in the other half.

Anhydrous calcium chloride absorbs water vapour from the air.

Then we could proceed as with the light/dark investigation or, because we can see both halves of the choice chamber, we could make an assessment of the kinetic response of woodlice. To do this, we mark the surface of the choice chamber with a grid of 1 cm by 1 cm squares. Then, place a woodlouse in each side of the chamber and proceed as follows:

- Count the number of squares the woodlouse covers in a 10-minute period (this is a measure of the total movement of the woodlouse).
- Count the number of turns made by the woodlouse (this is a measure of how frequently the woodlouse changes direction).
- Repeat the experiments with other woodlice to obtain an average.

How does *Daphnia* respond to a directional light stimulus?

Daphnia is a small aquatic crustacean. There are many species of *Daphnia*, and there are also other genera of invertebrates that have a similar structure.

The aquatic crustacean *Daphnia*

Nearly all species of *Daphnia* feed on phytoplankton — tiny unicellular photosynthetic organisms. To maximise their photosynthesis, these organisms are found mainly in the surface layers of oceans and of inland waterways, where the light intensity is greatest. A positive phototactic response ensures that *Daphnia* move towards the area of greatest light intensity, i.e. into the same area as the phytoplankton so that they can feed effectively.

Box 7.2 Investigating phototaxis in *Daphnia*

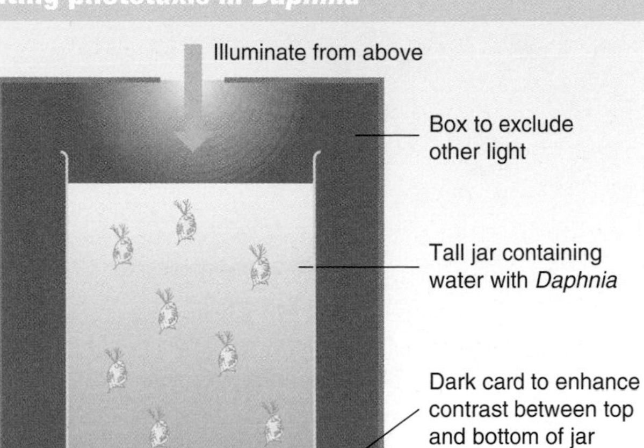

Figure 7.3 Investigating phototaxis in *Daphnia*

- Place a number of *Daphnia* in a tall jar containing pond water.
- Stand the jar on a dark card (this enhances the contrast in light intensity between the top and bottom of the jar during the investigation).
- Place an opaque box (e.g. a cardboard box) over the jar to exclude light. The box should have a hole in the top (Figure 7.3).
- Illuminate from above. This will create a gradient of light intensity in the water — highest near the surface and lowest near the bottom.
- Leave for a few minutes and observe how the *Daphnia* redistribute themselves.

How do humans detect internal stimuli?

There are many internal stimuli to which humans can respond. These include changes in:

- core body temperature
- plasma glucose concentration
- blood pressure
- plasma carbon dioxide concentration

Changes in blood pressure and plasma carbon dioxide concentration are important in the regulation of heart function.

Sensors that monitor each are located close to the heart in:

- the aortic arch
- the carotid artery

The sensors that detect changes in blood pressure are **baroreceptors**; those that detect changes in plasma carbon dioxide concentration are **chemoreceptors**. The locations of both are shown in Figure 7.4.

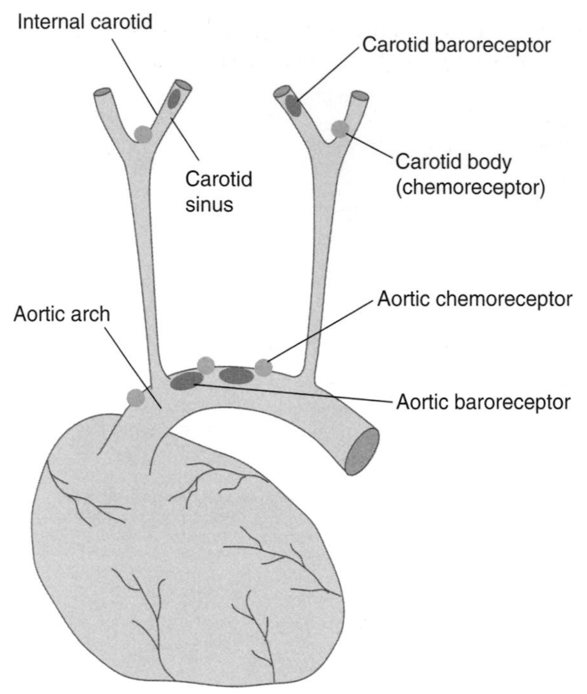

Figure 7.4 The location of the baroreceptors and chemoreceptors close to the heart

How do humans detect external stimuli?

The external stimuli to which humans respond are:

- light
- sound
- chemical stimuli (taste and smell)
- physical stimulation of the skin (light touch through to heavy pressure)
- heat

Each stimulus has a different receptor. For example, taste buds in the tongue respond to chemicals in solution, rods and cones in the eye respond to light and a range of receptors in the skin respond to different levels of physical pressure.

Each receptor is specific and responds only to one particular stimulus. Receptors are **energy transducers**; a particular form of energy in the environment (light, heat, physical pressure) is used to generate an electrochemical potential that initiates a nerve impulse. This is covered in more detail in Chapter 8.

We shall look at human receptors to just two external stimuli — pressure and light.

How do humans detect changes in pressure on the skin?

There are several different pressure receptors in the skin. Some, near the surface, are sensitive to very light touch. Others, deeper in the skin, are sensitive to more intense pressure. One of these is the **Pacinian corpuscle**. Besides being found deep in the skin, Pacinian corpuscles also occur in some joints and tendons.

Pacinian corpuscles consist of a 'naked nerve ending' (a nerve ending with no myelin sheath, surrounded by 20–60 lamellae made from connective tissue.

The lamellae are separated by a gel (Figure 7.5).

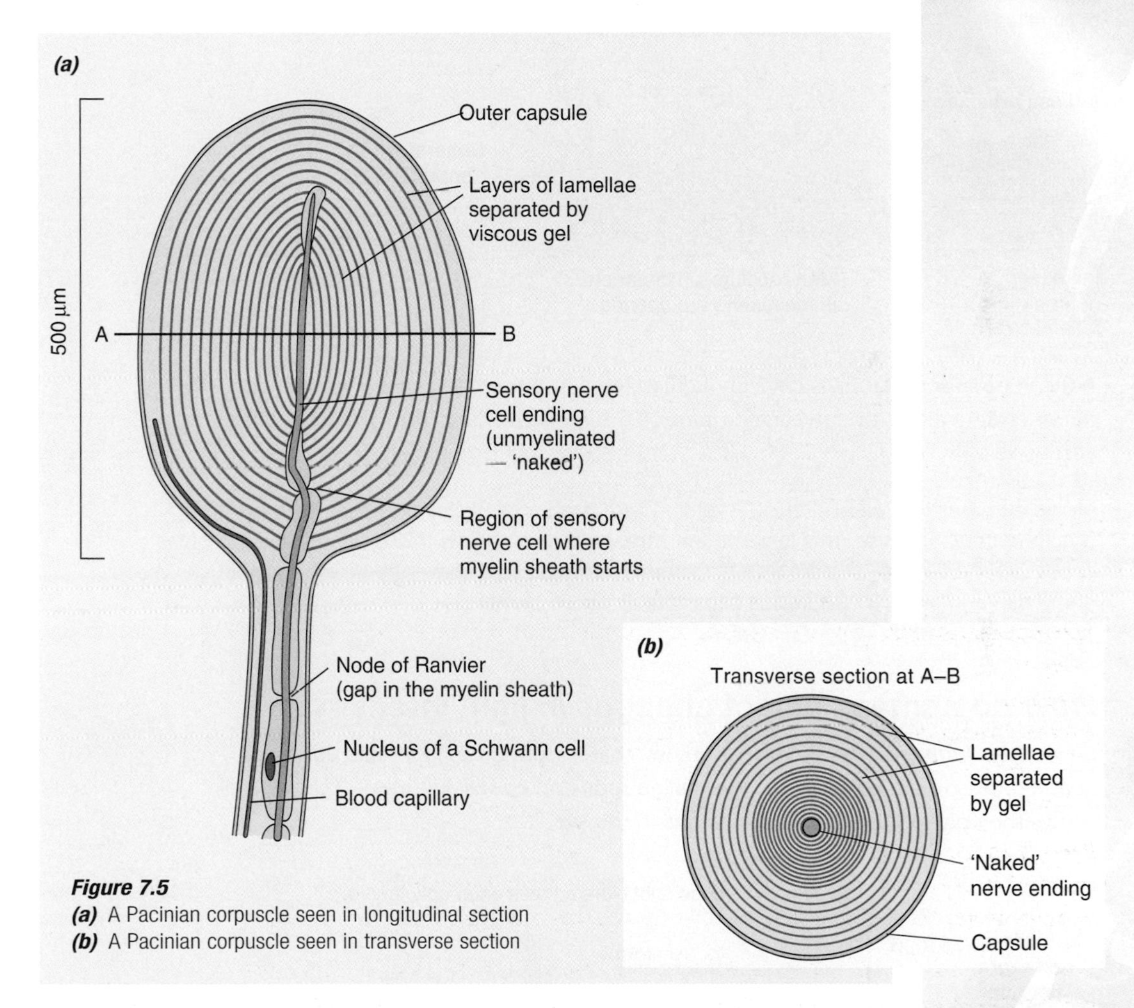

Figure 7.5
(a) A Pacinian corpuscle seen in longitudinal section
(b) A Pacinian corpuscle seen in transverse section

The lamellar structure of the corpuscle is important in ensuring that only quite firm pressure stimulates the naked nerve ending. Light pressure deforms the lamellae slightly, but most of this pressure is then absorbed by the gel.

The membrane of the naked nerve ending contains pressure-sensitive sodium ion channels. As pressure on the nerve ending increases, more and more of these ion channels open, allowing positively charged sodium ions to enter. This creates a change in the balance of charge on the two sides of the membrane. This is called the **generator potential**. A greater pressure deforms the Pacinian corpuscle more and opens more ion channels, producing a larger generator potential (Figure 7.6).

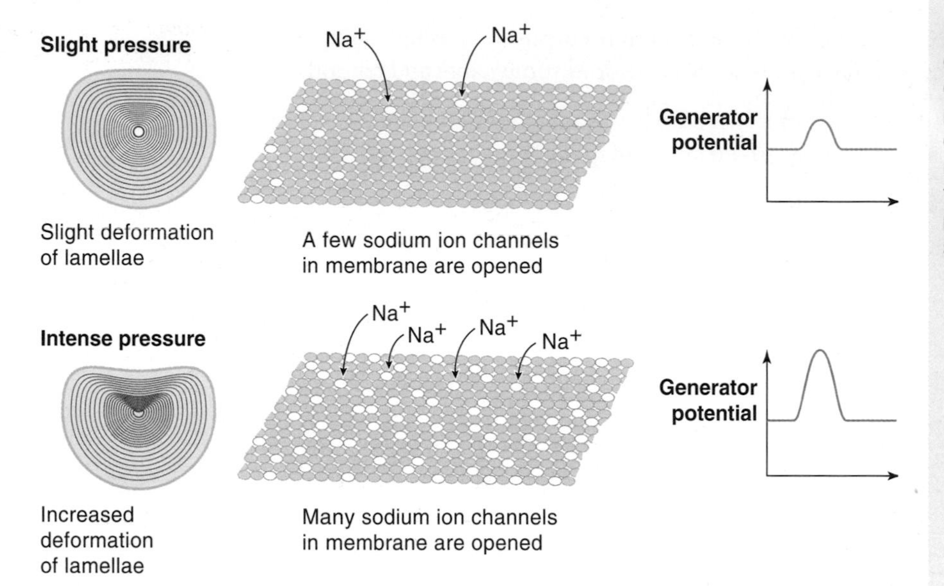

Figure 7.6 Effect of pressure on generator potential. The size of the generator potential produced by a Pacinian corpuscle depends on the number of sodium ion channels that open; this depends on the pressure being applied to the skin.

If the generator potential is large enough, a nerve impulse passes from the naked nerve ending along the myelinated part of the nerve leading from the Pacinian corpuscle.

Nerve impulses are 'all-or-nothing' events. There is a certain level of stimulation needed to initiate a nerve impulse. This is called the **threshold** level; below this level, no nerve impulse results. All nerve impulses are also the same. A stimulus just above the threshold produces the same nerve impulse as a stimulus considerably above the threshold. A larger stimulus could, however, result in *more* nerve impulses.

How do humans detect changes in light intensity?

Obviously, with our eyes! The receptors that respond to light are specially adapted cells in the retina of the eye, called **rods** and **cones**. Figures 7.7 and 7.8 shows the position of the rods and cones in the eye.

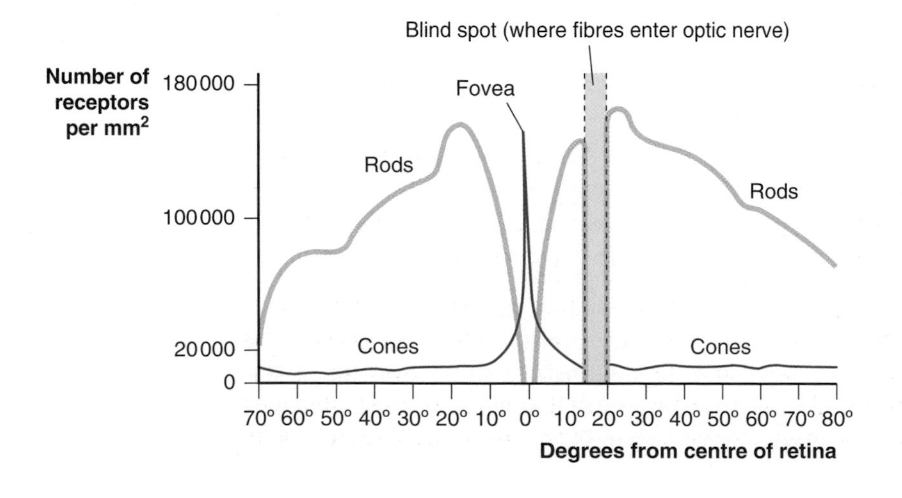

Figure 7.7 The distribution of the rods and cones in the eye

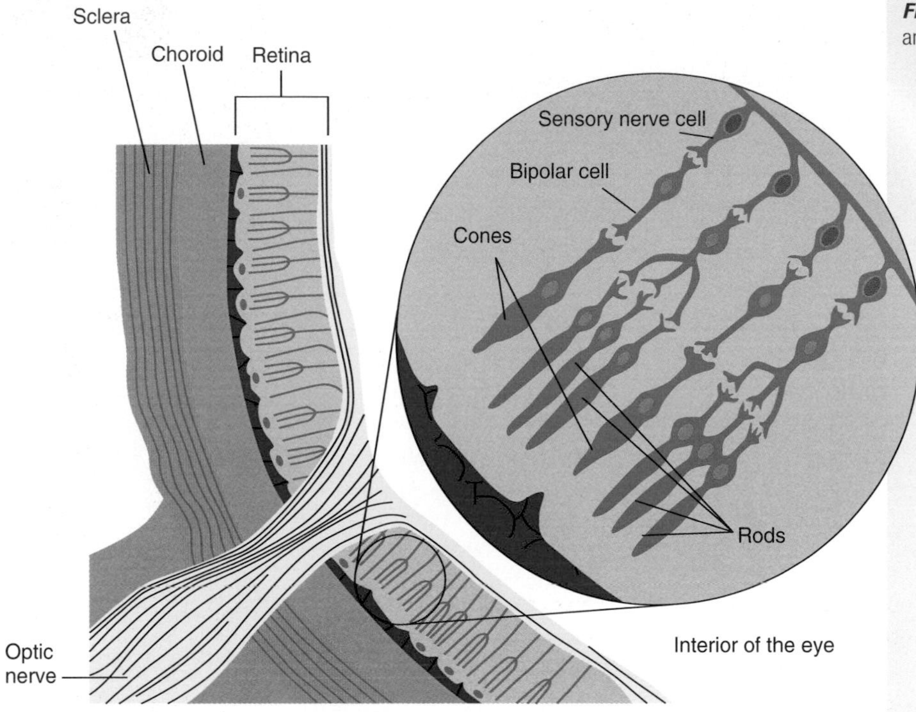

Figure 7.8 Location of rods and cones within the retina

Cones and rods in the retina synapse with (are linked to) nerve cells by **bipolar cells**. The cones and rods differ in **visual acuity** (the degree of detail in which the object is perceived) and in **sensitivity** (the intensity of light required to produce a sufficiently large generator potential to initiate a nerve impulse) because of the way in which they are connected to the bipolar cells.

Each cone is connected to a separate bipolar cell that synapses with a single nerve cell. Several rods, however, are connected to just one bipolar cell. This is called **retinal convergence**. These connections affect the sensitivity and acuity of the rods and cones.

The sensitivity is, effectively, the lowest intensity of stimulation that results in a nerve impulse. To do this, the rods or cones must produce a generator potential large enough to exceed the threshold level of stimulation of the nerve cell. Several rods connect to just one bipolar cell. Therefore, although the individual generator potentials fall short of the threshold, their *combined* generator potentials may exceed the threshold. This effect is called **summation**. Summation is not possible in cones because each cone is connected separately to a single bipolar cell. As a result, sensitivity is greater in rods than in cones.

Visual acuity is the detail in which an object is perceived. This depends on how many individual components there are in the image, rather like pixels on a screen. More pixels mean more definition. If 1 million cones were stimulated, this would result in 1 million nerve impulses going to the brain from which the brain could construct an image. However, if 1 million rods were stimulated, then, because of retinal convergence, this would result in fewer impulses to the brain from which it could construct the image. It would therefore be less detailed. Visual acuity is greater in cones.

The differences in sensitivity and visual acuity are summarised in Table 7.1.

Table 7.1 Sensitivity and visual acuity

Property	Cones	Rods
Sensitivity	Low: light energy transduced by a single cone must produce a generator potential large enough to exceed the threshold needed for a nerve impulse. In low light intensities this is unlikely.	High: in low light intensities, generator potentials from several rods can combine and so the threshold is more likely to be exceeded and a nerve impulse initiated. This phenomenon is called summation. It is possible because several rods are linked to (or converge on) one neurone (via bipolar cells). This is called retinal convergence.
Acuity	High: each cone is connected to a single bipolar cell, so in high light intensities each cone stimulated represents a separate part of the image which can be seen in detail.	Low: several rods are connected to the same bipolar cell, so the individual parts of the image represented by each rod are merged into one — they are indistinguishable and detail is poor.

The greatest concentration of cones is found at the fovea in the centre of the retina. Looking straight at an object focuses light from it onto the fovea, enabling the object to be seen in great detail when the light intensity is high. For example, to read this print, you must look straight at it. This results in the rays of light from the print being focused on the **fovea**. This gives the greatest possible acuity and allows you to make out the letters and your brain to construct the words. If you look just fractionally away from the print, you will be aware of letters and words, but you will not be able to read them. The area of the retina where the light rays now fall has too few cones to give the necessary acuity.

The greatest concentration of rods is about 20° away from the fovea. In very low light intensities, looking slightly to the side of an object causes the light rays to fall on this area of the retina. Summation by the rods allows better perception than if light fell on the fovea. For example, if on a clear night you find the constellation of stars called the Pleiades (or the seven sisters) and look straight at it, you will be able to make out a few quite bright stars.

If you look just to one side of the constellation, it will appear brighter and you will be able to make out more stars. This is because the light from the dimmer stars is too low in intensity to be detected by the cones when you look straight at the constellation. By looking slightly to one side, the light rays are focused on an area of the retina with a high concentration of rods. Summation allows the threshold to be exceeded so you can see the extra stars.

7

The Pleiades

Summary

Plant responses

- Tropic responses by plants help to maintain different parts of the plant in the most favourable environmental conditions.
- Plant shoots are positively phototropic (they grow towards light); this response exposes the leaves to the maximum light intensity, which enhances photosynthesis.
- Plant roots are positively gravitropic (they grow towards gravity); this response ensures that the roots always grow into the soil from which they absorb water and mineral ions.

Responses in simple animals

- Simple animals exhibit taxes and kineses.
- A taxis involves a directional response to the differential intensity of a stimulus — for example, animals may move towards (or away from) the strongest light or the most humid atmosphere.
- A kinesis involves a change in the rate of movement as a result of a change in intensity of a stimulus. An increase in the rate of movement when exposed to a stimulus associated with hostile conditions makes it more likely that the animal will move out of the hostile conditions.

Responses in humans

- Human receptors are energy transducers that respond to specific stimuli.
- The Pacinian corpuscle, is a receptor located deep in the dermis of the skin that responds to changes in pressure. An increase in pressure deforms the lamellae of the Pacinian corpuscle, which causes:
  - pressure-sensitive sodium ion channels in the membrane of the naked nerve ending to open
  - positively charged sodium ions to enter the naked nerve ending, altering the balance of charge on either side of the membrane
  - a generator potential to be produced
  - a nerve impulse to pass along the myelinated part of the nerve cell leading from the naked nerve ending

- A nerve impulse is only initiated if the generator potential is larger than the threshold level of stimulation of the nerve cell.
- Rods and cones are receptors in the retina of the eye that are sensitive to light.
- Rods show retinal convergence; several rods are connected to one bipolar cell.
- Cones are connected to bipolar cells on a 1:1 basis.
- Because of retinal convergence, rods have higher sensitivity but lower visual acuity than cones.

Questions

Multiple-choice

1 Positive phototropism is:
A growth towards gravity
B growth towards light
C growth away from light
D growth away from gravity

2 A taxis could involve:
A a change in the rate of movement as a result of a change in the intensity of a stimulus
B a change in the direction of movement as a result of a change in the intensity of a stimulus
C a change in the direction of movement as a result of a change in the direction of a stimulus
D a change in the rate of movement as a result of a change in the direction of a stimulus

3 A kinetic response to unfavourable stimuli makes it more likely that:
A the organism will remain in the unfavourable conditions due to decreased movement
B the organism will remain in the unfavourable conditions due to increased movement
C the organism will leave the unfavourable conditions due to increased movement
D the organism will leave the unfavourable conditions due to decreased movement

4 In the human eye, compared with cones, rods have:
A greater sensitivity and greater visual acuity
B greater sensitivity but less visual acuity
C less sensitivity and greater visual acuity
D less sensitivity and less visual acuity

5 The Pacinian corpuscle is able to produce generator potentials of different magnitudes because:
A light pressure causes many sodium ion channels to open and heavy pressure causes few sodium ion channels to open
B light pressure causes many sodium ion channels to open and heavy pressure causes many sodium ion channels to open

C light pressure causes few sodium ion channels to open and heavy pressure causes few sodium ion channels to open

D light pressure causes few sodium ion channels to open and heavy pressure causes many sodium ion channels to open

Examination-style

1 Sense cells, such as Pacinian corpuscles in the skin, are biological energy transducers. They transduce one form of energy from the environment into the electrochemical energy of a generator potential. Nerve impulses may be initiated as a result of these generator potentials.

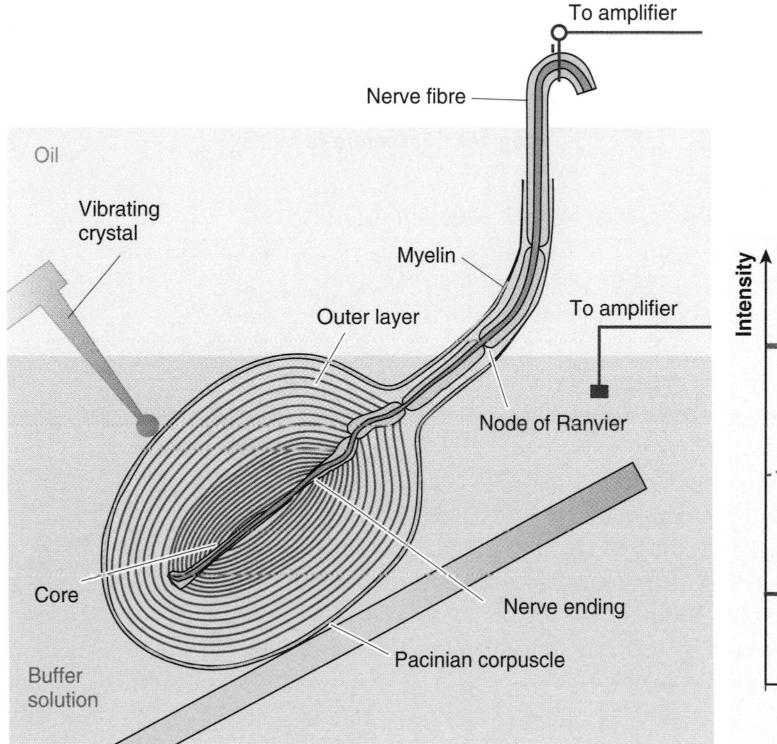

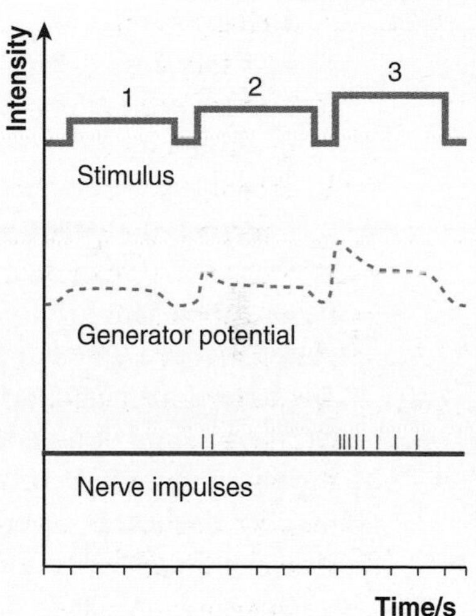

The left-hand diagram shows the apparatus used in an investigation into the mode of action of Pacinian corpuscles. The vibrating crystal is connected to a variable power pack. The right-hand diagram summarises the results obtained.

(a) How does the Pacinian corpuscle convert the vibrations of the crystal into a generator potential? *(3 marks)*

(b) Suggest how the variations in the stimulus shown in the right-hand diagram could be generated using the above equipment. *(2 marks)*

(c) Use your knowledge of the 'all or nothing' nature of nerve impulses to suggest an explanation for the results obtained. *(3 marks)*

Total: 8 marks

2 The diagram shows the distribution of rods and cones across the retina of the human eye.

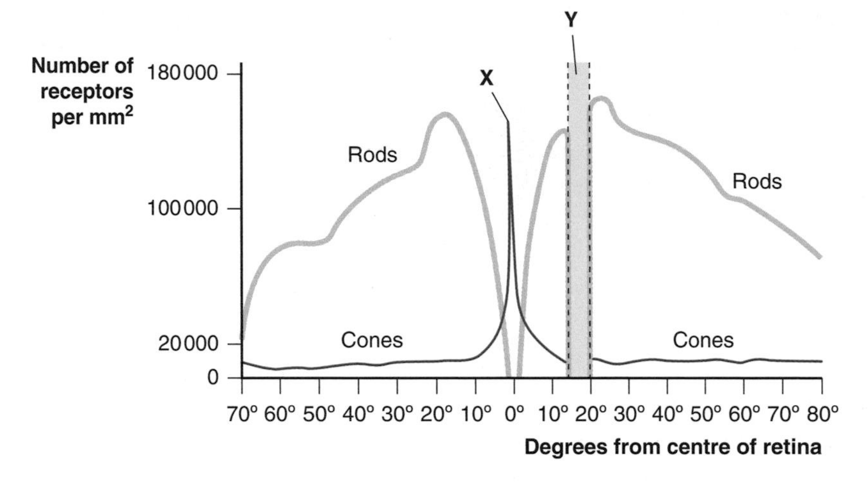

(a) Describe and explain the difference in sensitivity of rods and cones. *(4 marks)*

(b) (i) Explain why images formed on point **X** of the retina can be seen in more detail than those formed anywhere else on the retina *(2 marks)*

(ii) Explain why, when an image is formed on spot **Y** of the retina, nothing is perceived by the brain. *(2 marks)*

Total: 8 marks

3 In an investigation into kinetic responses in woodlice, the woodlice were placed one at a time into a chamber in which the conditions could be altered. The base of the chamber was marked out in 1 cm squares. The woodlice were left for a period of time and the number of squares they covered during that time was recorded. Then, a kinetic value was calculated by dividing the number of squares covered by the time. A high kinetic value indicates more movement than a low kinetic value.

(a) One set of results is shown in the table. The conditions were cold and dry.

Investigation	Squares covered	Time	Kinetic value
1	122	300	0.407
2	206	300	
3	62	180	0.344
4	42	180	0.233
5	97	180	0.539
6	76	180	0.422

Mean kinetic value = 0.439

(i) What is a kinesis? *(2 marks)*

(ii) Calculate the missing kinetic value. Show your working. *(2 marks)*

(b) The graph summarises the results from several investigations.

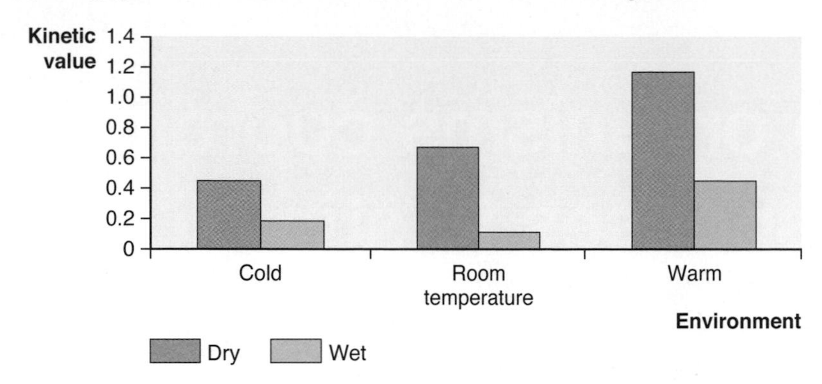

(i) Suggest **one** way in which the results lack precision and suggest how this could be improved. *(2 marks)*

(ii) Use information from the graph and your knowledge of kineses to suggest the conditions preferred by woodlice. Explain your answer. *(4 marks)*

Total 10 marks

4 There are two types of cell in the retina: rods and cones. The graph shows the intensity of light of different wavelengths that is required to stimulate the two types of cell sufficiently for the generator potentials produced to produce nerve impulses in the nerve cells to which they are connected.

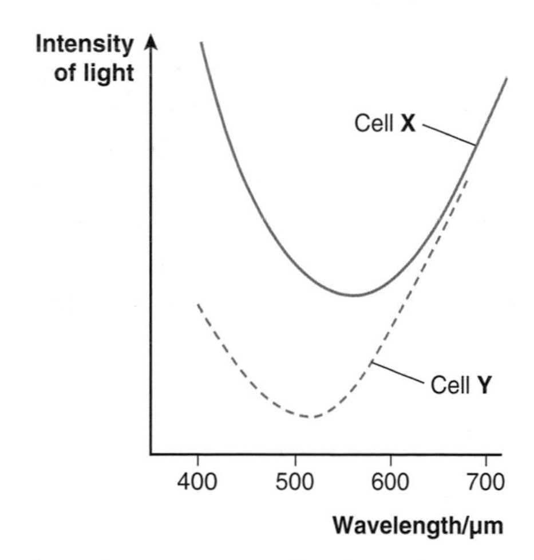

(a) (i) Describe how the sensitivity of cell **X** changes at different wavelengths. *(2 marks)*

(ii) Which type of cell is cell **Y**? Explain your answer. *(3 marks)*

(b) (i) To what type of nerve cell are cells **X** and **Y** connected? *(1 mark)*

(ii) Explain what is meant by 'threshold stimulation'. *(2 marks)*

(c) Cones have a greater visual acuity than rods. Explain why. *(4 marks)*

Total: 12 marks

Chapter 8

How do organisms coordinate their responses to stimuli?

This chapter covers:
- the structure of human nerve cells
- how a nerve impulse is initiated and propagated along a nerve cell
- how the structure of a nerve cell influences the speed at which it conducts nerve impulses
- how a nerve impulse crosses the synapse between nerve cells
- how nerve cells are organised into a reflex arc to produce a reflex action
- how human hormones produce responses
- the roles of histamine and prostaglandins
- how plant growth substances produce tropic responses

We saw in Chapter 7 that, for a stimulus to produce a response by an organism, some kind of linking or coordinating system is needed. In this chapter we shall look at the different ways in which organisms coordinate their responses.

In humans, some responses rely on nervous coordination, whereas others are coordinated by hormones.

Plant growth substances coordinate the directional growth responses shown by plants to the directional stimuli of light and gravity.

How are nerve impulses passed through the human nervous system?

How are nerve cells arranged in our nervous system?

The nervous system is composed of billions of cells. Most of these cells are involved in transmitting 'information' and are called **neurones**. These are organised into larger structures, including:
- nerves
- the spinal cord
- the brain

The brain and the spinal cord form the **central nervous system (CNS)**; the nerves form the **peripheral nervous system (PNS)**. Figure 8.1 shows this organisation.

Neurones can be classified by shape, but for our purposes we shall classify them according to function:

- **sensory neurones** — these carry **nerve impulses** from sense cells into the CNS (Figure 8.2a)
- **motor neurones** — these carry impulses from the CNS to effectors such as muscles (Figure 8.2b)
- **inter-neurones** or **relay neurones** — these carry impulses from a sensory neurone to a motor neurone, often in **reflex arcs**, within the CNS

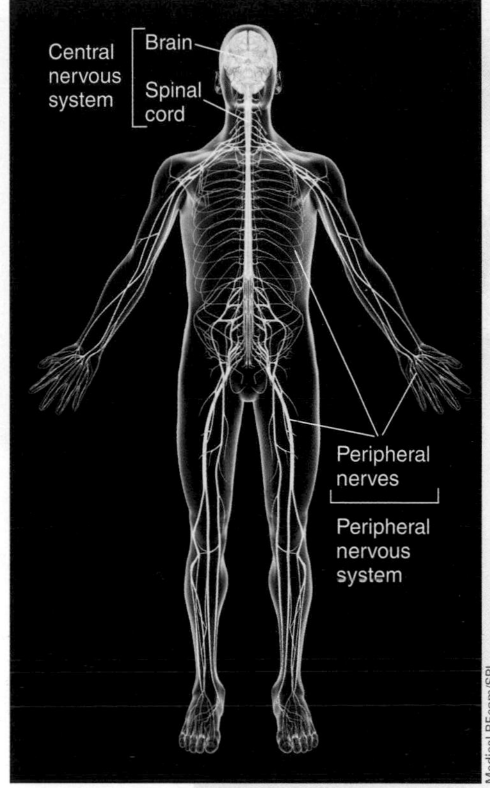

Figure 8.1 The nervous system.

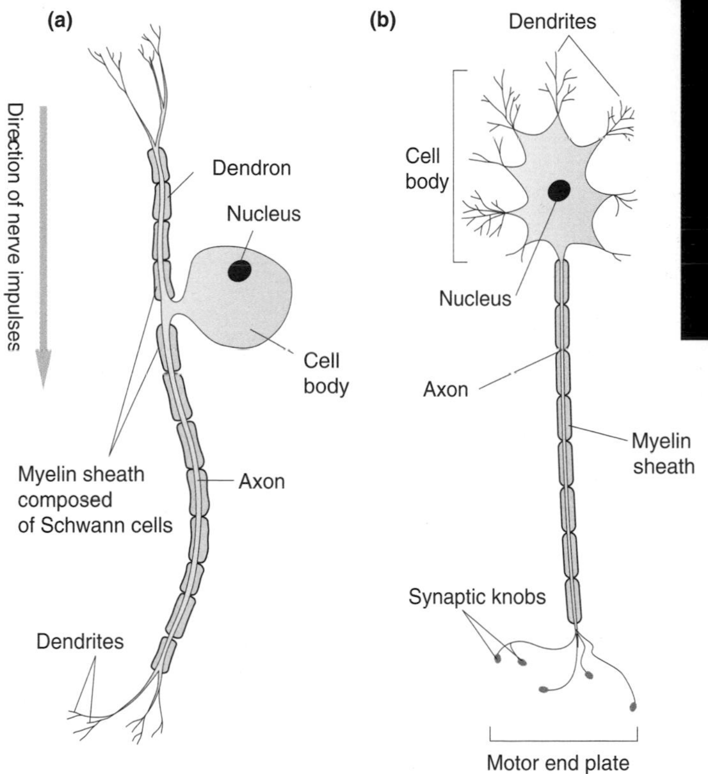

Figure 8.2 ***(a)*** Sensory and ***(b)*** motor neurones. Impulses pass along the axons of neurones. The myelin sheath is important in ensuring that the impulses pass quickly along the axon without any deterioration.

Box 8.1 The importance of myelin

The axons of neurones acquire their myelin courtesy of another nerve cell. These other cells are:

- in the CNS, oligodendrocytes
- in the PNS, Schwann cells

The dendrites of the oligodendrocytes wrap themselves around the axons of neurones forming a sheath that contains myelin (Figure 8.3).

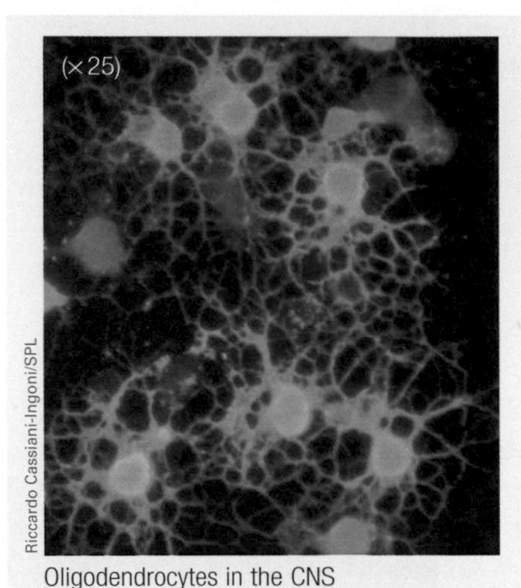

Oligodendrocytes in the CNS

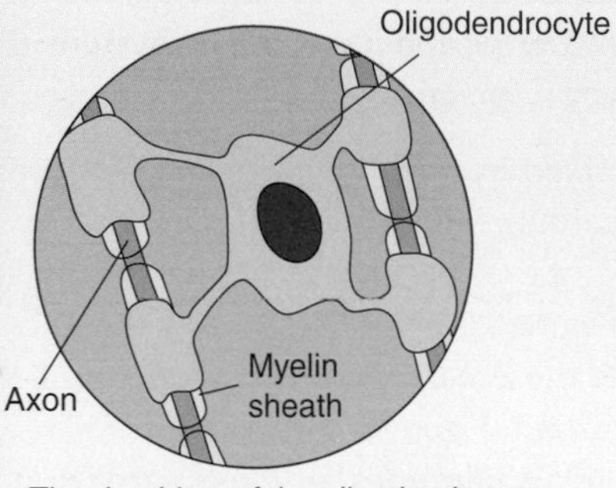

The dendrites of the oligodendrocyte wrap themselves around the axon forming a multi-layer myelin sheath

Figure 8.3 Myelin sheath formation in the central nervous system

In the peripheral nervous system, each Schwann cell forms a myelin sheath on only one axon, as shown in Figure 8.4.

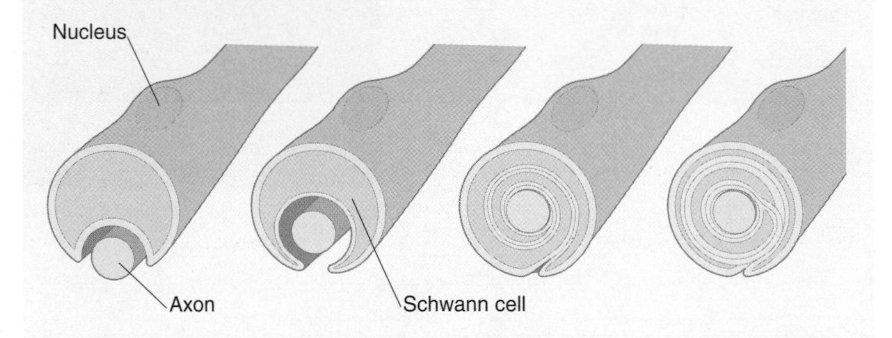

Figure 8.4 Myelin sheath formation in the peripheral nervous system. Each Schwann cell wraps itself round one axon, forming a multi-layer myelin sheath.

Myelin effectively insulates each axon and allows nerve impulses to pass along the axon efficiently. In some diseases, for example multiple sclerosis, myelin is progressively lost from axons. This is called demyelination.

Demyelinated axons conduct impulses much more slowly — the impulse may never reach the end of the axon. This results in a decrease in the ability to coordinate actions.

How does the structure of an axon membrane allow a nerve impulse to pass?

First, you must appreciate that a nerve impulse is not like an electric current in which electrons flow through wires. It is more like a Mexican wave in which people in a stadium stand and wave in turn. The people do not move around the stadium, but the wave goes the whole way round.

A nerve impulse results from a change in the electrical balance of the membrane of an axon being passed along the axon. This change in electrical balance is called an **action potential**.

Action potentials can be initiated and transmitted along neurones because of the specialised plasma membrane of the axons of these cells. When a nerve impulse is not being conducted, an electrical potential difference is maintained between the inside and outside of the axon membrane. This is called the **resting potential** and typically, at rest, the inside of the axon is about 70 mV (millivolts) more negative than the outside.

This is often written as –70 mV.

The resting potential is maintained by:

- large anions, for example negatively charged protein molecules inside the axon
- passive diffusion of sodium and potassium ions across the membrane through ion channels; sodium ions diffuse in more slowly than potassium ions diffuse out
- active transport of sodium and potassium ions across the membrane by transport proteins; sodium ions are pumped out faster than potassium ions are pumped in (Figure 8.5)

Sodium ions and potassium ions each carry one positive charge. More positively charged ions moving out than there are moving in means that there is a build up of positive ions outside the axon. Put another way, the inside of the axon is less positive (more negative) than the outside.

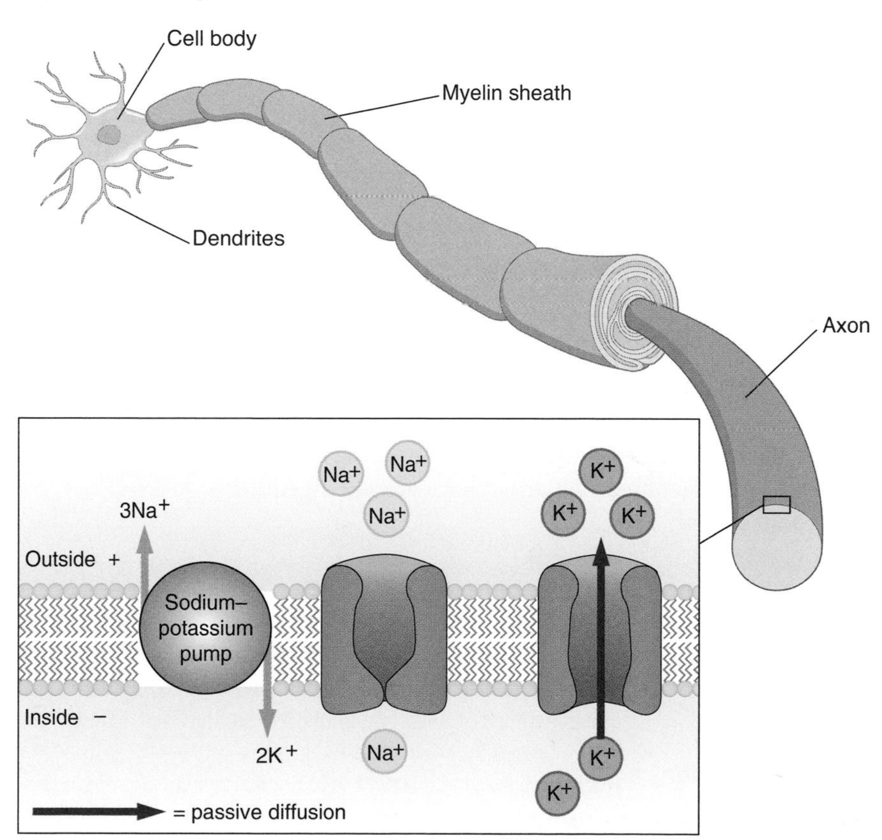

Figure 8.5 The membrane of an axon

To maintain the resting potential, positive ions are moved out of the axon at a greater rate than they are moved in.

The term 'resting potential' is a little misleading because it implies that the membrane is doing nothing. In fact, it is using ATP to actively maintain the 'resting potential'.

How is an action potential initiated?

Because the inside of an axon membrane in the 'resting' condition has a potential of –70 mV (with respect to the outside), we say that the membrane is **polarised**.

If the membrane of a neurone is stimulated, then the situation changes and an action potential may result.

Besides the ion channels that are open as the resting potential is maintained, there are also **gated channels** that are firmly shut during maintenance of the resting potential. On stimulation, some gated sodium ion channels in the membrane open and allow positively charged sodium ions to enter.

During this period, similar gated potassium ion channels remain firmly shut.

This reduces the resting potential. If it is reduced to –55 mV, thousands more 'voltage-activated' sodium ion channels open, allowing sodium ions to 'flood' in. These positive ions quickly reduce the resting potential and even allow the inside of the membrane to become positively charged (with respect to the outside — Figure 8.6).

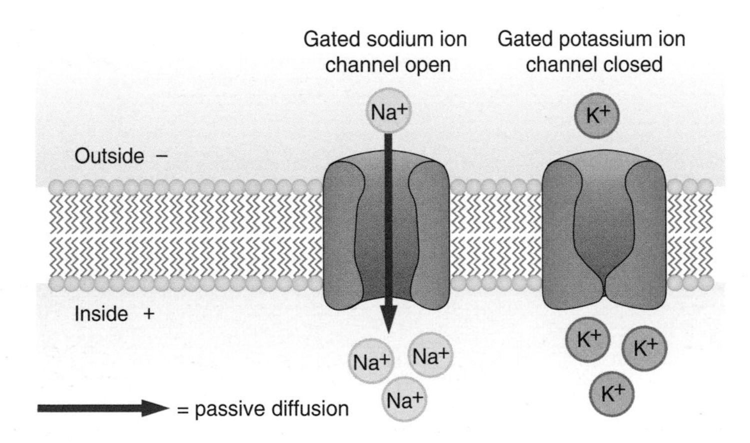

Figure 8.6 During depolarisation, gated sodium ion channels open allowing sodium ions to enter the axon. Gated potassium ion channels remain closed.

The size of the action potential does not vary: it is always +40 mV.

There either *is* an action potential or there *isn't*. It is an **all-or-nothing response**. If the critical potential of –55 mV is not reached, the voltage-activated gates do not open, sodium ions cannot flood in and there is no action potential. The critical value of –55 mV is known as the **threshold value**.

Once an action potential has been generated, it is propagated along the axon because it stimulates depolarisation in successive regions of the axon membrane.

What happens next?

The axon membrane must return to the 'resting' condition — it must be **repolarised**. If this did not happen, it would not be possible to initiate another action potential.

Box 8.2 How do we detect the intensity of a stimulus?

If the size of an action potential doesn't vary, how do we know, for example, the difference between a loud noise and a quiet sound, or between a bright light and a dim light? The links between detection of a stimulus and generation of an action potential vary between different receptors but, in general, the following hold true:

- An increase in intensity of the stimulus results in a *larger* generator potential in the receptor cell (an increase in *amplitude* of the generator potential).
- An increased generator potential results in *more* action potentials per second (an increase in *frequency* of the action potentials).

So, a loud noise results in more action potentials per second along the auditory nerve than a quiet sound does and bright light results in more action potentials per second along the optic nerve than a dim light does.

Repolarisation begins as soon as the action potential is reached. The gated sodium ion channels close and gated potassium ion channels open. Positively charged potassium ions flood out. This makes the inside of the membrane once again 70 mV negative with respect to the outside and another action potential can be generated.

In practice, the potassium ions flood out just a little too quickly and the membrane potential becomes lower than the resting potential; perhaps to −80 mV. This 'undershoot' is called **hyperpolarisation**.

The events of depolarisation and repolarisation are shown in Figure 8.7.

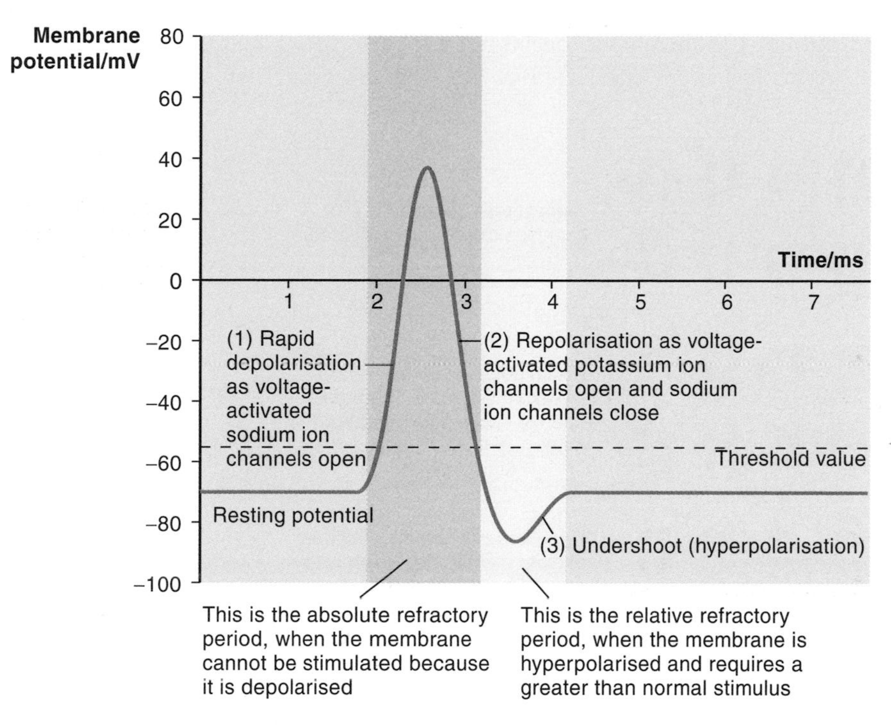

It is important to realise that this graph shows changes that occur *over a short period of time* at *one place* in the axon.

Figure 8.7 Changes that occur as an axon membrane is depolarised and then repolarised

There is a period of time following the initiation of one action potential during which another action potential cannot be generated. This is called the **absolute refractory period**. It includes:

- the period of depolarisation of the axon; another action potential cannot be initiated before the current one is achieved
- the period of repolarisation; the gated sodium ion channels are not only shut but also inactivated (think of them as 'locked'), so they cannot open during this phase and sodium ions cannot enter to initiate another action potential

It is also more difficult, but not impossible, to initiate another action potential during the period of hyperpolarisation. Because the membrane potential is more negative than the normal resting potential, it requires a larger generator potential to open more sodium ion channels so that the threshold can be reached. However, because it is not impossible to initiate another action potential during this period, it is called the **relative refractory period**.

The duration of the refractory period limits how many action potentials can be initiated per second. This, in turn, limits the number of nerve impulses that can be transmitted along a neurone per second.

How does the action potential pass along an axon?

When an action potential is generated at one end of an axon, the depolarisation affects the axon membrane just ahead of it; this region now begins to depolarise and generate an action potential in the new region. The action potential in the new region now affects the axon membrane just ahead of *it* and *this* begins to depolarise...and so on along the axon (Figure 8.8).

Meanwhile, once an area of the membrane begins to repolarise, the gated sodium ion channels in this region are inactivated as well as being shut. They

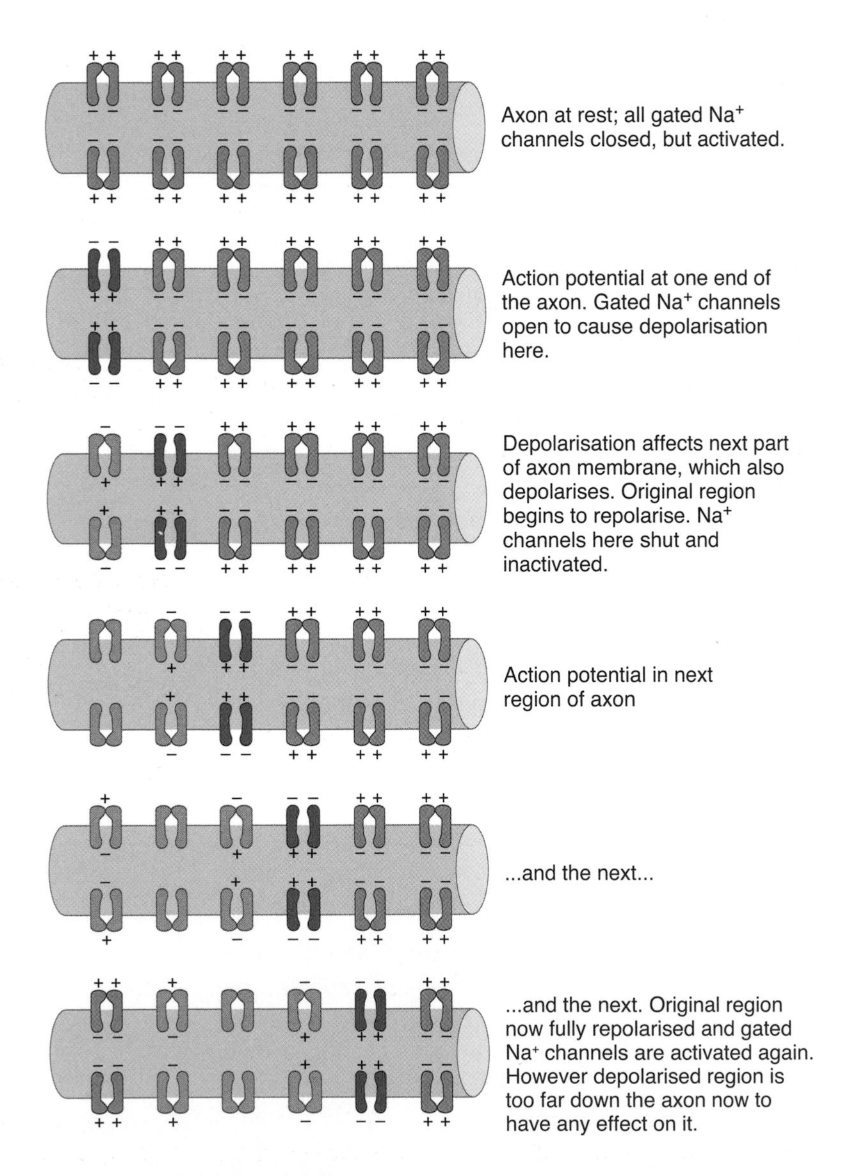

Figure 8.8 Propagation of an action potential. An action potential can only move in one direction along an axon. Inactivation of gated sodium ion channels following depolarisation prevents it from travelling backwards.

become activated again once the membrane is repolarised fully. By this time, the action potential has moved a considerable way down the axon and cannot affect the area where the action potential originated. This explains why an action potential can only travel in one direction along an axon. It cannot move backwards to where the action potential originated because the gated sodium ion channels are shut and inactivated in this region, so depolarisation leading to an action potential is impossible.

The speed of transmission of an action potential is affected by several factors. These include:

- temperature — like all physiological processes, an increase in temperature increases the rate of the process, up to the point at which enzymes and transport proteins begin to denature
- diameter of the neurone — neurones with a large diameter conduct impulses faster than those with a narrow diameter
- myelination — myelinated neurones conduct impulses faster than non-myelinated neurones

The method of propagation of action potentials shown above is what happens in neurones that have no myelin sheath. Those that are myelinated propagate action potentials by a method called **saltatory conduction**.

'Saltatory' is derived from the Latin verb *saltare*, which means to jump.

In a non-myelinated neurone, each portion of the axon depolarises in turn and the action potential is propagated along the entire axon. In myelinated neurones, the action potential is only generated at the nodes of Ranvier — 'gaps' in the myelin sheath. The nerve impulse 'jumps' from node to node and so is conducted *much* faster than in non-myelinated neurones (Figure 8.9).

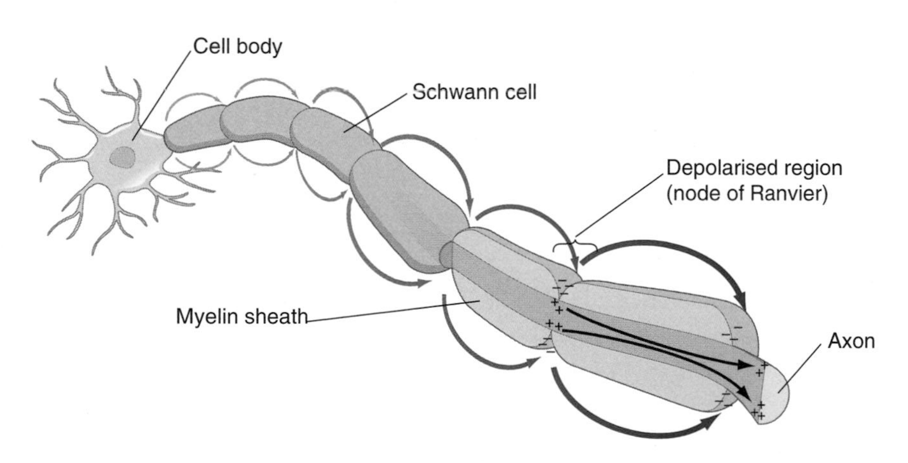

Figure 8.9 Saltatory conduction. In saltatory conduction, the action potential 'jumps' from one node of Ranvier to the next.

How do nerve impulses pass from one neurone to another?

Neurones do not touch each other. There is a small space between the dendrites of one neurone and the dendrites of other neurones. The region that includes a dendrite of one neurone and that of another is called a **synapse**.

Note that the synapse is not just the gap between the two dendrites; this is the synaptic cleft.

At a synapse, only a minute distance separates the membranes of two neurones. Chemical transmitters (**neurotransmitters**) are released from **secretory vesicles** in the **pre-synaptic neurone**. These cross the **synaptic cleft** and bind to **receptors** on the membrane of the **post-synaptic neurone**. The neurotransmitter at many synapses in the CNS and PNS is **acetylcholine**; these synapses are called **cholinergic** synapses (Figure 8.10).

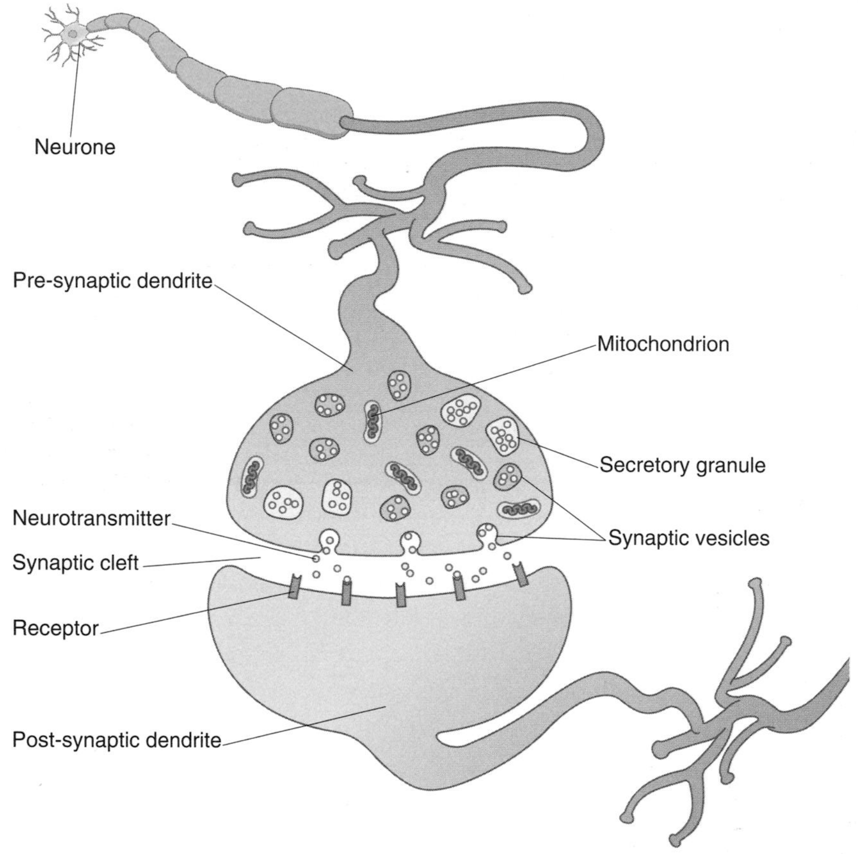

Figure 8.10 Structure of a cholinergic synapse

Notice that the presynaptic dendrite contains mitochondria. These produce the ATP that is needed to synthesise the acetylcholine (neurotransmitter) molecules.

When an action potential arrives at the pre-synaptic membrane, it causes the following to take place:

- Ion channels open that allow calcium ions to enter the neurone.
- This causes the secretory vesicles to move to, and fuse with, the pre-synaptic membrane and release acetylcholine (neurotransmitter) into the synaptic cleft.
- Acetylcholine binds to receptor proteins in the post-synaptic membrane.
- Once bound, the neurotransmitter causes gated sodium ion channels in the post-synaptic membrane to open.
- This leads to the generation of an action potential, depolarising the membrane in that region; the action potential is then propagated along the post-synaptic neurone.
- The neurotransmitter is then hydrolysed by the enzyme acetylcholinesterase and the products are reabsorbed into the pre-synaptic neurone.
- Once reabsorbed, they are used to resynthesise more acetylcholine.

It is important that the neurotransmitter is not allowed to remain for too long in the receptors in the post-synaptic neurone as it could cause repeated generation of an action potential.

Box 8.3 One way transmission at synapses

The structure of a synapse ensures that the nerve impulse can only pass from the pre-synaptic neurone to the post-synaptic neurone. This is because:

- only the pre-synaptic neurone has synaptic vesicles containing neurotransmitter
- only the post-synaptic membrane has receptor proteins
- there is a diffusion gradient of neurotransmitter from pre- to post-synaptic membrane

What has just been described is what happens at an **excitatory synapse**. However, synapses can be excitatory or **inhibitory**; that is, they can promote the initiation of an action potential in the post-synaptic neurone (excitatory synapse) or they can make it less likely to occur (inhibitory synapse). The process of synaptic transmission is essentially the same at excitatory and inhibitory synapses, except that the neurotransmitter involved is different and so is the effect on the post-synaptic membrane. The events are summarised in the flow chart in Figure 8.11.

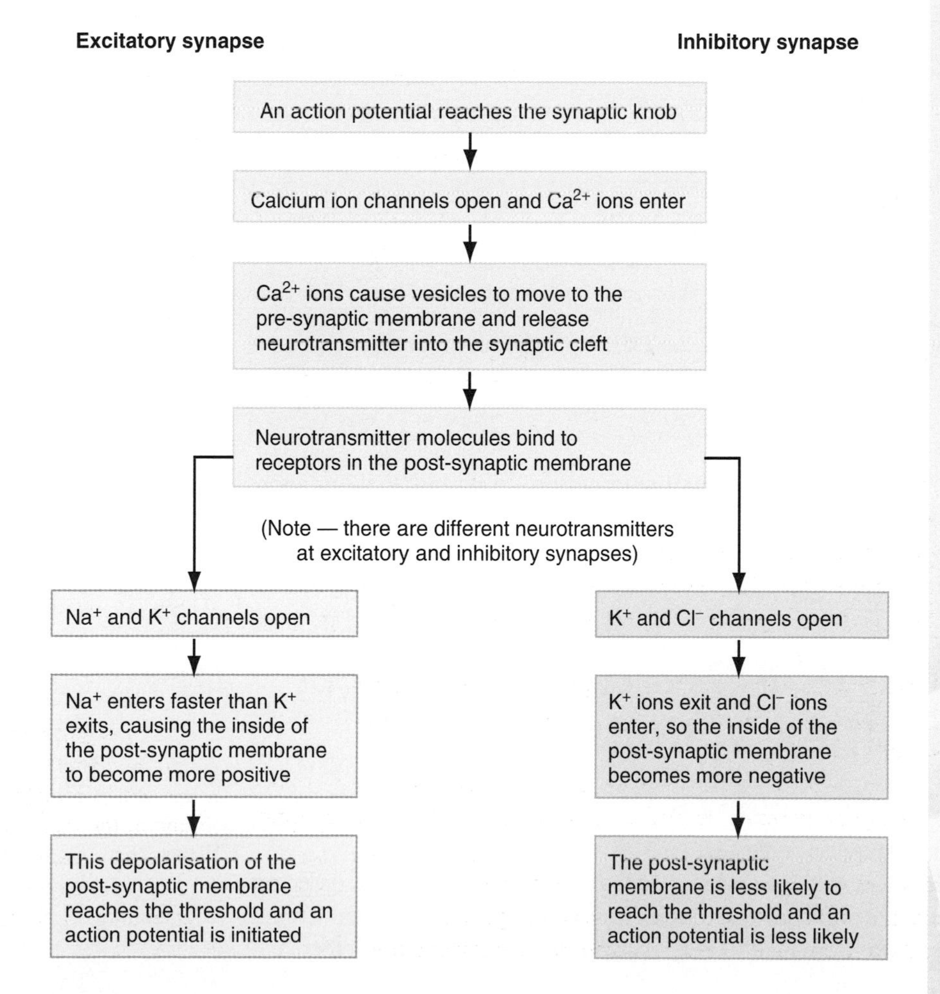

Figure 8.11 The differences between an excitatory and an inhibitory synapse

Usually, many neurones (rather than just two) synapse together. Some of the synapses might be excitatory, others might be inhibitory. The effect on the post-synaptic cell is the result of the **summation** of their individual effects. This type of summation is called **spatial summation** (Figure 8.12).

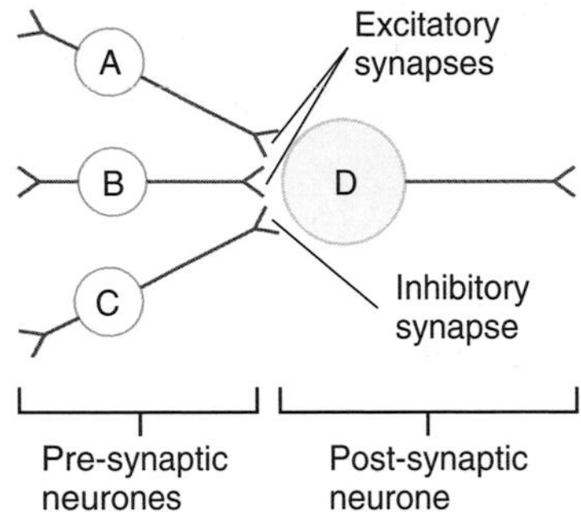

Figure 8.12 How summation operates at synapses

Action potential in pre-synaptic neurones	Action potential in neurone D	Reasons
A	✗	Sub-threshold stimulation — voltage-activated sodium channels are not opened
B	✗	Sub-threshold stimulation — voltage-activated sodium channels are not opened
A + B	✓	The summation of effects of neuro-transmitter of both excitatory synapses exceeds the threshold
A + C	✗	The summation of effects of excitatory and inhibitory synapses results in little change in the membrane potential — the threshold is not reached

Another type of summation is **temporal summation**. In temporal summation, several impulses arrive *at the same synapse* in rapid succession. Each depolarises the membrane of the post-synaptic neurone a little more until threshold is reached and an action potential is generated.

◀ Summation at synapses allows some control over whether or not an action potential is initiated in a post-synaptic neurone. This control is called modulation.

How are neurones organised to produce reflex actions?

What kind of reflex actions are there?

There are two main types of reflex action:

- **Somatic reflexes** involve our special senses (eyes, ears, pressure detectors etc.) and produce responses by muscles. They include the 'knee-jerk' reflex and the 'withdrawal-from-heat' reflex. Many of these reflexes are protective.

- **Autonomic reflexes** involve sensors in internal organs and produce responses in internal organs. They include those that control heart rate and breathing rate.

To understand how these two types of reflex action operate, we must look again at the structure of the nervous system. We have already seen that our nervous system is divided physically into two major components:
- the **central nervous system** (CNS), comprising the brain and spinal cord
- the **peripheral nervous system** (PNS), comprising the cranial and spinal nerves, each containing many hundreds of sensory and motor neurones

However, we can also divide our nervous system functionally into:
- the **somatic nervous system** (SNS), which integrates information from the special senses to produce responses in skeletal muscles
- the **autonomic nervous system** (ANS), which integrates information from receptors in internal organs and produces responses in the same or other organs or glands (Figure 8.13)

Both the functional divisions are present in each of the physical parts of the nervous system.

The autonomic nervous system is further subdivided into:
- the **sensory branch** (or division), which transmits sensory nerve impulses into the central nervous system
- the **sympathetic branch** (or division), which transmits impulses from the central nervous system to the organs, generally preparing the body for 'fight or flight' — for example by increasing cardiac output and pulmonary ventilation
- the **parasympathetic branch** (or division), which acts antagonistically to the sympathetic branch and prepares the body for 'rest and repair', decreasing cardiac output and pulmonary ventilation

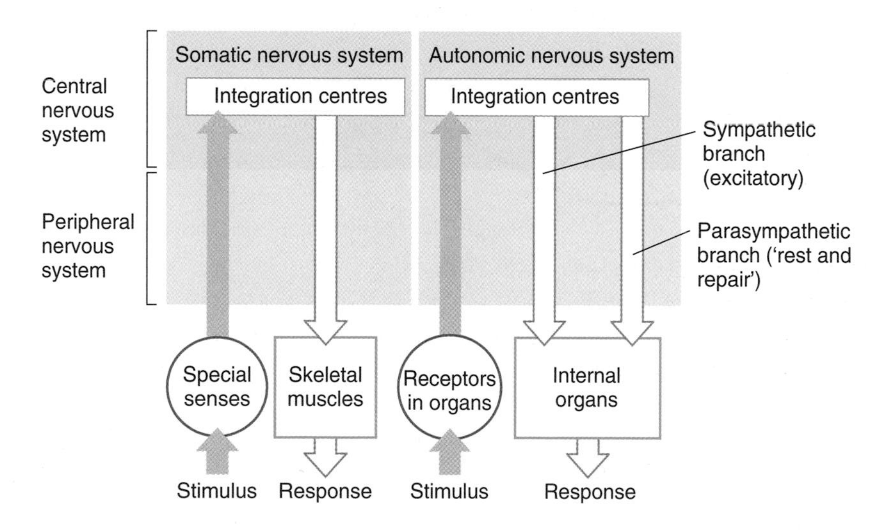

Figure 8.13 The main components of the human nervous system

How are somatic reflex actions produced?

Neurones are sometimes organised into simple **reflex arcs**. In a reflex arc, a few (usually two or three) neurones synapse with each other in sequence so that an impulse passing along the first *must* pass along the others in the arc. The consequence of this is that stimulating the sense cell that initiates the action potential in the first neurone always produces the same response.

The general layout of a three-neurone reflex arc is shown in Figure 8.14.

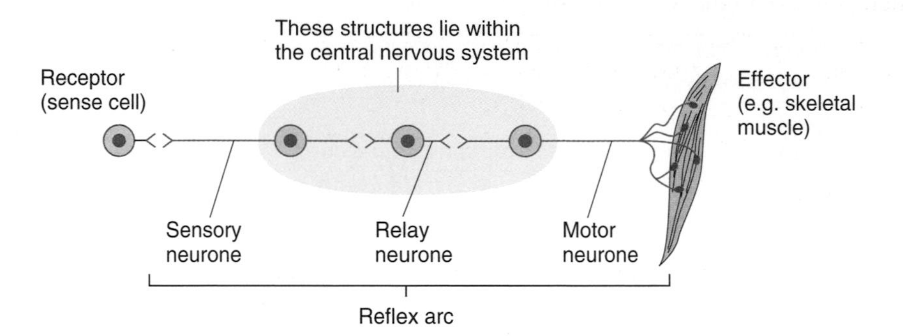

Figure 8.14 The main structures in a somatic reflex arc

Reflex arcs in which synapses with the relay neurone occur in the brain control **cranial reflexes**; those in which the synapses occur in the spinal cord control **spinal reflexes**. Withdrawing your hand from a hot object is an example of a spinal reflex.

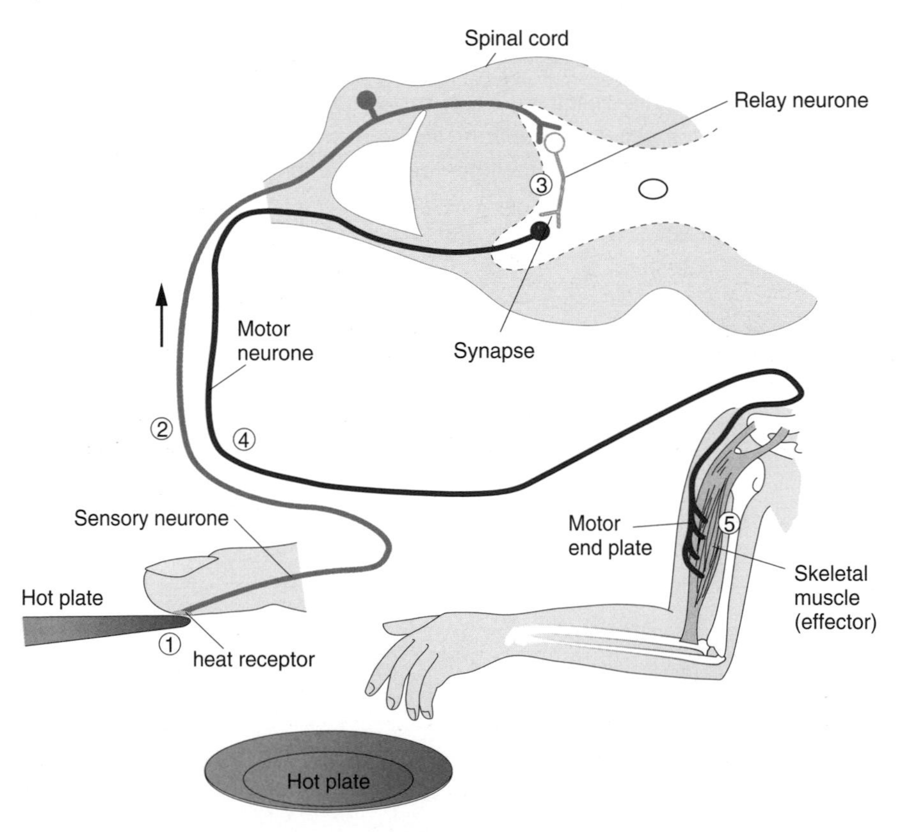

Figure 8.15 The withdrawal-from-heat reflex

The sequence of events in the withdrawal-from-heat reflex (Figure 8.15) is as follows:

(1) The heat receptor is stimulated by the hot plate and a generator potential results.

(2) The generator potential initiates an action potential in the sensory neurone, which is transmitted along the neurone.

(3) At the synapse with the relay neurone, neurotransmitter is released and initiates an action potential in the relay neurone.
(4) This is repeated at the synapse between the relay neurone and the motor neurone and an action potential is transmitted along the motor neurone.
(5) At the motor end plate, the action potential stimulates the contraction of the skeletal muscle, causing the automatic withdrawal of the hand from the hot plate.

How is our heart rate controlled?

The control of heart rate is an example of autonomic reflex control (Figure 8.16). It involves two reflex arcs:

- one that involves the sympathetic branch of the ANS and increases heart rate
- one that involves the parasympathetic branch of the ANS and decreases heart rate

Notice the antagonistic action of the sympathetic and parasympathetic branches.

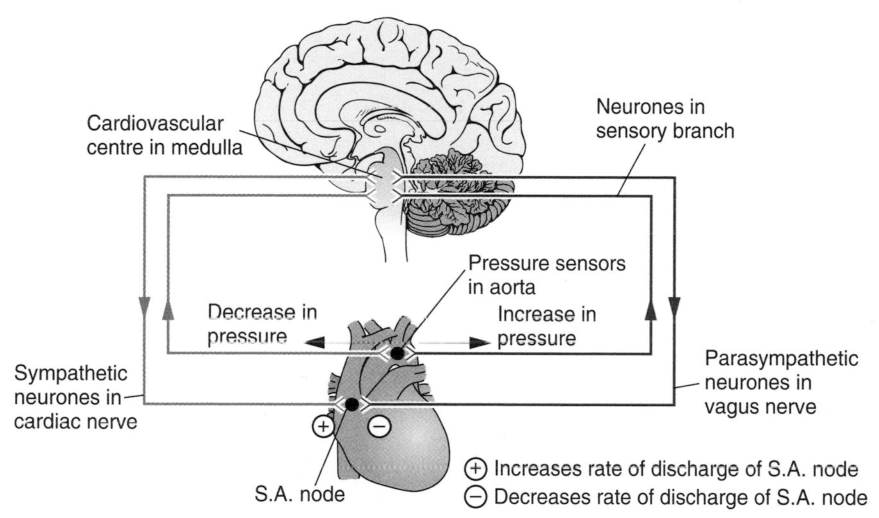

Figure 8.16 How reflex arcs in the ANS control heart rate

Sympathetic control	Parasympathetic control
Initiated by a decrease in blood pressure in the aorta	Initiated by an increase in blood pressure in the aorta
Impulses pass to the cardiovascular centre in sensory neurones and then to the sino-atrial (S.A.) node in the heart via neurones in the cardiac nerve	Impulses pass to the cardiovascular centre in sensory neurones and then to the sino-atrial node in the heart via neurones in the vagus nerve
Noradrenaline is released from endings of sympathetic neurones in the sino-atrial node, which increases activity of the node and so increases heart rate	Acetylcholine is released from endings of parasympathetic neurones in the sino-atrial node, which reduces activity of the node and so reduces heart rate
Other branches of the cardiac nerve affect the ventricle muscle and increase stroke volume	Other branches of the vagus nerve affect the ventricle muscle and reduce stroke volume

How do human hormones control metabolic processes?

What are hormones?

Hormones are chemicals produced by **endocrine glands**. They target particular cells in other parts of the body which have specific **receptor proteins** on or in them. The tertiary structure of a receptor protein has a binding site that has a shape that is complementary to only one hormone. The cells with the appropriate receptors are called **target cells** (Figure 8.17); the organ in which they are found is the **target organ**.

Collectively, all the endocrine glands form the **endocrine system** (Figure 8.18).

Secreting cell
Target cell
Blood vessel
Hormone molecule
Non-target cells
Target cell

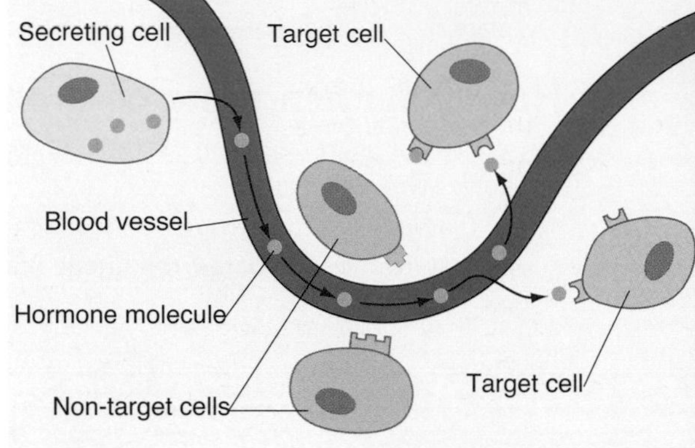

Figure 8.17 How hormones influence cells. Hormones are able to influence some cells and not others because there are different receptors on different cells

▲ Remember that the shapes are complementary, like an egg cup and an egg. Two objects with the same shape (such as two eggs) cannot fit together.

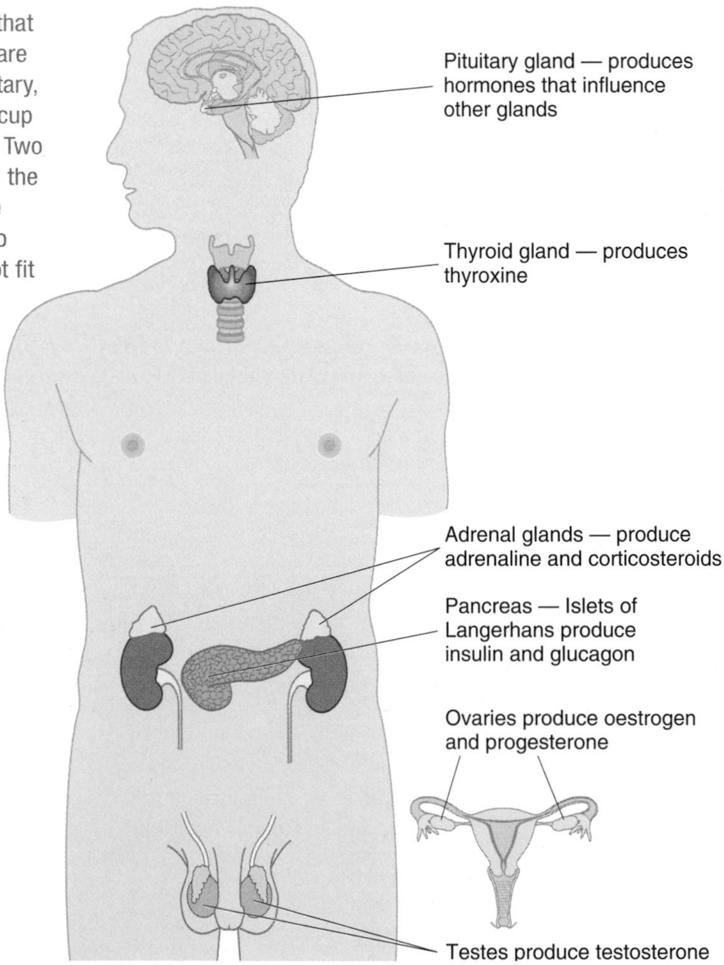

Pituitary gland — produces hormones that influence other glands
Thyroid gland — produces thyroxine
Adrenal glands — produce adrenaline and corticosteroids
Pancreas — Islets of Langerhans produce insulin and glucagon
Ovaries produce oestrogen and progesterone
Testes produce testosterone

Figure 8.18 The glands of the endocrine system and the major hormones they produce

How do hormones act?

Hormones activate processes in cells in one of two main ways.

- Steroid hormones (for example the male and female sex hormones) are small, lipid-soluble molecules that can pass freely through plasma membranes and bind with receptors in the cytoplasm. The receptor–hormone complex then moves into the nucleus, and binds with and activates specific genes (Figure 8.19).
- Non-steroid hormones (for example insulin and glucagon) bind to receptors on the surface of the cell. This activates a molecule in the membrane called a 'G-protein', which activates specific enzymes that produce specific metabolic effects (Figure 8.20).

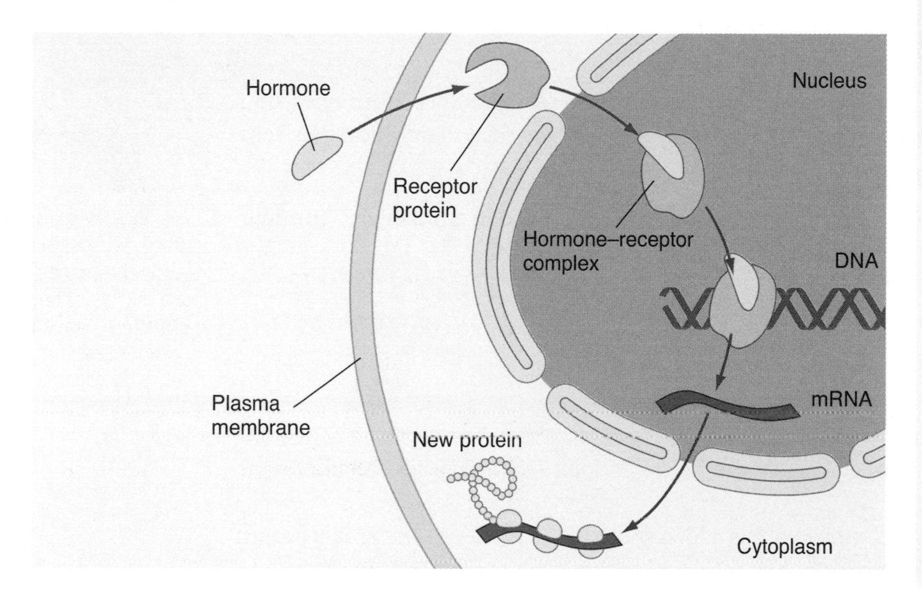

Figure 8.19 Action of steroid hormones. Steroid hormones are able to activate specific genes by binding with receptors that move into the nucleus and bind with specific genes.

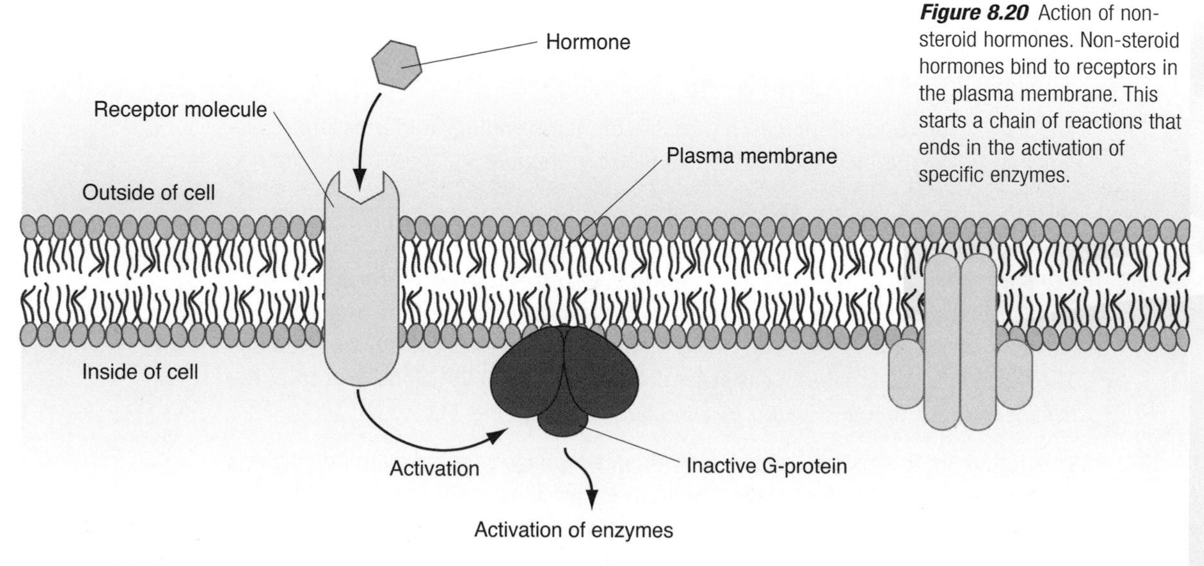

Figure 8.20 Action of non-steroid hormones. Non-steroid hormones bind to receptors in the plasma membrane. This starts a chain of reactions that ends in the activation of specific enzymes.

Box 8.4 How is hormonal control different from nervous control?

The action of hormones is different from that of nerves in several ways. Compared with nervous control, hormonal control of responses is:

- slower, because the hormone is secreted into, and travels in, the bloodstream to reach its target cells
- longer lasting, because the effect persists for as long as the hormone remains bound to its receptor in or on the target cell
- more general in their effects, because hormones often have target cells in different regions of the body — some hormones affect nearly all cells

Prostaglandins and histamine

Prostaglandins and **histamine** are 'chemical messengers' like hormones but, unlike hormones, they act locally on the cells that produce them and other cells in the immediate area.

They are involved in producing the inflammatory response to injury. The four characteristics of an inflammatory response are:

- redness
- warmth
- pain
- swelling

Prostaglandins are mediators of the first three of these characteristics. They are released from damaged cells at the site of an injury and have several important effects including:

- vasodilation of arterioles — this allows more blood to flow to the area (causing the redness and warmth) and more phagocytic white blood cells to migrate to the area
- promoting clotting of the blood — this minimises blood loss and the entry of microorganisms

The pain is the result of increased pressure on nerve endings and pressure receptors caused by the increased volume of blood in the area.

Histamine has a slightly different role in the inflammatory response. It is not produced by damaged cells, but by **mast cells** in the area. Histamine acts on the capillary walls allowing these to dilate and also to become more 'leaky' than usual. This allows some of the plasma, including some protein molecules, to leave the blood. The increased volume of liquid in the tissues causes the swelling. The 'leakiness' also makes it easier for phagocytes to exit the blood and enter the tissues in order to ingest dead cells and bacteria (Figure 8.21).

The liquid in the tissues, together with dead phagocytes containing bacteria and/or dead cells that they have engulfed, eventually escape from the inflamed site as 'pus'.

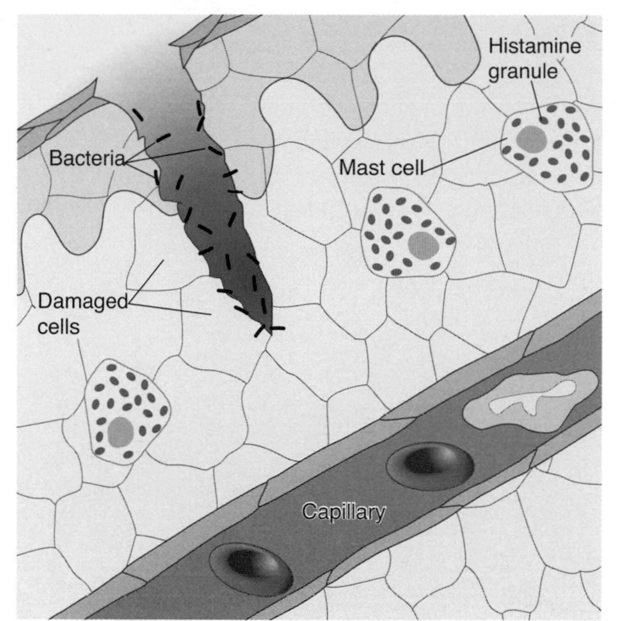

(a) Injury to the skin damages cells and may allow bacteria to enter

(b) The damaged cells release prostaglandins and stimulate mast cells to release histamine

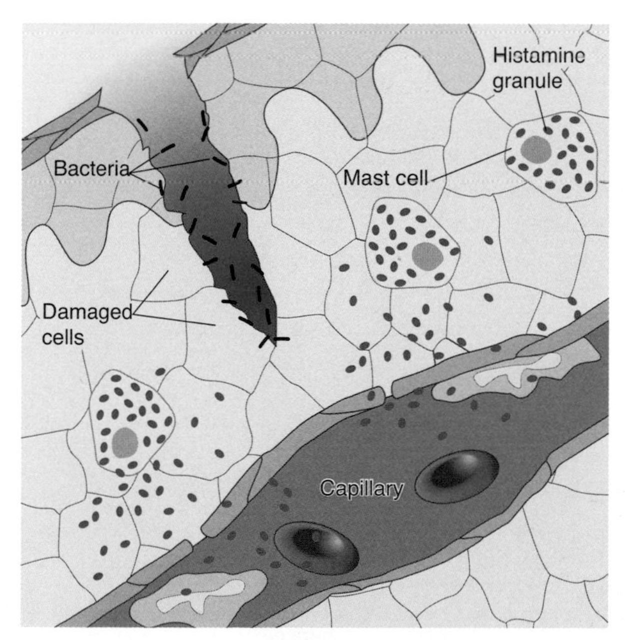

(c) Histamine causes the capillaries to dilate and become leakier...

(d) ...allowing phagocytes to escape into the tissues where they engulf dead cells and bacteria

Figure 8.21 The role of histamines in the inflammatory response

Do plants have hormones?

Well do they?

This is a tricky one to answer. They certainly have 'chemical messengers', but they aren't secreted by 'glands' and don't travel in a transport system of any kind. So, biologists prefer to call them plant growth regulators. They include:

- auxins
- gibberellins
- abscisic acid
- cytokinins
- ethene

The auxins are the best known plant growth regulators and were the first to be discovered.

The name auxin comes from the Greek verb *auxein* meaning to grow.

There are several auxins, but they are all similar, chemically, to indole acetic acid (IAA — Figure 8.22).

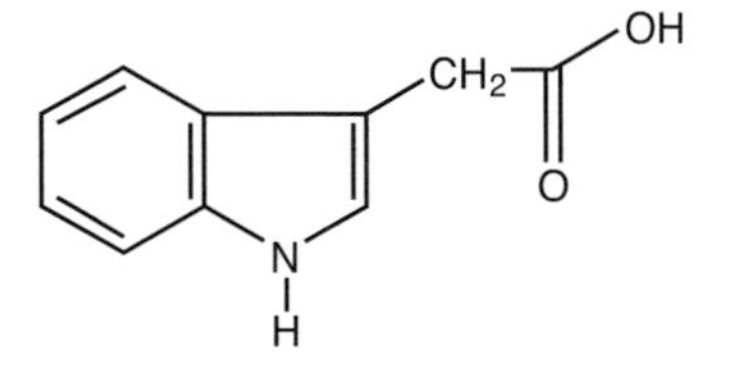

Figure 8.22 The structure of a molecule of indole acetic acid

HOW SCIENCE WORKS

Box 8.5 The discovery of the action of auxins

One of the first researchers into the action of auxins was Charles Darwin, although he is known less for this than for his theory of natural selection. Charles and his son Francis carried out several experiments in which they illuminated plant coleoptiles from one side. The results of these experiments are summarised in Figure 8.23.

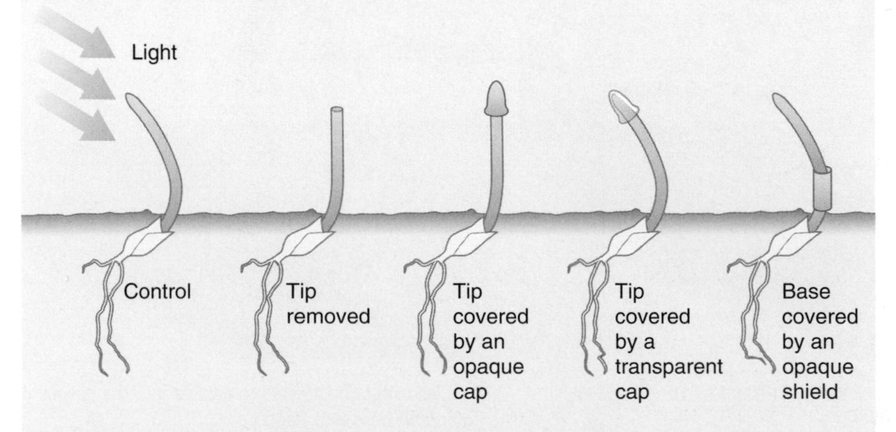

Figure 8.23 The results of experiments carried out by Charles and Francis Darwin in 1880

The Darwins concluded that detection of the stimulus occurs in the tip of the coleoptile and a region some distance behind the tip brings about the response.

A coleoptile is not a stem as such. It is the curved sheath that covers the emerging shoot of a seedling of a cereal.

They suggested that there must be some communication between the two, probably chemical in nature.

In 1913, Boysen-Jensen took the research further and obtained strong evidence that a chemical messenger is involved. He cut the tips of coleoptiles and treated them in one of two ways:

- He placed the tip on a block of gelatine (which contains water and allows chemicals to diffuse through it) and replaced it on the coleoptile.
- He placed the tip on a sheet of mica (which is a hard, impervious substance) and replaced it on the coleoptile.

His results are summarised in Figure 8.24.

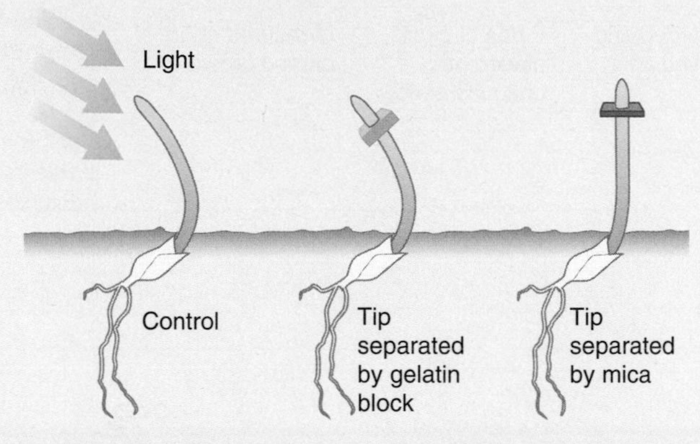

Figure 8.24 Results of experiments carried out by Boysen-Jensen in 1913

In 1919, Paal extended the work further by removing the tip and then replacing it either centrally or asymmetrically on the coleoptile. His results are shown in Figure 8.25.

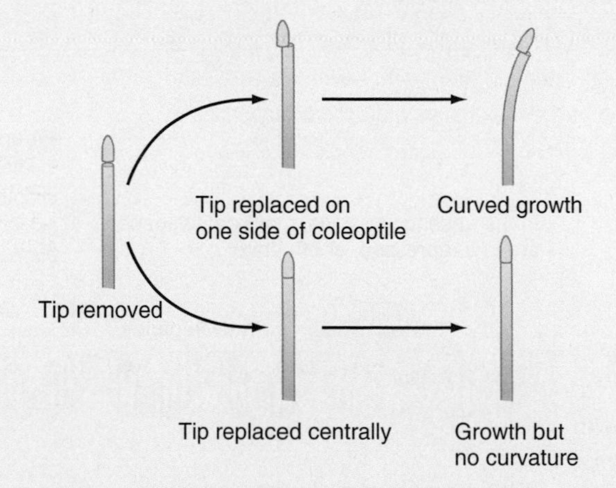

Figure 8.25 The results of experiments carried out by Paal in 1919

Paal's results seemed to give evidence that curved growth was due to the chemical accumulating on one side of the stem.

In 1926, a Dutch biologist called Fritz Went confirmed that the tropic response is coordinated by a chemical and that the extent of the response depends on the concentration of the chemical. He cut tips off coleoptiles and stood them on agar (gel) blocks for a period of time to allow the auxin to diffuse into the agar. He then replaced the blocks asymmetrically (as in Paal's experiment) and measured the angle of curvature. By using different numbers of tips for the same time, he varied the concentration of auxin in the agar. His results are summarised in Figure 8.26.

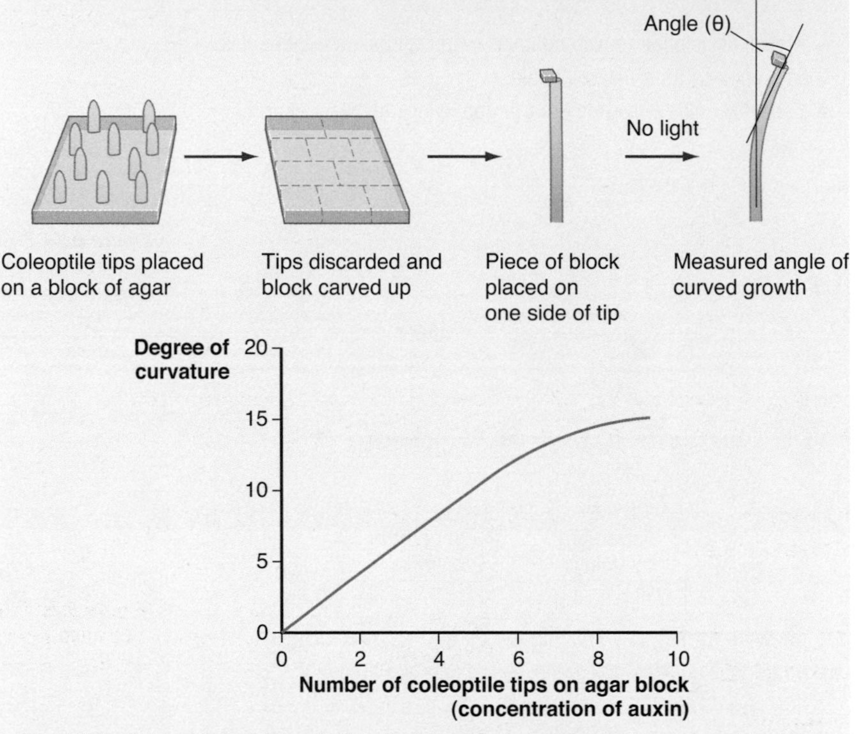

Figure 8.26
(a) The experiments carried out by Fritz Went

(b) The results from Went's experiments using different numbers of agar blocks

Since then, the auxin has been isolated and purified and shown to produce the same effects. In this sequence of investigations over a period of 60 years, biologists used the results of previous investigations and extended them with new experiments to provide more rigorous evidence to support the hypothesis that tropic responses in plants are brought about by chemicals.

How do auxins produce the phototropic response?

The way in which auxins, including IAA, act has only recently been discovered. We now believe that they act on growth genes, turning them on and stimulating both cell division and cell elongation (Figure 8.27).

The growth towards light from one side is a result of the auxin being redistributed to the shaded side of the shoot (Figure 8.28).

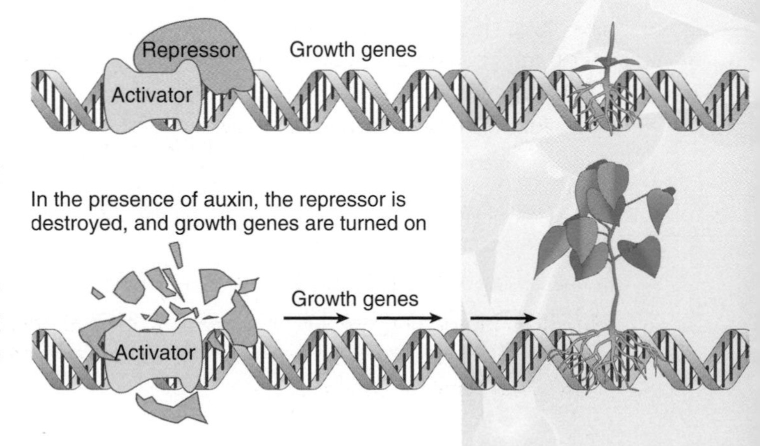

Figure 8.27 Auxins act by inactivating repressor substances that block the activation of growth genes in plants

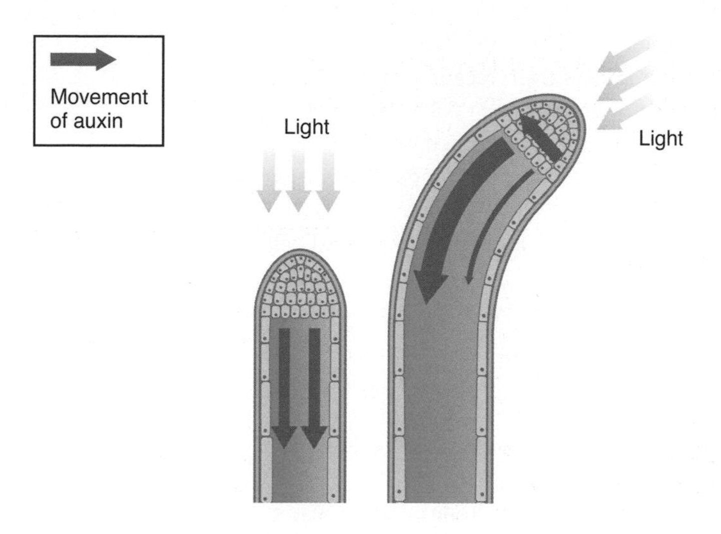

Figure 8.28 The phototropic response. Auxins move away from the lit side of a shoot and accumulate in the shaded side. This side grows faster as a result, causing a curvature towards the light.

What about the gravitropic response?

This is a similar story, but with an important difference. Auxins inhibit growth in roots. It is the absence, or low concentrations of auxins that bring about growth here.

If a seedling is placed horizontally and left, the root grows downwards and the shoot grows upwards (Figure 8.29).

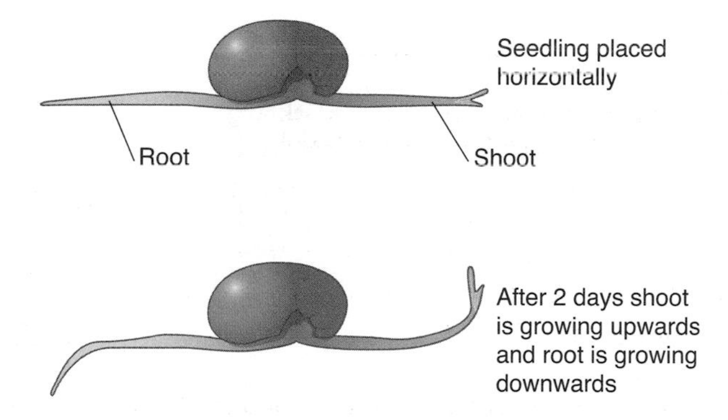

Figure 8.29 Investigating how a root and shoot respond to gravity

The explanation is that, in both root and shoot, the auxin is redistributed to the lower side. In the shoot, auxin stimulates growth, so cells on the lower side grow faster than those on the upper side, causing upward curvature. In the root, auxin inhibits growth so cells on the lower side grow more slowly than those on the upper side, causing downward curvature.

Summary

Nervous control

- The human nervous system comprises the central nervous system (the brain and spinal cord) and the peripheral nervous system (nerves leading to and from the central nervous system).

- There are three types of neurone:
  - sensory neurones that carry impulses from receptors to the CNS
  - motor neurones that carry impulses from the CNS to effectors
  - inter-neurones (relay neurones) that link other neurones
- A nerve impulse is a change in the membrane potential that sweeps along an axon.
- The myelin sheath around the axon insulates the axon and allows action potentials to be propagated more quickly.
- The resting potential of an axon membrane is about −70 mV. This is maintained by:
  - the presence of large anions inside the membrane
  - the sodium–potassium pump
  - sodium ions diffusing into the axon more slowly than potassium ions diffuse out
- The imbalance of charge across the axon membrane means that it is polarised.
- When stimulated, the axon membrane may generate an action potential and become depolarised. For this to happen:
  - sufficient gated sodium ion channels must open to raise the membrane potential to the threshold level (−55 mV)
  - at threshold, voltage-gated sodium ion channels open and sodium ions flood in, changing the membrane potential to +40 mV; the membrane is now depolarised
- To repolarise the membrane:
  - the gated sodium ion channels close and potassium ion channels open
  - potassium ions flood out taking the membrane potential to just below −70 mV (it is hyperpolarised) before it is restored to the normal polarised state
- The transmission of nerve impulses occurs faster:
  - if the axon is myelinated
  - if the temperature is raised
  - in a wider axon than in a narrower axon
- In myelinated axons, transmission of nerve impulses occurs by saltatory conduction as action potentials 'jump' to successive nodes of Ranvier.
- Nerve impulses are transmitted across synapses by neurotransmitters.
- When an action potential arrives at an excitatory synapse:
  - calcium ion channels open, allowing calcium ions into the axon
  - vesicles containing neurotransmitter fuse with the pre-synaptic membrane and release neurotransmitter into the synaptic cleft
  - the neurotransmitter diffuses across the cleft and binds with receptor proteins on the post-synaptic membrane
  - this binding opens gated sodium ion channels, which allows the generation of an action potential in the post-synaptic membrane
- Synapses can be inhibitory or excitatory; summation of impulses allows modulation of nerve impulses.
- In the somatic reflex arc for withdrawal from heat:
  - the heat receptor is stimulated by the hot plate

- an action potential is transmitted along the sensory neurone
- an action potential is initiated in the relay neurone and then the motor neurone
- neurotransmitters carry the impulse across the synapses between sensory neurone and relay neurone and between relay neurone and motor neurone

- In the reflex control of heart rate, impulses passing along sympathetic motor neurones increase the heart rate and impulses passing along parasympathetic motor neurones decrease the heart rate.

Hormonal control

- Hormones are produced in endocrine glands and travel in the bloodstream to produce their effect in a target organ elsewhere in the body.
- Cells in target organs have receptor proteins that have a shape complementary to that of the hormone molecule. These receptor proteins may be on the plasma membrane or in the nucleus.
- Compared with nervous control, hormonal control is slower, longer lasting and more general in its effect.
- Histamine and prostaglandins are hormone-like substances that produce their effects in the cells that secrete them or in nearby cells. They are important mediators in the inflammatory response to injury.

Plant growth regulators

- Auxins are the most widely known plant growth regulators. They are involved in producing the phototropic and gravitropic responses of shoots and roots:
  - in shoots, auxins stimulate growth and division of cells
  - in roots, auxins inhibit growth and division of cells.
- Redistribution of auxins in response to a directional stimulus produces directional growth responses:
  - in shoots, auxins become concentrated on the side of a shoot away from the light, causing increased growth here and curvature towards the light
  - in roots placed horizontally, auxins become concentrated on the lower surface of the root, inhibiting growth here and causing the curvature towards gravity

Questions

Multiple-choice

1 The central nervous system comprises:
  A the brain and spinal cord
  B the brain and cranial nerves
  C the spinal cord and spinal nerves
  D the cranial nerves and spinal nerves

2 The myelin sheath around the axons of neurones:
  A insulates the axons from other impulses
  B allows saltatory conduction, decreasing the speed of conduction
  C both A and B
  D neither A nor B

The diagram below of an action potential relates to both questions 3 and 4.

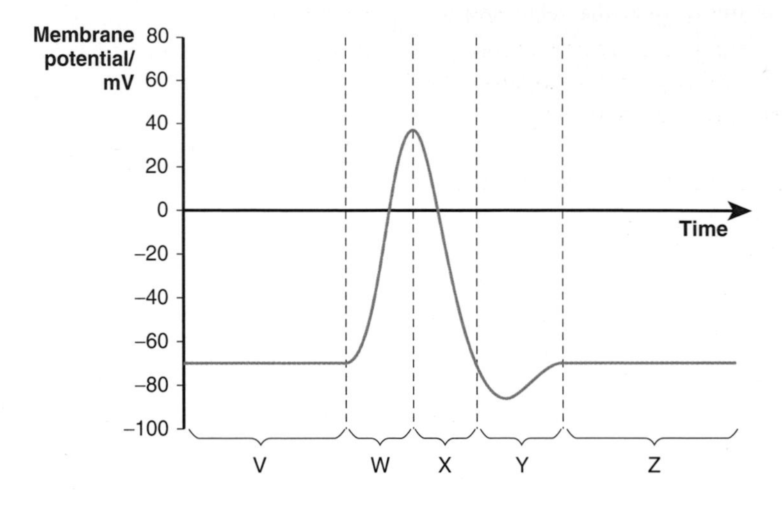

3 In the diagram, the period(s) labelled:

A W, X and Y represent the refractory period

B V represents the resting potential of the membrane

C Y represents a period of hyperpolarisaton

D all of the above

4 In the diagram, the period labelled W represents:

A a period when gated potassium ion channels open and gated sodium ion channels close

B a period when both gated potassium ion channels and gated sodium ion channels open

C a period when both gated potassium ion channels and gated sodium ion channels close

D a period when gated sodium ion channels open and gated potassium ion channels close

5 Excitatory and inhibitory synapses differ in that:

A different neurotransmitters are involved

B the effect on the post-synaptic membrane is different

C both A and B

D neither A nor B

6 Compared with nervous control, hormonal control is:

A slower, longer lasting and more specific in its effect

B faster, longer lasting and more general in its effect

C faster, longer lasting and more specific in its effect

D slower, longer lasting and more general in its effect

7 Hormones are able to target certain organs because the cells in the organ have protein receptors that are:

A the same shape as the hormone molecule

B the same charge as the hormone molecule

C the opposite charge to the hormone molecule

D a complementary shape to the hormone molecule

8 Plant shoots are:
  A positively gravitropic and negatively phototropic
  B positively gravitropic and positively phototropic
  C negatively gravitropic and negatively phototropic
  D negatively gravitropic and positively phototropic

9 In shoots, auxins cause:
  A increased growth of cells and increased cell division
  B increased growth of cells but decreased cell division
  C decreased growth of cells but increased cell division
  D decreased growth of cells and decreased cell division

10 The downward curvature of a root placed horizontally is due to:
  A an accumulation of auxin on the lower surface of the root
  B stimulation of growth by the auxin
  C increased growth of cells where the auxin accumulates
  D none of the above

Examination-style

1 Nerve impulses are propagated along the axons of neurones as a series of action potentials. The diagram shows the changes in membrane potential, sodium ion (Na^+) conductance and potassium ion (K^+) conductance of an axon membrane as an action potential is generated.

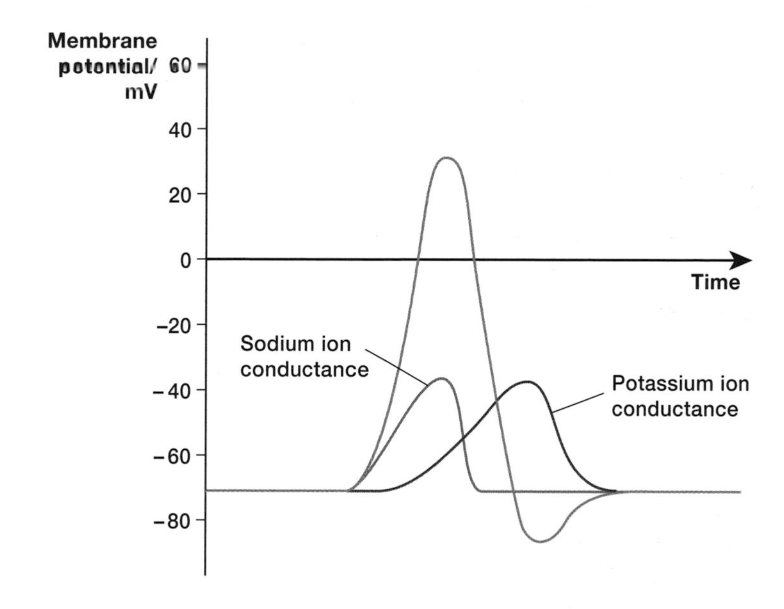

(a) Describe the evidence in the graphs that suggests that depolarisation is caused by an influx of sodium ions, while repolarisation is caused largely by the exit of potassium ions. *(3 marks)*

(b) Explain the role of the refractory period in the transmission of nerve impulses. *(2 marks)*

Total: 5 marks

2 The diagram and the table beneath it show how several sensory neurones can influence a single motor neurone.

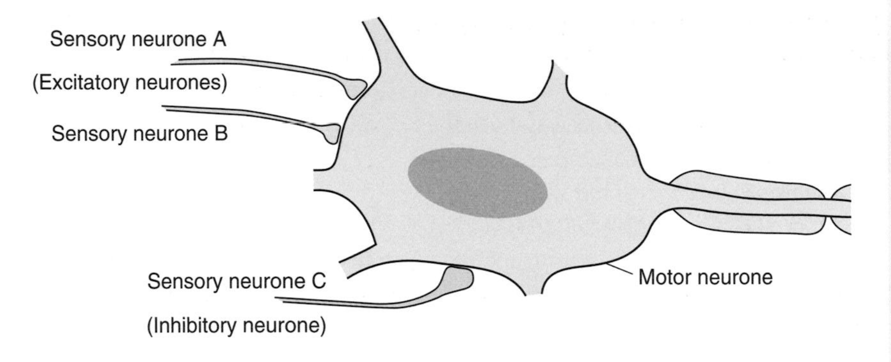

Neurone	Action potential				
Sensory neurone A (excitatory)	✗	✓	✗	✓	✓
Sensory neurone B (excitatory)	✗	✗	✓	✓	✓
Sensory neurone C (inhibitory)	✗	✗	✗	✗	✓
Motor neurone	✗	✗	✗	✓	✗

(a) Explain why a nerve impulse can only cross a synapse in one direction. *(4 marks)*

(b) Explain the results shown in the table. *(3 marks)*

Total: 7 marks

3 The graph shows the effect of different concentrations of an auxin on the growth of shoots and buds. Buds grow into side shoots.

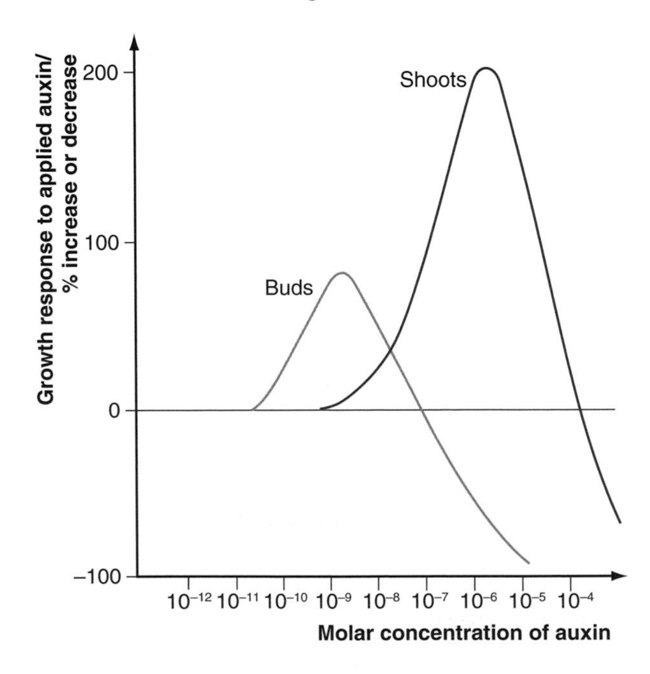

(a) What is auxin? *(2 marks)*

(b) Describe the effect of different concentrations of auxin on the growth of shoots. *(4 marks)*

(c) Suggest why auxin inhibits the growth of side branches in growing shoots. *(3 marks)*

Total: 9 marks

4 In the inflammatory response to physical injury, damaged cells release prostaglandins. Other cells release histamines.

(a) (i) Name the cells that release histamines. *(1 mark)*

(ii) Describe the roles of prostaglandins and histamines in the inflammatory response. *(4 marks)*

(b) The main stages in the synthesis of prostaglandins are shown in the flow chart.

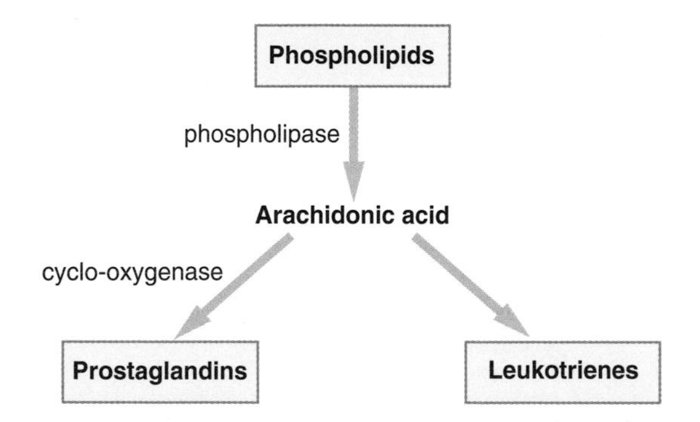

Leukotrienes are important in attracting phagocytic white blood cells to an injured area.

Suggest the benefit of anti-inflammatory drugs that act by blocking the action of cyclo-oxygenase rather than phospholipase. *(3 marks)*

Total: 8 marks

5 Some areas of the brain use the neurotransmitter dopamine. One of these areas is called the **nucleus acumbens**.

(a) The diagram shows the structure of a synapse from this area.

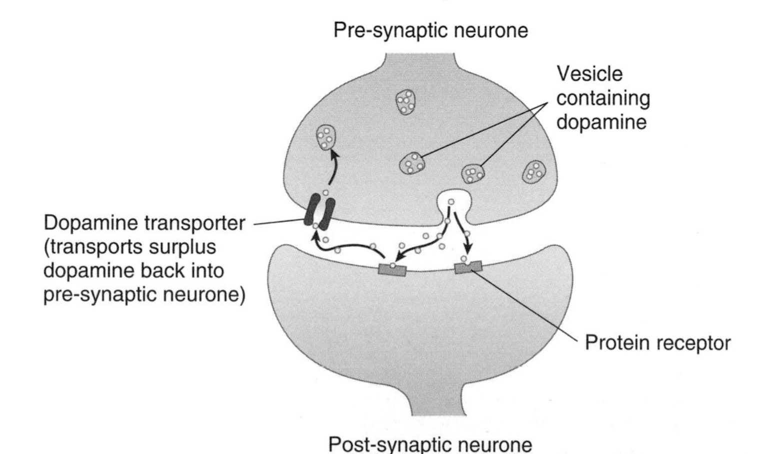

(i) Give two pieces of evidence, visible in the diagram, which show that transmission can take place in only one direction at this synapse. *(2 marks)*

(ii) Explain how the binding of dopamine to the receptors in the post-synaptic membrane results in a nerve impulse passing along the post-synaptic neurone. *(5 marks)*

(b) In the nucleus acumbens, when pleasurable events are occurring, large amounts of dopamine are released at the synapses, resulting in large numbers of nerve impulses. The diagram shows a synapse in the nucleus acumbens when cocaine is present.

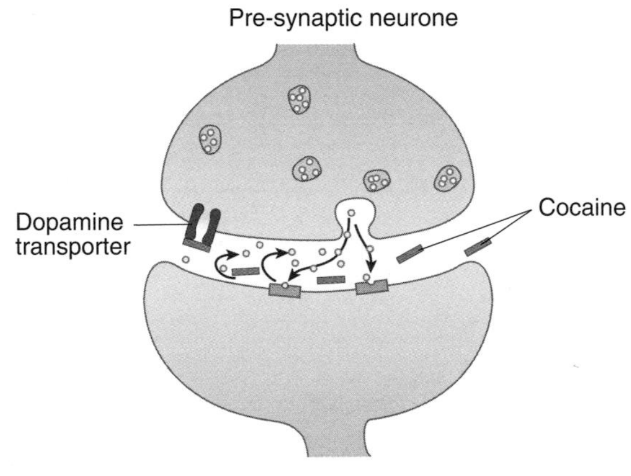

(i) Use the diagram and your own knowledge of synaptic transmission to explain why many cocaine users experience pleasurable 'highs'. *(5 marks)*

(ii) Suggest why, when the cocaine leaves the brain, users feel depressed. *(3 marks)*

Total: 15 marks

Chapter 9

How do animals make things happen *and* keep conditions the same?

This chapter covers:

- the macroscopic and microscopic structure of muscles
- the way in which energy is transferred within muscles
- the different types of muscle fibre and their proportions in different types of athlete
- the way in which temperature is gained and lost in ectotherms and in endotherms
- the way in which ectotherms and endotherms control their core body temperatures
- the way in which plasma glucose concentration can change
- the way in which humans control the plasma glucose concentration
- diabetes — the consequence of not being able to control plasma glucose concentration effectively
- the concepts of homeostasis and negative feedback control
- the way in which the female reproductive cycle is controlled

In Chapter 8, we saw that reflex arcs involving neurones in the somatic nervous system usually produce a response by skeletal muscle. This might be lifting an eyebrow, smiling, frowning, removing one's hand from a hot plate or just about any unconscious physical movement. By contrast, reflex arcs involving neurones of the autonomic nervous system usually produce responses in internal organs or glands. Responses by internal organs are often adaptations to new conditions — for example, an increase in heart rate to meet an increased energy demand by muscles or an increase in blood flow to the skin to help the body lose more heat because the core body temperature has increased.

In this chapter we shall see how skeletal muscle can exert a force to move all or part of the body in response to nerve impulses. We shall see how the autonomic nervous system helps to maintain the core body temperature within narrow limits and how hormones help to maintain the plasma glucose concentration and also bring about the changes that occur in the oestrous cycle of female mammals.

Effective coordination by nerves and hormones allows us to maintain some conditions constant while changing others.

How do skeletal muscles cause movement?

How are skeletal muscles arranged in the body?

Skeletal muscles contract and pull structures; they cannot expand and push. For this reason, muscles occur in **antagonistic pairs**. One muscle moves a structure in one direction; the other muscle of the pair moves it in the opposite direction (Figure 9.1).

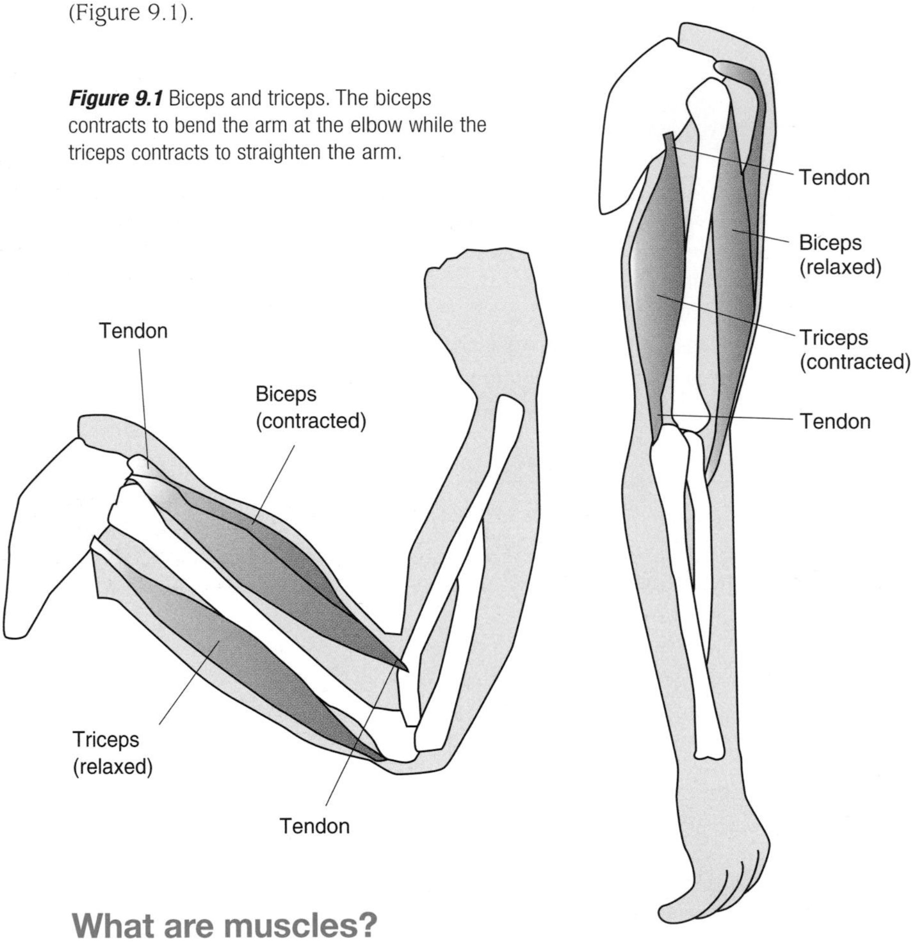

Figure 9.1 Biceps and triceps. The biceps contracts to bend the arm at the elbow while the triceps contracts to straighten the arm.

What are muscles?

Muscles such as the biceps, triceps and rectus muscles are organs. They contain several tissues, each important to the functioning of the muscle. Each muscle contains:

- muscle tissue, to provide the force for contraction
- blood vessels, to carry the blood
- blood, bringing oxygen and glucose for respiration
- sense cells to detect the degree of 'stretch' in the muscle
- sensory neurones to carry impulses from the stretch receptors
- motor neurones, to initiate contraction of the muscle
- connective tissue to 'pack' the groups of **muscle fibres**

◀ Each artery and vein is an organ made of connective tissue, smooth muscle, elastic tissue and epithelial tissue

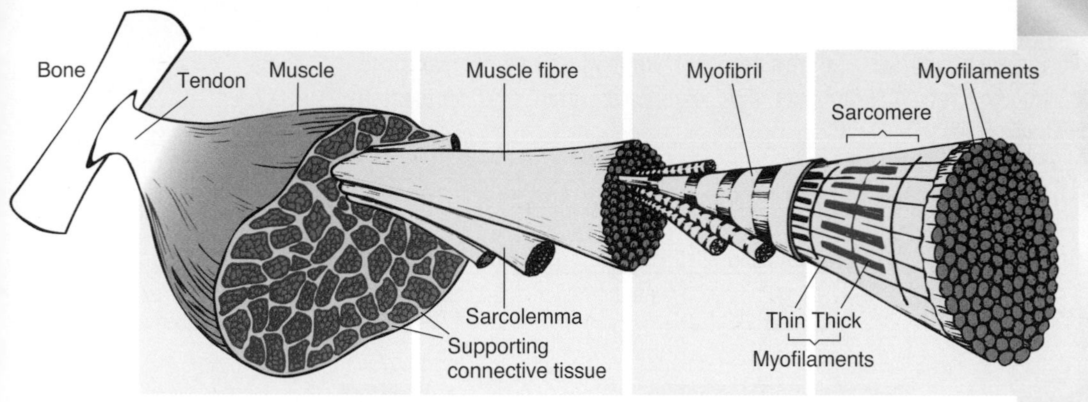

Figure 9.2 Structure of a muscle. A muscle is made from bundles of fibres. Each fibre contains many **myofibrils** made from thick and thin **myofilaments**. A single 'contracting unit' of a myofibril is called a **sarcomere**.

How do muscles contract?

To make the whole muscle contract, all the individual myofibrils must contract. Each myofibril is made largely from the fibrous proteins **actin** and **myosin**, arranged in repeating units called sarcomeres. Myosin forms the thick filaments of the myofibril and actin forms the thin filaments (Figure 9.2).

The generally accepted theory of muscle contraction is called the **sliding filament theory**. According to this theory, filaments of actin and myosin slide over one another, shortening each sarcomere. As a result, each myofilament becomes shorter and so the whole muscle becomes shorter.

Figure 9.3 shows how the filaments are arranged in a single, relaxed sarcomere.

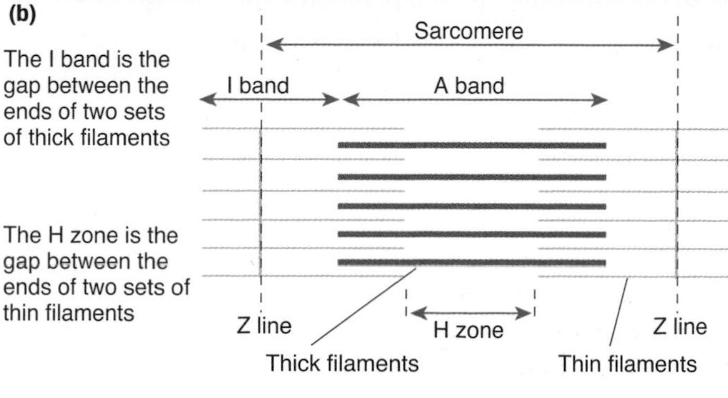

Figure 9.3 A relaxed sarcomere: **(a)** electron micrograph; **(b)** diagrammatic representation

When nerve impulses initiate a contraction, the appearance of the sarcomere changes as the two sets of filaments slide over each other and shorten the sarcomere (Figure 9.4).

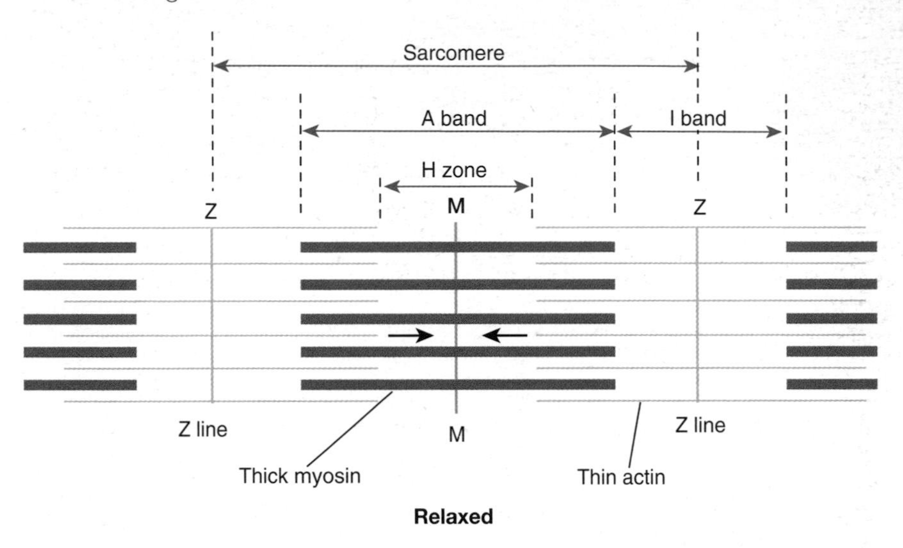

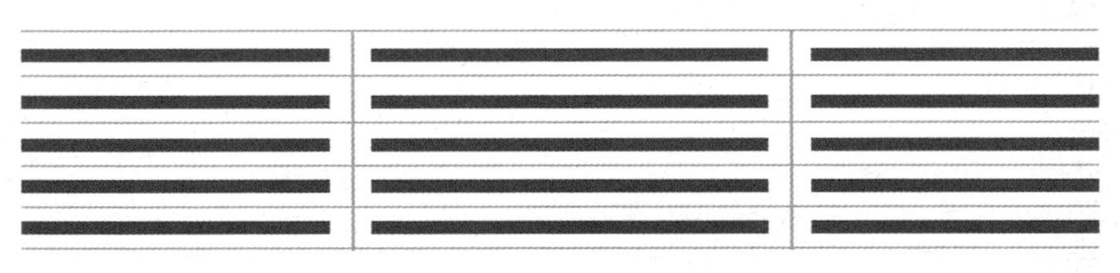

Figure 9.4 Contraction of sarcomere. When the filaments slide over each other, the I band and H zone become reduced as the whole sarcomere shortens.

The mechanism for the filaments sliding over each other is linked to the structure of the two types of filament.

The ends of the myosin filaments are bulbous and are called **myosin heads**. They can bind with **binding sites** on the actin filaments. On binding, the heads tilt, moving the actin filaments. The myosin heads are then released from the binding sites and return to their original positions. They then bind to other sites further along the actin filament and repeat the tilting and movement of the actin.

When the muscle is relaxed:

- molecules of **tropomyosin** block the binding sites, preventing the myosin heads from binding
- each myosin head is in a 'resting position' and is bound to a molecule of ATP; hydrolysis of this ATP supplies the energy needed for the movement of the myosin head

Neurones synapse with myofibrils at **neuromuscular junctions**. Here, nerve impulses initiate contraction when they release neurotransmitters into the junction (Figure 9.5).

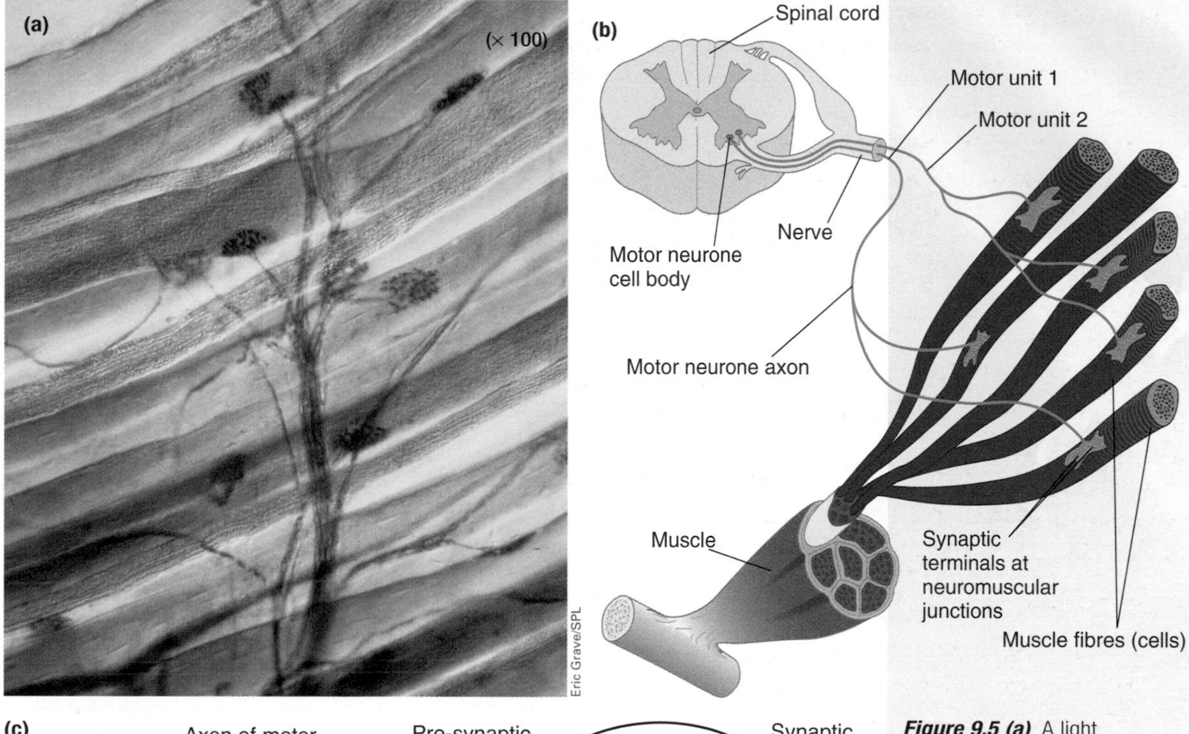

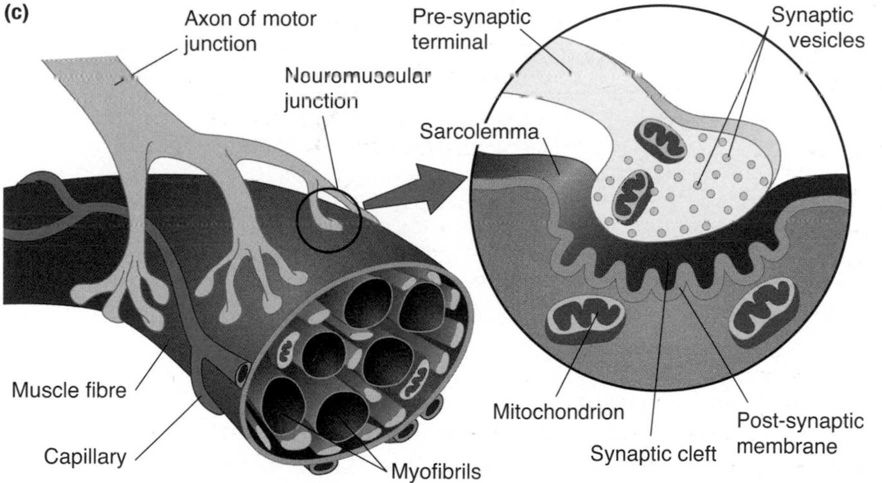

Figure 9.5 (a) A light micrograph showing several neuromuscular junctions.
(b) Neuromuscular junctions are where the synaptic terminals of motor neurones connect to muscles.
(c) The structure of a neuro-muscular junction.

When nerve impulses arrive at a neuromuscular junction, they cause contraction in the following way:

- **Calcium ions** are released around the actin molecules.
- These ions cause the tropomyosin molecules to move and expose the binding sites on the actin molecules.
- At the same time, the ATP molecules are hydrolysed to ADP and inorganic phosphate. The energy released is transferred to the myosin heads.
- The inorganic phosphate is released and the myosin heads, with ADP attached, bind to the exposed binding sites on the actin molecules. (This is sometimes referred to as **cross-bridge formation**.)

- ADP is released. This allows the energy stored in the myosin heads to move the heads and with them the actin filaments.
- Another ATP molecule attaches to each myosin head and, provided nervous stimulation continues to cause calcium ions to be released, the cycle continues; if not, the muscle returns to the relaxed state (Figure 9.6).

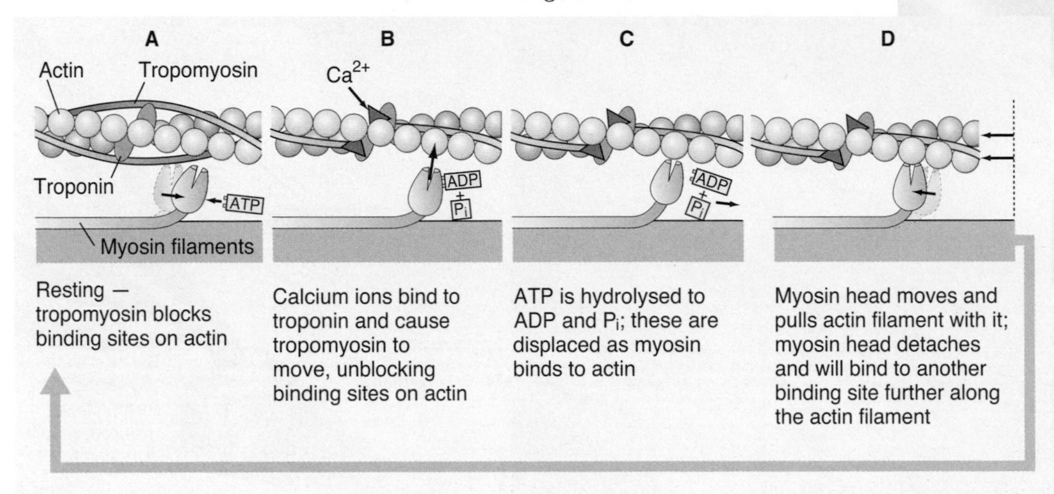

Figure 9.6 Sliding filament theory

How do muscles transfer the energy they need to contract?

To contract the sarcomeres repeatedly, ATP must be available continually. It is supplied by:

- aerobic respiration
- anaerobic respiration
- the **phosphocreatine–ATP system**

Molecules of phosphocreatine (CP) store energy in a similar way to ATP. Energy is used to attach a phosphate group to a creatine molecule. When phosphocreatine is hydrolysed, the energy is released again. This energy can be used to synthesise ATP. Ultimately, all the energy for the formation of ATP comes from the breakdown of food molecules in aerobic or anaerobic respiration (Figure 9.7).

During exercise, energy is released in the following ways:

- At the start of exercise, the small store of ATP held in muscles is hydrolysed to ADP and inorganic phosphate; this is used up quickly.
- CP is then hydrolysed to creatine and phosphate, which releases the energy needed for the resynthesis of ATP. However, this releases sufficient energy for the first few seconds of exercise only.
- For longer periods of exercise, ATP is resynthesised by aerobic or anaerobic respiration.
- Energy release by aerobic respiration requires an efficient supply of oxygenated blood. It takes some time for the circulatory system to adapt, so initially most energy is supplied anaerobically. The proportion of energy released aerobically increases with time.

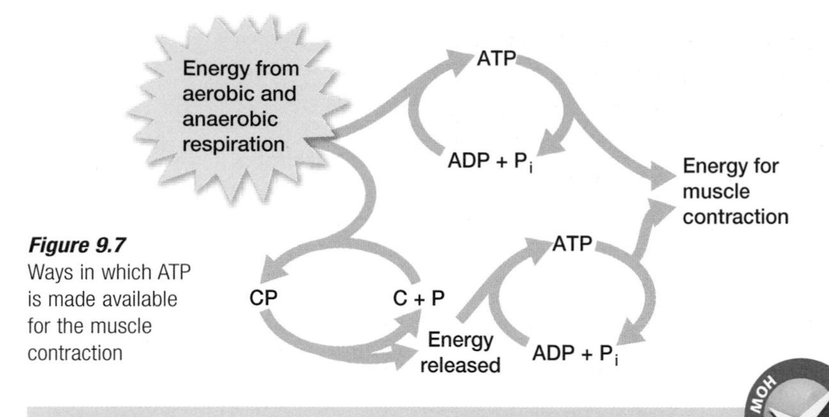

Figure 9.7
Ways in which ATP is made available for the muscle contraction

Box 9.1 Lactic acid and muscle fatigue

HOW SCIENCE WORKS

When muscle cells respire anaerobically, they produce lactate (part of the lactic acid molecule). It had been supposed for a long time that the accumulation of lactate was the cause of the fatigue in muscles when they continue to contract for long periods of time.

However, more recent research suggests that things are not that simple. It is not the lactate that causes the fatigue, although lactate is probably linked to the pain that is felt during prolonged intense exercise.

Biologists now think that as anaerobic respiration progresses, a build-up of hydrogen ions interferes with the mechanism controlling the release and reabsorption of calcium ions. The calcium ions are not reabsorbed properly, which means that the muscles do not relax properly and this makes the next contraction that little bit less effective. This leads to muscle fatigue.

Lactate can be respired aerobically once there is sufficient oxygen, or it can be used to resynthesise glycogen in the liver.

What makes natural sprinters and natural marathon runners?

The answer to this is in the type of muscle fibres that they have in their muscles. There are two types:

- **slow-twitch fibres**, which release energy principally by aerobic respiration
- **fast-twitch fibres**, which release energy principally by anaerobic respiration

The skeletal muscles of endurance athletes (such as marathon runners) contain a high proportion of slow-twitch fibres. These fibres are adapted to releasing energy aerobically over a sustained period. Slow-twitch fibres have:

- many mitochondria in each fibre
- high concentrations of the enzymes that regulate the Krebs cycle
- a more extensive capillary network than fast fibres (allowing more oxygen to be delivered)
- a high concentration of **myoglobin** (a pigment similar to haemoglobin, but which stores oxygen); because of this they appear darker than other fibres (Figure 9.8)

The reactions of the Krebs cycle and electron transport chain take place in the mitochondria and produce most of the ATP formed in aerobic respiration.

- a relatively slow contraction rate with less force than fast-twitch fibres (which is therefore sustainable for longer periods)
- a high resistance to fatigue

The skeletal muscles of athletes involved in 'explosive' events (such as sprinters and weight-lifters) contain a high proportion of fast-twitch fibres, which are adapted to releasing energy anaerobically over short periods of time. Fast-twitch fibres:

- have fewer mitochondria than slow fibres
- have high concentrations of the enzymes that control glycolysis
- have a higher concentration of ATPase (the enzyme that hydrolyses ATP to ADP and inorganic phosphate) than slow fibres; this ensures that a lot of ATP can be hydrolysed in a short period, so releasing energy quickly
- have a lower resistance to fatigue than slow-twitch fibres

Fast-twitch fibres contract more rapidly and with more force than slow fibres, but they cannot sustain this because the rapid increase in hydrogen ion concentration during anaerobic respiration induces fatigue.

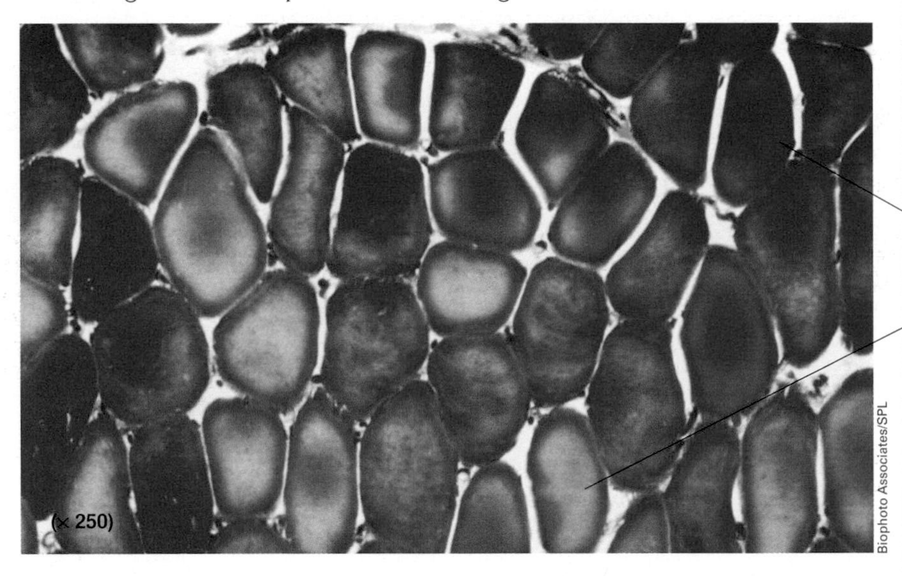

Figure 9.8 Slow-twitch fibres appear darker than fast-twitch fibres because of their higher concentration of myoglobin

How do organisms 'keep things the same'?

What is the benefit of keeping things the same?

Maintaining a constant internal environment is correctly referred to as **homeostasis**.

Homeostasis literally means 'steady-standing'. As a result of the many homeostatic mechanisms that operate in our bodies, our cells are bathed in a fluid of near constant temperature, pH and water potential, with other factors also being kept constant.

This unchanging internal environment allows enzymes to function with optimum efficiency, regardless of the external conditions. Homeostasis gives an organism a degree of independence from its surroundings.

How do homeostatic mechanisms operate?

Most systems that keep a factor constant in the body operate on the principle of **negative feedback**. In negative feedback systems, the factor being monitored has a 'pre-programmed set value' that is the norm. Sensors detect any change from this preset value and switch on mechanisms that correct the situation.

A room thermostat linked to a central heating system is an example of negative feedback in a non-living system. It usually operates in the following way:

- The temperature falls too far below the preset temperature.
- A sensor detects this fall.
- The sensor switches on the heater.
- The temperature is raised — probably to a little above the preset value.
- The sensor detects this change and switches off the heater.

However, there is a slight problem with a system like this. There is no system to bring the temperature back *down* to normal once it is too high. We just have to wait. It would be more effective if there was another system to, say, blow cold air into the room to cool it down once it became too hot. Most biological negative feedback systems have this 'dual mechanism' feature and their mode of action is summarised in Figure 9.9.

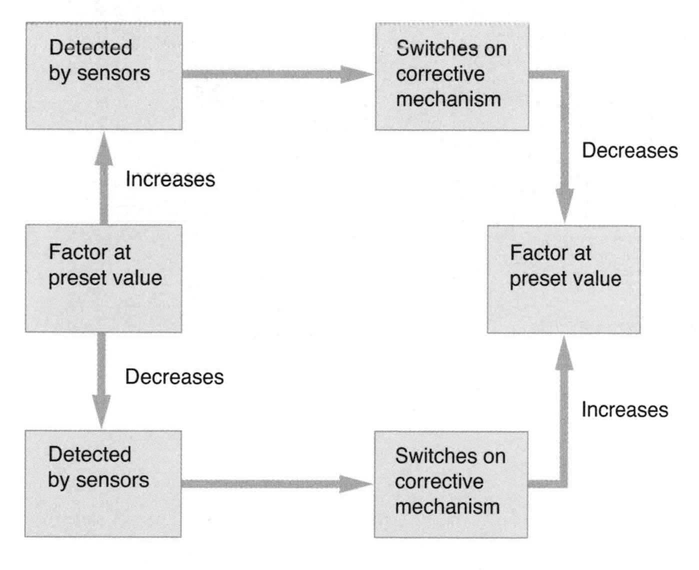

Figure 9.9 Negative feedback. Having separate systems for controlling deviations from the norm in different directions gives a greater degree of control than if there was only one system.

How do organisms control their core body temperatures?

Humans and other mammals are **endotherms**. They maintain a constant internal temperature, despite changing environmental conditions, by using physiological processes and metabolic energy. They can also use behavioural techniques, such as seeking shade or putting on extra clothes.

Endotherms are sometimes called warm-blooded. This is inaccurate and you should avoid using it in examinations.

Ectotherms, such as reptiles, have a limited capacity to regulate internal body temperature, which therefore fluctuates with that of the surroundings. After losing heat overnight, they seek warm environments, particularly direct sunlight, early in the morning. Once the desired body temperature is reached, reptiles alternate periods in the sun and shade to keep it at that level.

Some reptiles have a limited ability to control their temperatures by physiological means. Fossil evidence suggests that some of the dinosaurs had partial control over the amount of blood passing to the surface of their bodies. However, most regulation of temperature in ectotherms uses behavioural techniques.

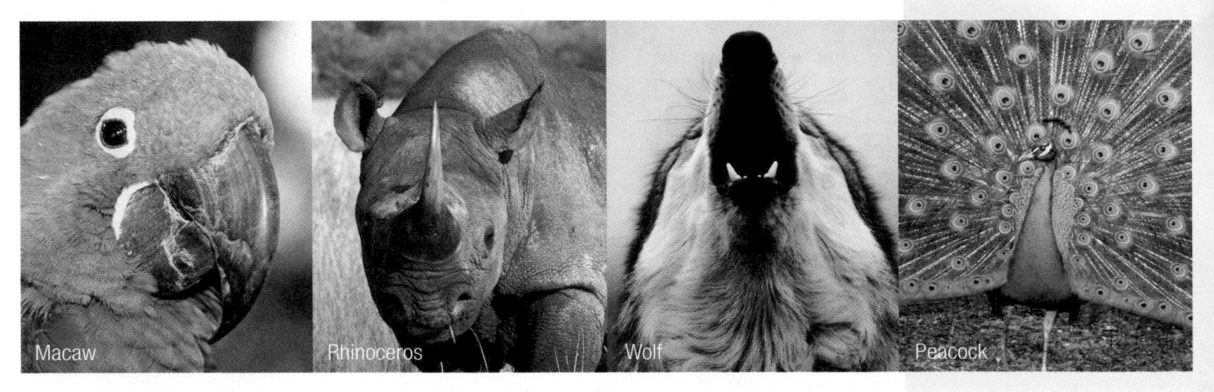

These animals are endotherms.

These animals are ectotherms.

Body temperature will remain constant if heat losses balance heat gains. Humans gain heat in the following ways:

- radiation from other objects — we can all feel the effects of solar radiation on a hot summer day
- eating warm food — the heat energy in the food is transferred to the body
- respiration — much of the energy released in respiration is released as heat; very active organs such as the liver and active skeletal muscle generate large amounts of heat

Humans lose heat by conduction, convection, radiation and evaporation. Table 9.1 describes the processes and gives some ways in which each method is used in humans.

Table 9.1 Heat loss in humans

Method of heat loss	Description of method	Involvement of method in temperature regulation in humans
Conduction	Heat energy is transferred through direct contact — heat energy moves to the cooler object	This usually plays only a small part in temperature regulation, but we can all feel heat loss by conduction when we take food out of a freezer
Convection	Heat energy is transferred by particles moving around carrying the heat with them	When air or water vapour next to the body is warmed, the particles move away from the body carrying the heat energy with them; this is more pronounced if it is windy — the 'wind-chill' factor
Radiation	Heat energy is lost by a body giving off 'rays' of heat energy; no particles are involved in transfer of the heat	Any hot body radiates heat; the hotter the body is, the more heat it radiates: as activity increases, more warm blood flows to muscles and skin; the increased heat in the skin results in more heat loss by radiation
Evaporation	When a liquid is converted into a vapour, heat energy is used	Water vaporised in breathing and sweating accounts for most heat loss in warm environments

Box 9.2 Heat loss from the skin involves all four methods of heat transfer

TopFoto

This man is losing heat as the sweat evaporates from his skin

The photograph shows clearly that heat is being lost by evaporation of the water in sweat. Other heat losses are also happening. The evaporated water is being carried away by convection. Look at the colour of the skin — any skin *that* red must be radiating quite a lot of heat! Finally, cold air particles touching the skin will gain heat by conduction and then carry the heat away by convection.

The control of core body temperature is the responsibility of the hypothalamus, a region of the brain that controls many homeostatic functions (Figure 9.10).

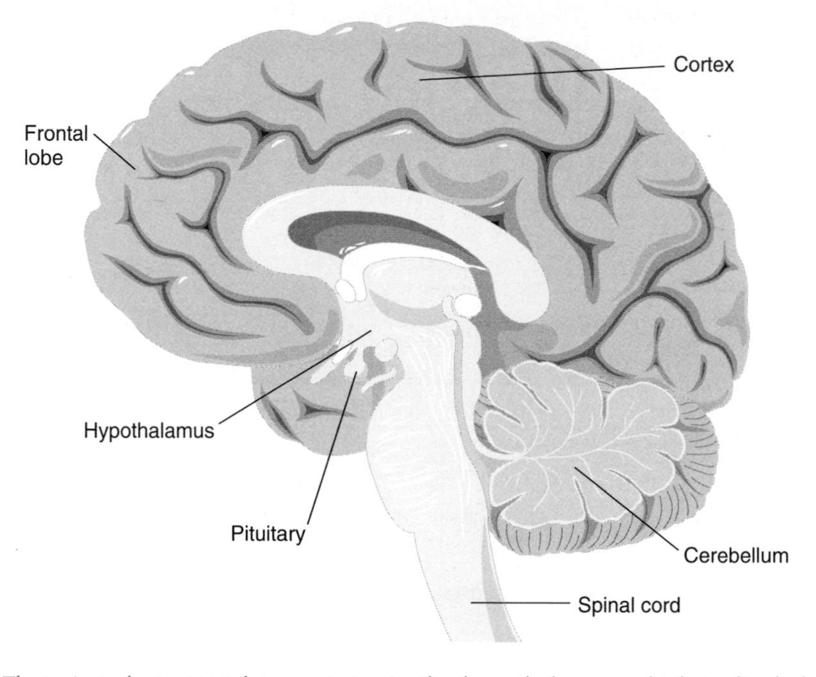

Figure 9.10 The position of the hypothalamus

There is a thermoregulatory centre in the hypothalamus, which is divided into two regions:

- the heat gain centre; this region 'switches on' all the measures that conserve heat and that increase heat generation
- the heat loss centre; this region 'switches on' all the measures that increase heat loss from the body and decrease heat generation

The two regions inhibit each other. As the heat gain centre becomes active (because the blood flowing through the hypothalamus is cooler than the norm) it inhibits the heat loss centre. The heat loss centre also inhibits the heat gain centre when it becomes active.

The hypothalamus controls body temperature through the autonomic nervous system. It receives information about body temperature from two main sources:

- temperature sensors in the skin detect changes in the temperature of the skin and therefore indirectly detect changes in the environmental temperature
- sensors in the hypothalamus itself detect changes in the temperature of the blood flowing through it, which reflects changes in the 'core' body temperature.

The hypothalamus acts as a kind of body 'thermostat'. It monitors body temperature and compares the actual temperature against a 'pre-set' value (37°C in humans). Any significant deviation from this is detected and appropriate responses are initiated. The responses to overheating are brought about by the heat loss centre of the hypothalamus while those to overcooling are brought about by the heat gain centre. Figure 9.11 summarises the role of the hypothalamus in controlling body temperature.

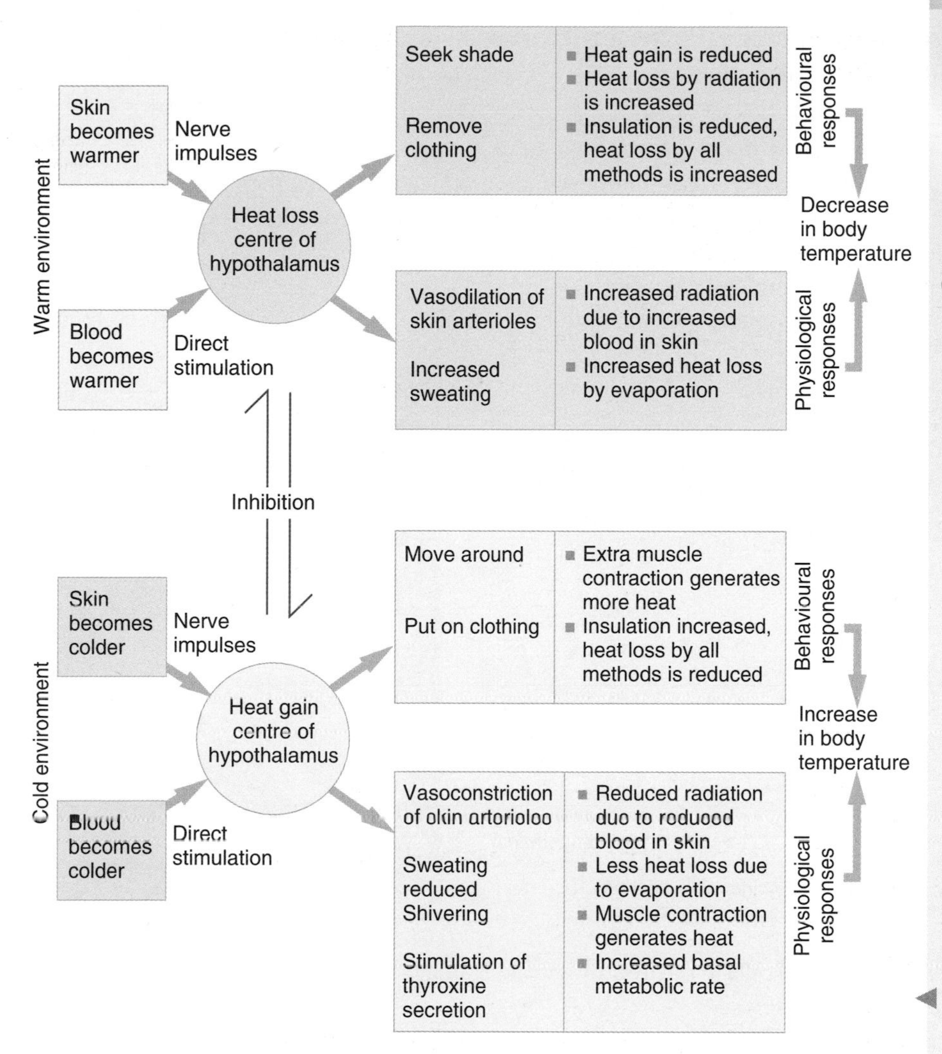

Figure 9.11 Mechanisms involved in regulating core body temperature in humans

Make sure that you never say that the capillaries move nearer to (or further away from) the surface of the skin. They stay firmly in one place all the time.

Basal metabolic rate is the rate of energy expenditure when a person is awake but resting, has not eaten for 12 hours and is comfortably warm.

Box 9.3 Controlling blood flow to the skin involves several mechanisms

Controlling the blood flow to capillaries near the surface of the skin involves three major mechanisms:

- Vasodilation (or vasoconstriction) of arterioles allows more (or less) blood to flow through them to the capillaries.
- Relaxation (or contraction) of the pre-capillary sphincters also allows more (or less) blood to flow to the capillaries.
- Constriction (or dilation) of the shunt vessel allows less (or more) blood to flow through it and so directs more (or less) blood to the capillaries.

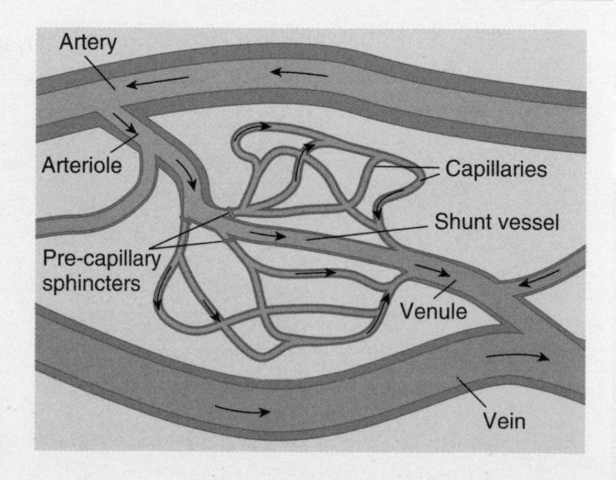

Figure 9.12 Microcirculation in the skin. The blood flow to the capillaries in the skin is influenced by the action of the arterioles, the pre-capillary sphincters and the shunt vessels.

How do we control our plasma glucose concentration?

The plasma glucose concentration of most people is normally between 75 mg and 125 mg per 100 cm^3. It is maintained in this range by the action of two hormones — insulin and glucagon. These hormones are produced in structures called **islets of Langerhans** in the pancreas (Figure 9.13). These structures contain two different types of secretory cell:

- α-cells, which secrete the hormone glucagon
- β-cells, which secrete the hormone insulin

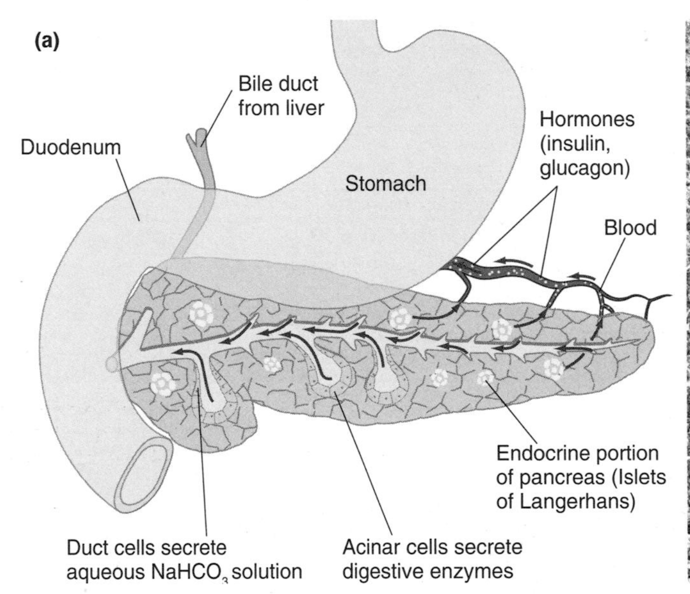

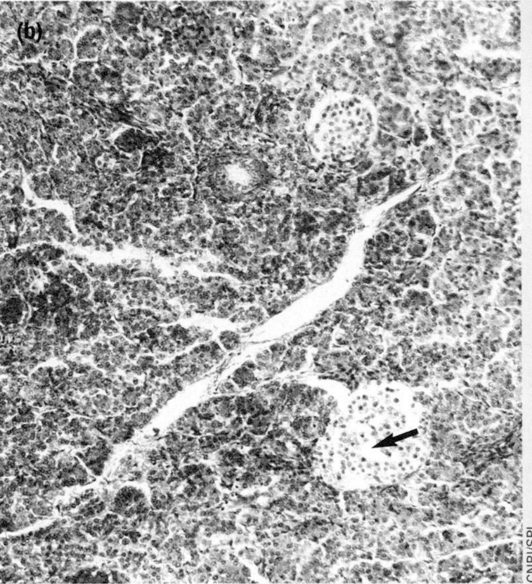

Figure 9.13 (a) Position of the pancreas and the location of the islets of Langerhans within the pancreas.
(b) A micrograph of pancreatic tissue showing islets of Langerhans. The arrow indicates β-cells, which secrete the hormone insulin.

Insulin and glucagon influence cells in the liver by activating enzymes concerned with carbohydrate metabolism:

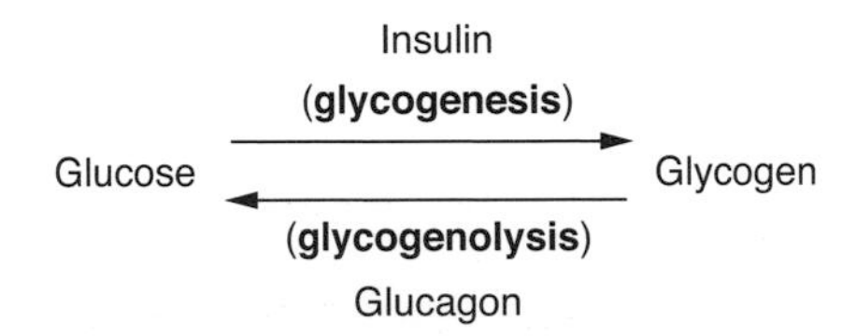

The term glycogenesis means 'making glycogen' (from glucose). Glycogenolysis means 'splitting glycogen' (into glucose).

When the plasma glucose concentration increases above normal, β-cells increase the production and secretion of insulin. This circulates in the bloodstream to the liver, where it causes increased uptake of glucose by the liver cells and activates enzymes that convert glucose to glycogen.

When the plasma glucose concentration falls below normal, α-cells increase their production and secretion of glucagon. This circulates in the bloodstream to the liver where it activates enzymes that convert glycogen to glucose. The glucose then enters the blood plasma, raising the concentration.

The flow chart in Figure 9.14 summarises the control of plasma glucose concentration by insulin and glucagon.

e Make sure that you know the difference between **glucagon** and **glycogen**. The words are very similar — too similar for an examiner to make any allowance and give you the benefit of the doubt for misspelling.

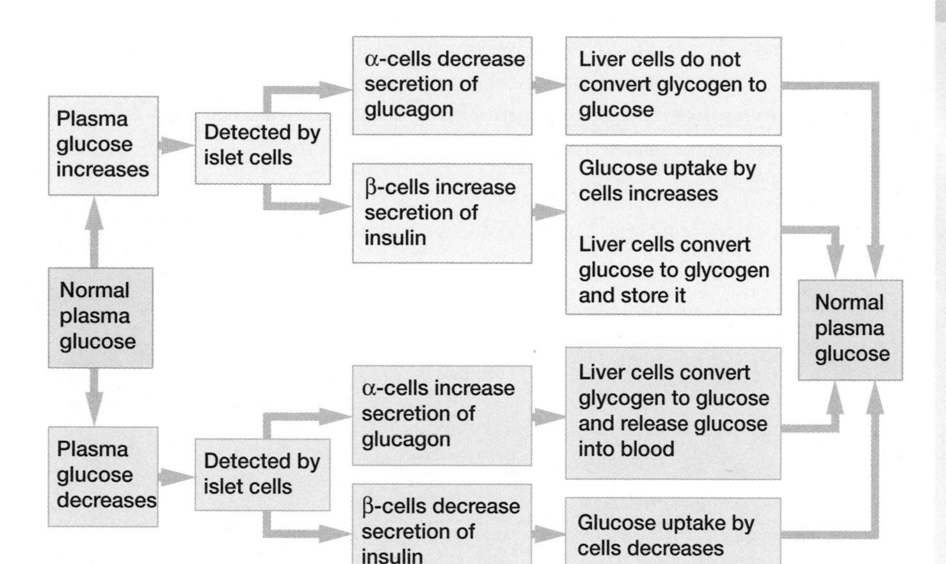

Figure 9.14 How humans control the concentration of glucose in the blood plasma

Both hormones bind to receptor proteins on the surface of liver cells and through a cascade system activate enzymes that catalyse the interconversion of glycogen and glucose. Figure 9.15 shows the cascade system that operates for insulin.

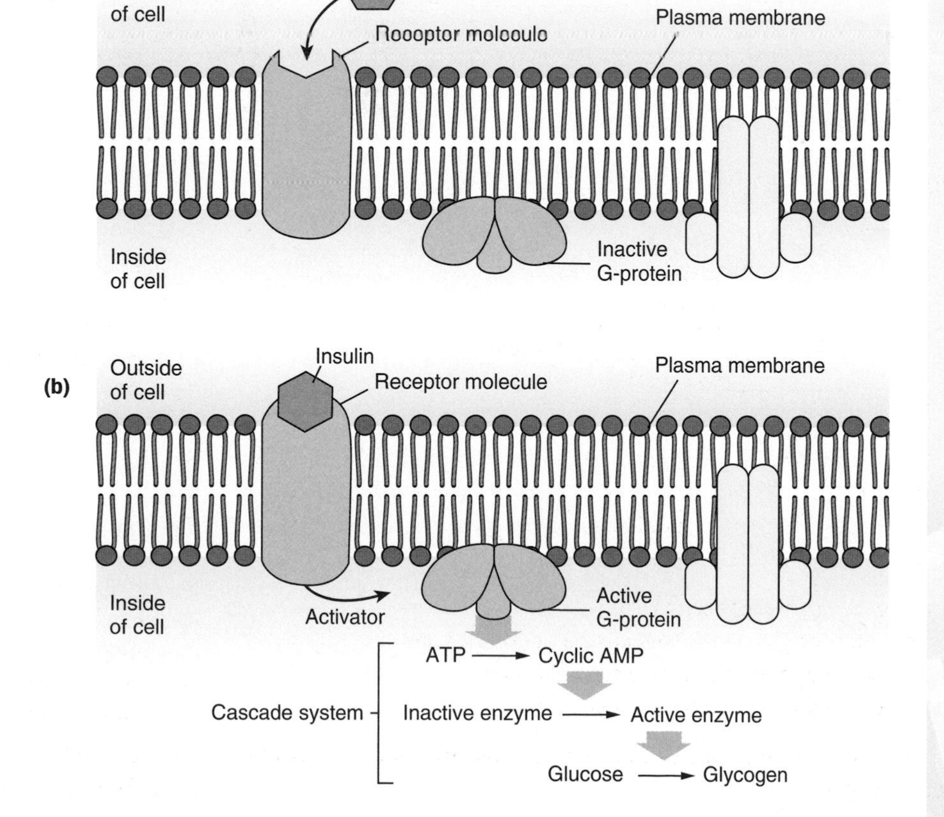

Figure 9.15 (a) Activation of enzymes by insulin. ***(b)*** The 'cascade' system.

At each stage of the cascade, the effect is amplified. Each insulin molecule that binds to the receptor protein stimulates the formation of several molecules of cyclic AMP (called the **second messenger**), which in turn stimulates the activation of several enzyme molecules. Each enzyme molecule has a turnover rate of many thousands of glucose molecules per second.

The cascade system for glucagon is very similar, but the receptor is different and the cascade activates a different enzyme.

Box 9.4 Other hormones also influence plasma glucose concentration

Insulin and glucagon are the hormones that maintain the concentration of glucose in the plasma within set limits. However, other hormones have a non-regulatory effect.

Plasma glucose concentration is also influenced by:

- **adrenaline**, which massively increases the breakdown of glycogen to glucose and the release of glucose into the bloodstream to allow increased energy release in the muscles, particularly during periods of stress or vigorous exercise
- **thyroxine**, which increases the basal metabolic rate, so an increased rate of energy release is required

In addition, a group of hormones called **corticosteroids** promote reactions that result in the synthesis of glucose from non-carbohydrate sources. This is called **gluconeogenesis** (Figure 9.16).

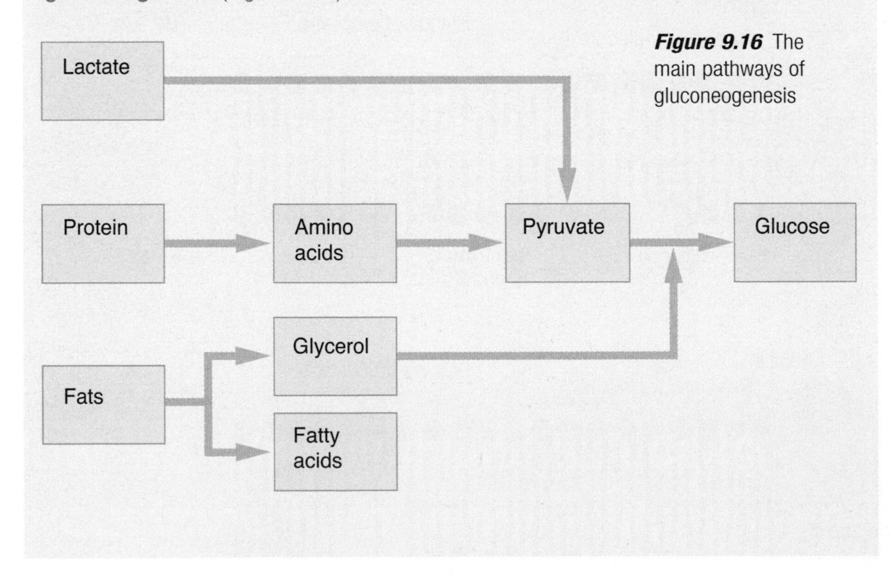

Figure 9.16 The main pathways of gluconeogenesis

What is diabetes?

There are two quite different diseases, both called diabetes — diabetes mellitus and diabetes insipidus. They have different causes, but a common symptom is the production of large amounts of dilute urine, which is what gives the disease its name.

Diabetes is derived from Ancient Greek and means 'through-passing' or 'the syphon', so you can see how a disease with the symptom of producing too much urine would get this name. Only later did scientists discover that there are two different causes.

Here we shall deal with diabetes mellitus, which is linked to an insulin deficiency.

Diabetes insipidus is the result of a deficiency of a hormone that causes reabsorption of water in the kidney.

There are two types of diabetes mellitus: type 1 diabetes and type 2 diabetes.

In type 1 diabetes, the immune system attacks the β-cells in the islets of Langerhans. As a result, they are unable to produce sufficient insulin and regulation of plasma glucose concentrations becomes erratic. This type of diabetes often appears during childhood. It used to be called early-onset diabetes.

In type 2 diabetes, the cause is different. The β-cells still produce some insulin (but less than normal), and the receptors on the liver cells, in particular, become insensitive to the hormone. The precise reason for this is not known. This condition tends to develop slowly as, gradually, more and more β-cells stop producing insulin and receptors become less and less sensitive to insulin.

It is not the sudden change that is often associated with type 1 diabetes and so the condition is often less severe.

Both type 1 and type 2 diabetes have some symptoms in common, including:

- the presence of glucose in the urine
- urinating frequently, particularly at night
- tiredness
- loss of weight
- abnormal thirst

These symptoms are linked to the inability of the body to convert glucose to glycogen, as shown in Figure 9.17.

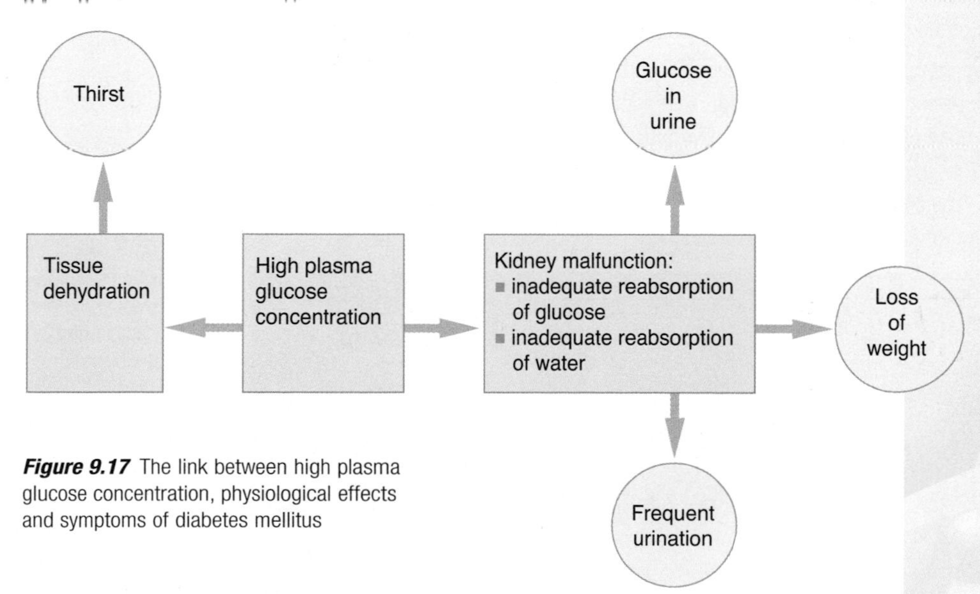

Figure 9.17 The link between high plasma glucose concentration, physiological effects and symptoms of diabetes mellitus

Symptoms of type 1 diabetes may also include:

- cramps
- constipation
- blurred vision
- recurrent skin infections

The treatment of type 1 diabetes always involves administering insulin to replace the insulin the body cannot make. Currently, this is almost always done by injections into the skin, although about 130 000 diabetics have their insulin administered by infusion pumps that deliver insulin continuously.

Treatment also involves following a carefully planned diet and taking regular exercise.

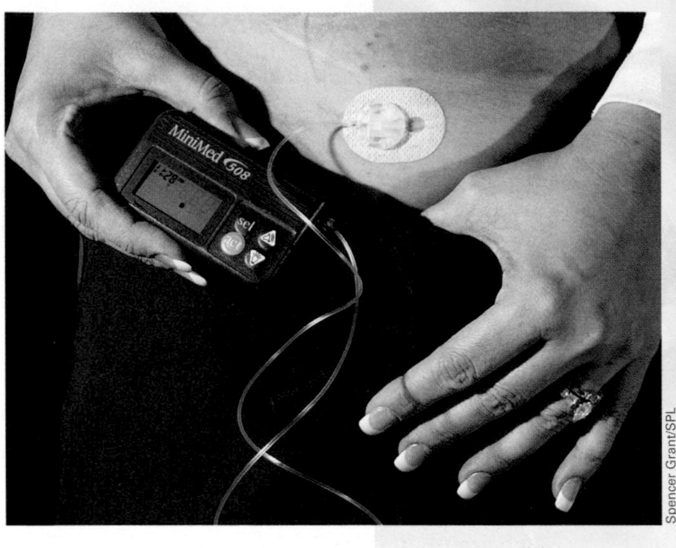

A continuous infusion pump being used by a diabetic

Type 2 diabetes is increasingly being treated by insulin injections, but it can also be treated by:

- a carefully planned diet
- an exercise program
- oral anti-diabetes medication that either:
  - reduces gluconeogenesis in the liver
  - enhances the production of insulin in the islets of Langerhans

Box 9.5 Potential dangers in managing diabetes

Diabetes results in high plasma glucose concentrations. If the treatment regime is not followed correctly, this can lead to a **hyperglycaemic** (high glucose) **coma**. It results from dehydration of the tissues, particularly the brain and can be treated by an injection of insulin.

A diabetic can also suffer a coma as a result of too little glucose — a **hypoglycaemic** (low glucose) **coma**. This can happen if the person mistakenly takes an extra dose of insulin or exercises without eating sufficient carbohydrate or misses a meal. The low glucose concentration leads to trembling, confusion, palpitations and seizures. The condition can be treated by an injection of glucagon or glucose given directly into the bloodstream.

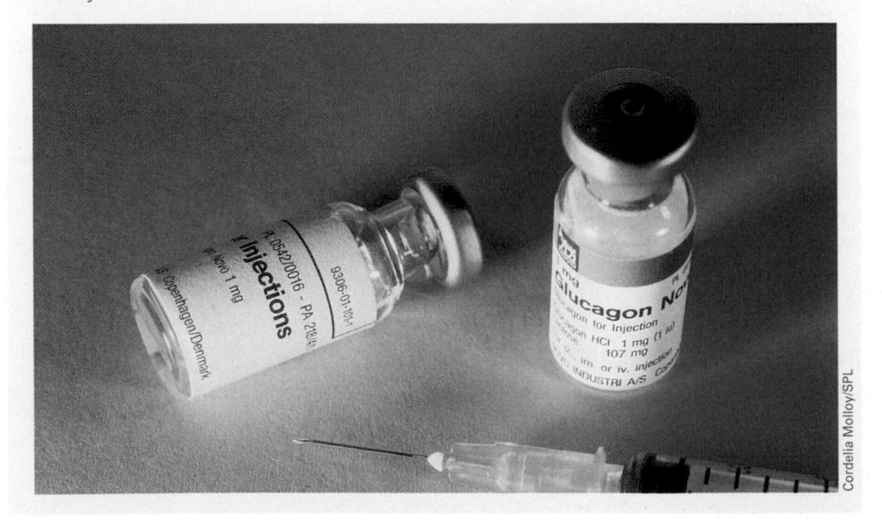

Glucagon can be injected to prevent coma from hypoglycaemia (too little glucose in the plasma)

Does positive feedback ever occur?

What is positive feedback?

In negative feedback, any deviation from a preset value is corrected and the preset value is restored. Some mechanisms in the body operate on a **positive feedback** system where any difference from the preset value is detected and then exaggerated — it is not returned to the preset value but made even more different from it.

Positive feedback sometimes occurs when the negative feedback systems fail to function properly. A good example of this is the progressive development of **hypothermia**.

We saw earlier that one of the responses to body temperature falling is to increase the metabolic rate. This results in more energy being released as heat. However, in some circumstances, this extra heat production is not enough to offset the heat losses and the core body temperature continues to decrease. Beyond a certain point, reduction in body temperature reduces the metabolic rate, and so less heat is produced. This means that the core body temperature now drops more quickly, so the metabolic rate reduces more quickly — and so on. This is positive feedback.

Positive feedback also occurs during childbirth. The hormone **oxytocin** from the pituitary gland causes contractions of the wall of the uterus. These, in turn, stimulate extra secretion of oxytocin which increases the strength of the contractions — and so on. This is summarised in the flow chart in Figure 9.18.

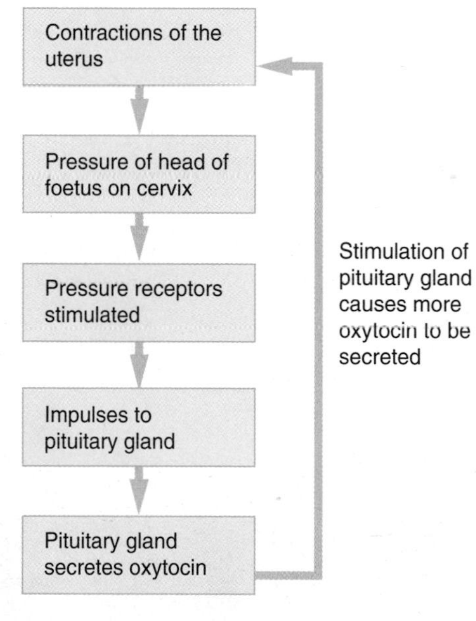

Figure 9.18

Positive and negative feedback mechanisms combine in the control of oestrus in mammals

The oestrus cycle in mammals (the menstrual cycle in humans) is controlled by several hormones. The pituitary gland (just beneath the brain) produces and secretes **follicle-stimulating hormone** (FSH) and **luteinising hormone** (LH), which are involved in controlling the cycle of events that takes place in the ovary (Figure 9.19).

As a result of stimulation by the pituitary hormones, cells in the follicles in the ovaries produce **oestrogen** and **progesterone**, which influence the events of the

Follicles are groups of cells in an ovary that contain a developing oocyte (egg cell).

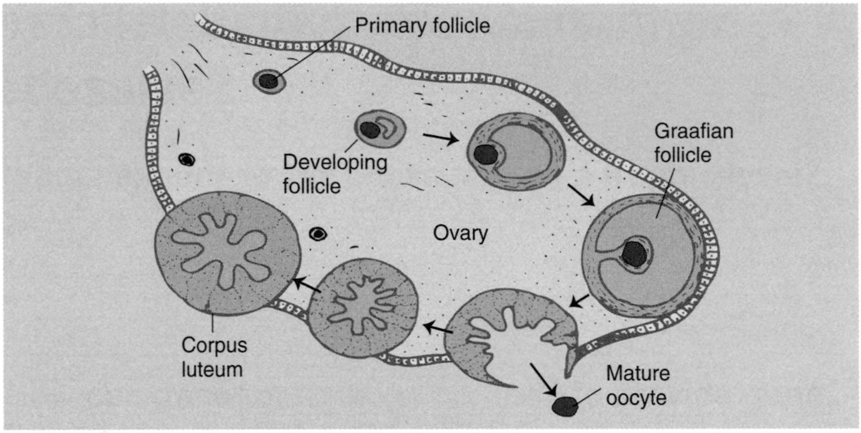

Figure 9.19 Stages in the development of a follicle

menstrual cycle. Negative and positive feedback loops control the production of these hormones.

An increase in the plasma concentration of FSH stimulates follicles to grow and to produce oestrogen. As the concentration of oestrogen increases in the blood plasma, it has two effects on the pituitary gland (Figure 9.20):

- it inhibits the secretion of FSH
- it stimulates the secretion of LH

As a result, the pituitary gland:

- reduces the secretion of FSH, so the plasma concentration of FSH falls
- increases the secretion of LH, so the plasma concentration of LH increases

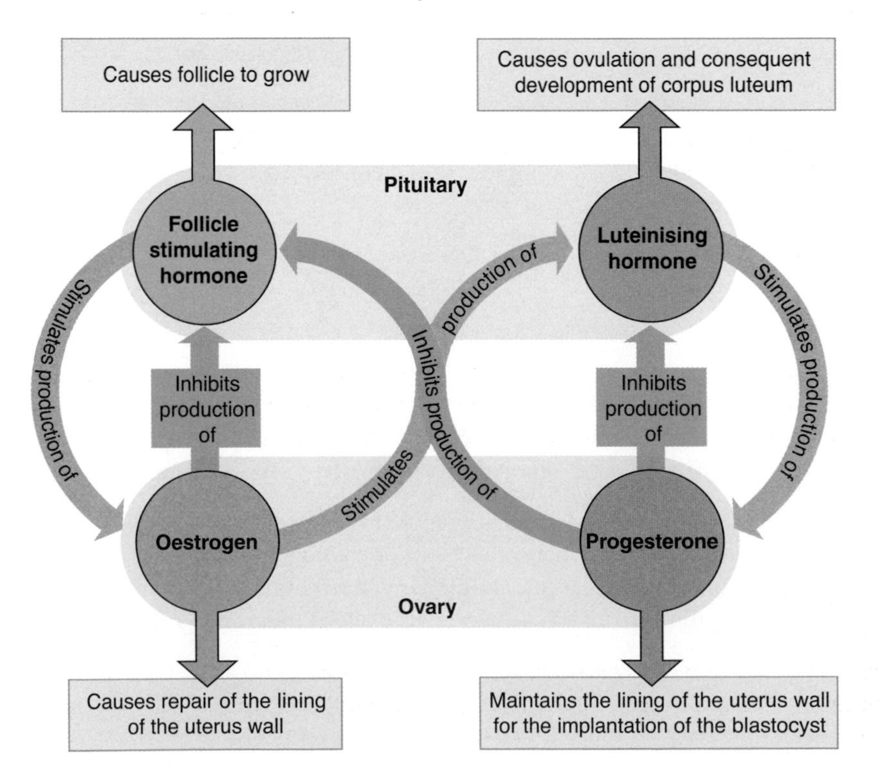

Figure 9.20 Feedback systems that control the secretion of pituitary and ovarian hormones

The levels of LH rise slowly at first and then sharply by day 13–14 of the cycle. This sharp increase in the concentration of LH triggers ovulation. LH also stimulates the corpus luteum (the remains of the follicle formed at ovulation) to secrete increased amounts of progesterone. The increased concentrations of progesterone in the plasma inhibit the secretion of both FSH and LH by the pituitary gland.

As levels of FSH and LH fall, the production of progesterone also decreases. This removes its inhibitory effect on the pituitary, which starts a new cycle by increasing the secretion of FSH.

These control mechanisms are summarised in Figures 9.20 and 9.21.

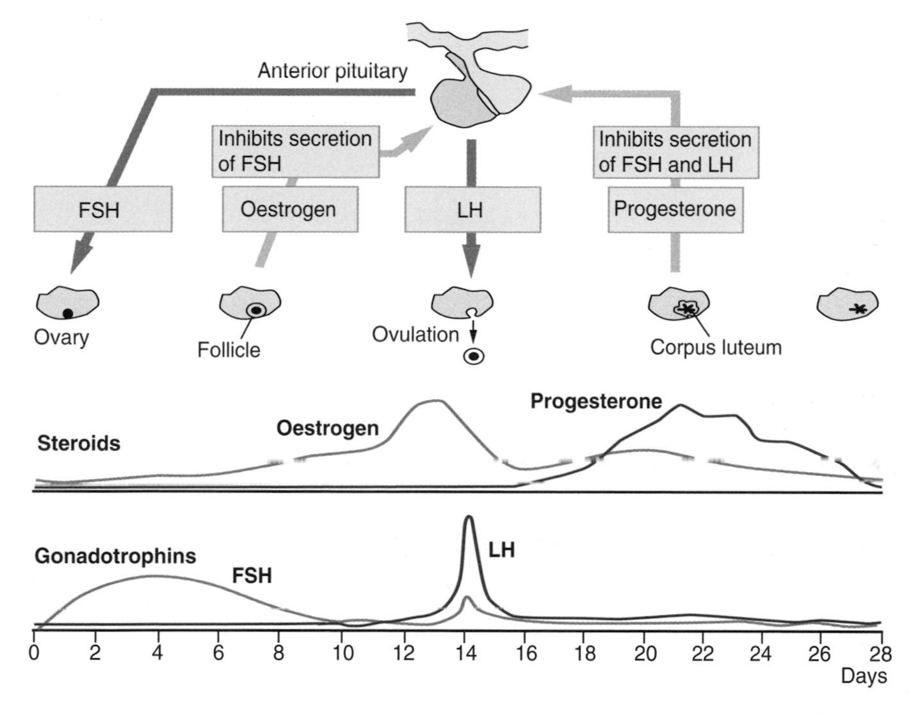

Figure 9.21 A summary of the control of the menstrual and ovarian cycles

All these controls operate using negative feedback loops. As the concentration of a particular hormone increases, it initiates a mechanism that reduces its level again.

The control of the secretion of oestrogen and progesterone also involves positive feedback loops. As the concentration of each of these hormones increases in the blood, it stimulates cells in the follicle to produce increased amounts of the hormone.

Box 9.6 Fertility in cattle

HOW SCIENCE WORKS

Knowledge of reproductive hormones has allowed farmers to breed cattle with increased milk yields.

If a farmer has one cow in the herd that produces significantly more milk than the others, it would be desirable to breed from this cow to produce a herd with a similar yield. Using conventional breeding methods, it would take many years to produce such a herd. However, with the techniques of **in vitro fertilisation** (**IVF**) and **embryo transplantation**, this can be achieved in a much shorter time.

The main stages in the process are as follows:

- Injections of FSH are given to the 'superior' cow to induce **multiple ovulation** (sometimes called 'super-ovulation'). The oocytes are collected and stored in a culture medium that has a similar composition to the fluid in the Fallopian tubes.
- Using IVF, the oocytes are fertilised using semen from a registered bull.
- The fertilised ova are allowed to develop into 4-cell or 8-cell embryos.
- The cells of these embryos are then split, returned to the culture medium and each begins to develop again. The development and splitting is repeated until sufficient embryos have been obtained.
- The embryos are screened for:
  – sex (only female embryos are used)
  – the genes needed to produce the high milk yield
- Each suitable embryo is then transplanted into the uterus of a surrogate cow. A herd of surrogates can carry embryos all derived from just one 'superior' cow. The calves that are born will develop into high-milk-yielding cattle.

The process is summarised in Figure 9.22.

Oocytes are produced by meiosis, so they are not genetically identical.

The cells obtained from one original embryo (no matter how many times splitting has occurred) are genetically identical because they were formed by mitosis.

Figure 9.22 IVF and embryo transplantation in cattle

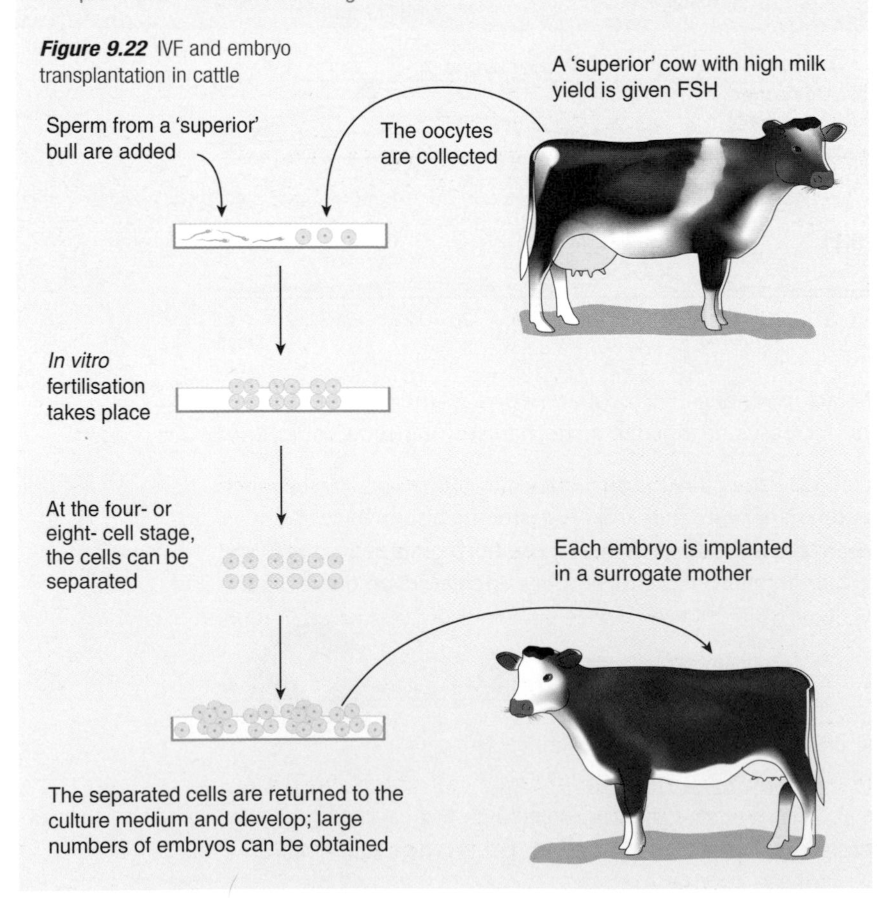

The surrogate 'mothers-to-be' are given daily injections of progesterone for a few days. This suspends their oestrous cycles by suppressing the secretion of the pituitary hormones FSH and LH. When the injections of progesterone are stopped, the inhibition is removed and all start a new cycle at the same time. This means that the transplantation of all the embryos can take place at the same time.

Box 9.7 Increasing milk yield with BST

Selective breeding is not the only method of increasing the milk yield of cattle. Dairy cattle can be given injections of BST (bovine somatotrophin). This is a growth hormone that has the added effect of increasing milk production and letdown in cattle. Regular injections of BST increase the milk yield significantly. Genetically engineered bacteria produce much of the BST used.

Typical results are shown in the table below.

	Feed/kg d^{-1}	Milk output/kg d^{-1}	Milk-to-feed ratio
Without BST	34.1	27.9	0.82
With BST	37.8	37.3	0.99

The milk-to-feed ratio is a measure of productivity. It measures how many kilograms of milk are produced from a kilogram of feed. This is a more important measure than just an increase in yield, since it might require an excessive amount of feed. The figures shown give a 21% increase in milk-to-feed ratio — a 21% increase in real productivity.

Summary

Muscles

- Muscles occur in antagonistic pairs because they can only exert a force when they contract.
- A muscle is an organ; it contains blood, nervous tissue, connective tissue and sense cells.
- Each muscle is made from bundles of fibres, each made from myofilaments that are made from myofibrils.
- A single contracting unit of a muscle is called a sarcomere; it contains overlapping filaments of actin and myosin. To contract the sarcomere (and therefore the muscle):
  - calcium ions are released and cause the tropomyosin molecules to move and expose the binding sites on the actin molecules
  - ATP molecules are hydrolysed to ADP and P_i; the energy released is transferred to the myosin heads

- P_i is released and the myosin heads bind to the exposed binding sites on the actin molecules
- ADP is released and the energy stored in the myosin heads moves them and, with them, the actin filaments

- Energy is released in muscles by:
  - aerobic respiration
  - anaerobic respiration
  - the phosphocreatine–ATP system
- Slow-twitch and fast-twitch muscle fibres have different properties:

Feature of fibre	Slow-twitch fibres	Fast-twitch fibres
Respiratory enzymes	Krebs cycle	Glycolysis
Capillary network	Extensive	Less extensive
Myoglobin concentration	High	Lower
Contractions	Slow, less forceful	Fast, more forceful
Fatigue resistance	High	Low

Homeostasis

- Homeostasis is the maintenance of a constant internal environment so that enzyme systems can function with optimum efficiency.
- Negative feedback systems maintain systems at a preset level by detecting deviations from that level and initiating corrective mechanisms to restore it.
- Negative feedback systems exist to control core body temperature and the concentration of plasma glucose.
- Ectotherms usually control their core body temperature by behavioural methods and it varies with the environment. Endotherms are able to maintain a more constant core body temperature because they also use physiological methods.
- In endotherms, core body temperature is controlled by the heat loss centre and the heat gain centre of the hypothalamus. They initiate responses by sending nerve impulses through the autonomic nervous system.
- The heating and cooling responses to stimulation by the hypothalamus are given in the following table:

Controlling mechanism	Heating response	Cooling response
Arterioles leading to skin	Constriction	Dilation
Radiation from skin	Decreased	Increased
Sweating	Decreased	Increased
Shivering	Increased	Decreased
Metabolic rate	Increased	Decreased

- Insulin and glucagon from the islets of Langerhans in the pancreas are both involved in the interconversion of glucose and glycogen in the liver.

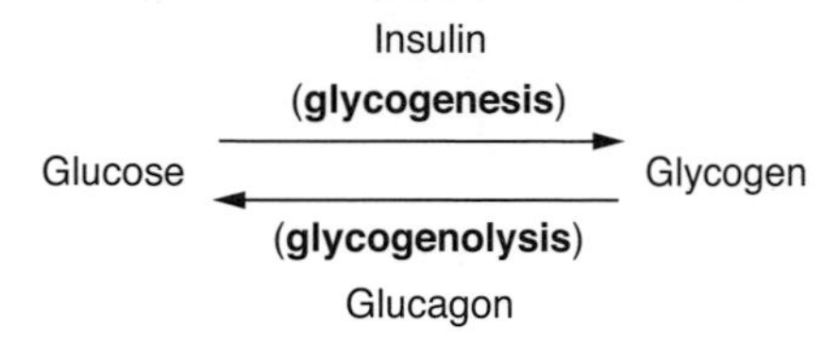

- The events in the control of plasma glucose concentration are summarised in the following table:

Stage in response	Glucose too high	Glucose too low
Cells active in pancreas	β-cells	α-cells
Hormone secreted	Insulin	Glucagon
Reaction in liver	Glucose → glycogen	Glycogen → glucose
Name of process	Glycogenesis	Glycogenolysis

- Other hormones involved with glucose are:
  - adrenaline, which causes the conversion of glycogen to glucose irrespective of the plasma glucose concentration
  - thyroxine, which raises the metabolic rate which increases the use of glucose
  - corticosteroid hormones, which increase gluconeogenesis
- There are two types of diabetes mellitus:
  - Type 1, in which the immune system attacks the β-cells in the islets of Langerhans
  - Type 2, in which the insulin receptors on cells become less sensitive to insulin; there is also a gradual reduction in insulin production
- Type 1 diabetes must be treated with insulin injections; type 2 can be treated by planning diet and exercise carefully.

Feedback

- In positive feedback, a deviation from a preset value is exaggerated.
- Positive feedback happens in hypothermia; normal negative feedback control breaks down and a positive feedback loop is set up between decreased metabolic rate and decreased core body temperature.
- Positive feedback control results in the progressively stronger contractions of the uterus during childbirth because of increased oxytocin secretion.
- A combination of negative feedback and positive feedback controls the levels of hormones involved in the menstrual cycle:
  - the interaction of the pituitary and ovarian hormones involves negative feedback
  - the secretion of the ovarian hormones sets up a positive feedback loop in which the increased concentration increases their secretion

Questions

Multiple-choice

1 When a sarcomere contracts:

A the H zone decreases and the I band increases
B the H zone increases and the I band increases
C the H zone increases and the I band decreases
D the H zone decreases and the I band decreases

2 Muscle myofibrils are made from:

A thick filaments of actin and thin filaments of myosin
B thin filaments of actin and thick filaments of myosin
C thin filaments of actin and thin filaments of myosin
D thick filaments of actin and thick filaments of myosin

3 Compared with fast-twitch fibres, slow-twitch fibres have:

A Krebs cycle enzymes and high fatigue resistance
B Krebs cycle enzymes and low fatigue resistance
C Glycolysis enzymes and low fatigue resistance
D Glycolysis enzymes and high fatigue resistance

4 Negative feedback mechanisms detect:

A a deviation from the norm and exaggerate it
B a return to the norm and make it deviate again
C a return to the norm and maintain it
D a deviation from the norm and return it to the norm

5 Ectotherms are different from endotherms in their control of core body temperature because they:

A cannot control core body temperature
B can use only physiological means to control core body temperature
C can use only behavioural means to control core body temperature
D can use both behavioural and physiological means to control core body temperature

6 Insulin, glucagon, adrenaline, thyroxine and corticosteroid hormones all affect plasma glucose concentration. The hormones that reduce plasma glucose concentration are:

A insulin and glucagon
B glucagon and adrenaline
C insulin and corticosteroids
D glucagon and corticosteroids

7 The two types of diabetes mellitus are different in that:

A type 1 is caused by the immune system destroying β-cells and can be treated by diet and exercise
B type 2 is caused by the immune system destroying β-cells and can be treated by diet and exercise
C type 2 is caused by lack of sensitivity of protein receptors and must be treated with insulin injections
D type 1 is caused by the immune system destroying β-cells and must be treated with insulin injections

8 Increased secretion of FSH causes:
- **A** a follicle to develop and increase its secretion of oestrogen
- **B** a follicle to regress and increase its secretion of oestrogen
- **C** a follicle to regress and increase its secretion of progesterone
- **D** a follicle to develop and increase its secretion of progesterone

9 In human females, increased secretion of LH causes:
- **A** ovulation and the secretion of oestrogen by the corpus luteum
- **B** ovulation and the secretion of progesterone by the corpus luteum
- **C** menstruation and the secretion of progesterone by the corpus luteum
- **D** menstruation and the secretion of oestrogen by the corpus luteum

10 Positive feedback systems:
- **A** detect a deviation from the norm and maintain it
- **B** detect a deviation from the norm and bring it back to the norm
- **C** detect a return to the norm and maintain it
- **D** detect a deviation from the norm and exaggerate it

Examination-style

1 Three groups of cyclists were fed different diets for three days prior to a cycling event:
- group 1 was fed a mixed diet
- group 2 was fed a low carbohydrate diet
- group 3 was fed a high carbohydrate diet

The graphs show the glycogen concentrations in the cyclists' muscles immediately before and after the event.

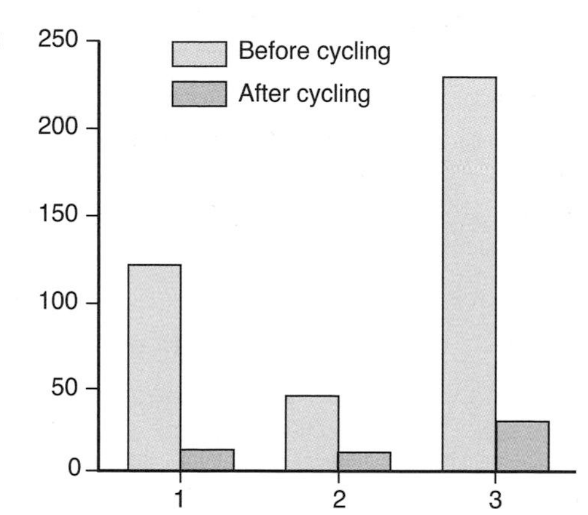

(a) Explain the difference in muscle glycogen concentration between groups 2 and 3 before the event. *(4 marks)*

(b) There was no difference in the concentration of muscle glycogen of groups 1 and 2 after the event, even though the concentrations were very different before the event.

(i) Explain why the muscle glycogen of both groups is reduced during the event. *(2 marks)*

(ii) Both groups were able to complete the event comfortably. Explain why. *(2 marks)*

Total: 8 marks

2 It is possible to estimate the levels of the steroid hormones oestrogen and progesterone in the blood by measuring their concentrations in saliva. The procedure is as follows. The patient:

- rinses out her mouth with water
- waits 5 minutes
- collects saliva in a test tube
- stores it for up to 1 week in a refrigerator (if necessary)
- posts it to the laboratory for analysis

(a) Suggest why:

(i) she first rinses out her mouth with water *(1 mark)*

(ii) she must wait 5 minutes before collecting the sample *(2 marks)*

(b) Explain one advantage of this procedure compared with having a doctor take a blood sample. *(1 mark)*

(c) The results from an analysis of one patient's saliva over 70 days are shown in the graph below.

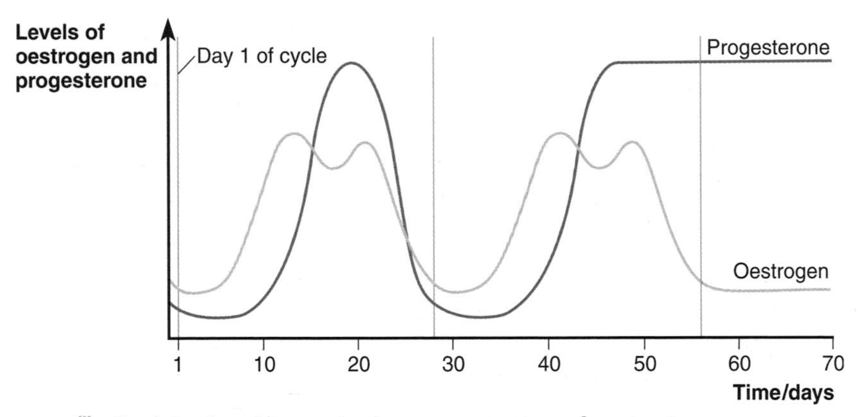

(i) Explain the changes in the concentration of oestrogen from day 7 to day 17. *(2 marks)*

(ii) Did this woman become pregnant during the 70-day period? Give evidence from the graph to support your answer. *(3 marks)*

Total: 9 marks

3 The graph shows the core body temperature and the rate of heat production in a human over a range of temperatures.

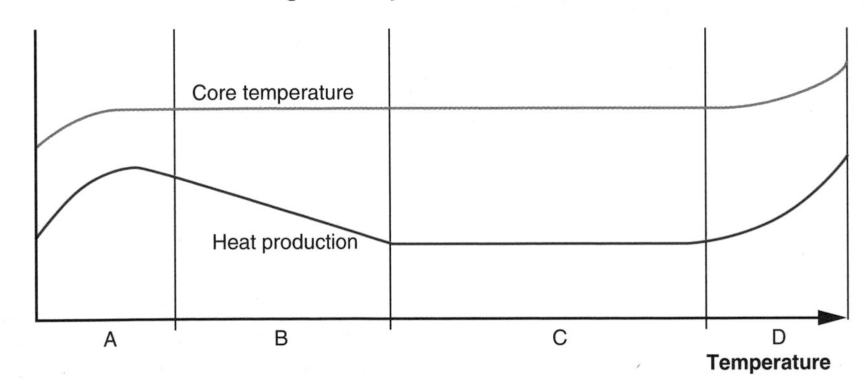

(a) Name the condition of the person in temperature range A. Explain how this condition is caused. *(3 marks)*

(b) The core temperature is constant over temperature range B.

(i) Explain the benefit to humans of a constant core body temperature. *(2 marks)*

(ii) Use the graph to explain how the core temperature is maintained in these conditions. *(3 marks)*

(c) Suggest how the heat loss systems in the body would respond to the conditions in temperature range D. *(3 marks)*

Total: 11 marks

4 The sliding-filament hypothesis is believed to offer the best explanation of muscle contraction.

(a) The diagram shows part of a relaxed myofibril.

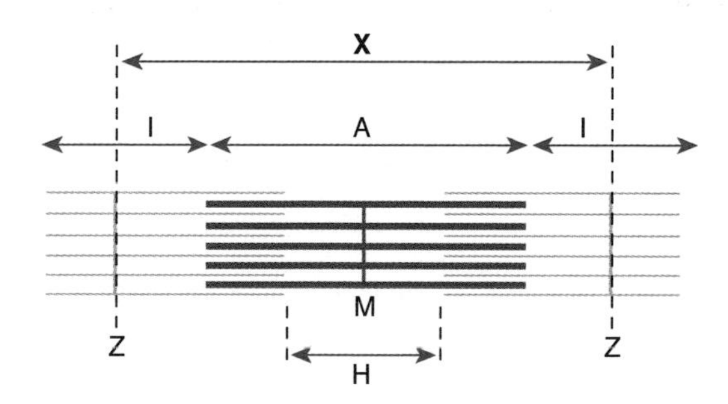

(i) Name the region labelled **X**. *(1 mark)*

(ii) Re-draw the diagram to show the appearance of the myofibril following contraction. *(3 marks)*

(b) Describe the roles of calcium ions and ATP in bringing about contraction of skeletal muscle. *(4 marks)*

(c) Give *two* differences between slow-twitch fibres and fast-twitch fibres. *(2 marks)*

Total: 10 marks

5 Within set limits, humans are able to maintain a plasma glucose concentration and high body temperature that both vary only slightly.

(a) (i) Explain the role of pancreatic hormones in maintaining the plasma glucose concentration within set limits. *(6 marks)*

(ii) Explain why these pancreatic hormones are able to target liver and skeletal muscle cells in particular. *(4 marks)*

(b) Explain the benefit of being able to maintain a constant, high body temperature. *(5 marks)*

Total: 15 marks

Chapter 10

How do cells synthesise proteins?

This chapter covers:
- a reminder of the structure of DNA
- the nature of the genetic code
- transcription of DNA to messenger RNA
- translation of the code in the messenger RNA into a sequence of amino acids
- the effects of mutations
- the way in which gene expression is controlled
- stem cells and their importance in plants and in animals

The discovery of the structure of DNA in 1953 by Watson and Crick allowed biologists to understand how chromosomes duplicate themselves prior to cell division and how DNA controls protein synthesis. Proteins are polymers of amino acids. Therefore, to control protein synthesis, DNA must be able to code for the amino acid sequences. One gene controls the synthesis of a protein; not all these proteins are enzymes.

◀ Not all genes code for proteins. Some are 'regulator genes' that switch on and off genes that do code for proteins.

The code for a protein that is specified by DNA has to be carried to the ribosomes so that they can assemble the amino acids in the correct sequence to form the appropriate protein. DNA remains in the nucleus at all times. The following events occur:
- The DNA code is rewritten in a molecule of **messenger RNA** (mRNA); this rewriting of the code is called **transcription**.
- The mRNA travels from the nucleus through pores in the nuclear envelope to the ribosomes.
- Free amino acids are carried from the cytoplasm to the ribosomes by molecules of **transfer RNA** (tRNA).
- The ribosome reads the mRNA code and assembles the amino acids into a protein; this is called **translation**.

Mutations may alter the sequence of bases in DNA within a gene and, as a result, may alter the structure of the protein coded for by the gene. This may be trivial, have serious consequences, or it could turn out to be beneficial.

What is DNA?

DNA is a huge molecule made from two **anti-parallel** strands of **nucleotides**. The nucleotides in one strand are paired with nucleotides in the other according to the base pairing rule. This states that:

- adenine-containing nucleotides are always opposite thymine-containing nucleotides
- cytosine-containing nucleotides are always opposite guanine-containing nucleotides (Figure 10.1)

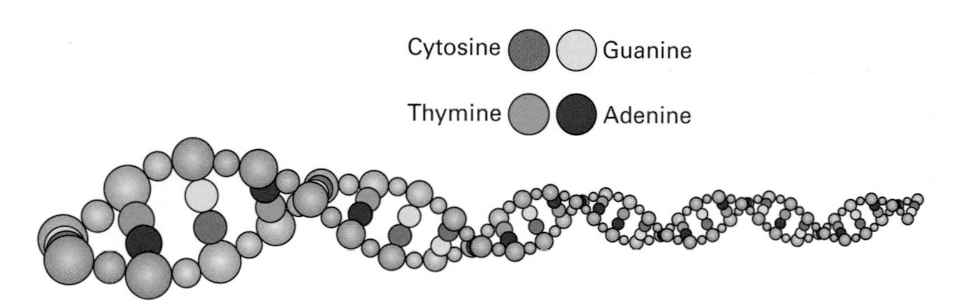

Figure 10.1 The DNA molecule is an anti-parallel double helix. Adenine on one strand is always opposite thymine on the other strand; cytosine is always opposite guanine.

DNA is a stable molecule at normal temperatures. The individual hydrogen bonds that hold the two strands together are quite weak, but the sheer number of them ensures that the two strands stay in position.

In eukaryotic cells, DNA is a linear molecule associated with histone proteins to form chromatin and is organised into chromosomes. The DNA of prokaryotic cells is similar in that it is a double helix with antiparallel strands, but different in that it is:

- much smaller
- not associated with histones (it is sometimes called naked DNA)
- a looped molecule (the ends are, in effect, joined)

How does DNA carry the code for proteins?

Where is the code carried?

Proteins are polymers of amino acids. Each protein has a unique tertiary structure because of the sequence of amino acids in its molecule.

To control protein synthesis, DNA must be able to specify this sequence of amino acids. Only one of the two strands carries the genetic code for proteins. This is the **coding strand** (sense strand). The other strand is the **non-coding strand** (anti-sense strand).

It is the sequence of nucleotides in the sense strand that carries the code for the amino acids, and, therefore, the protein (Figure 10.2). Every DNA nucleotide comprises an organic base, a pentose sugar (deoxyribose) and a phosphate

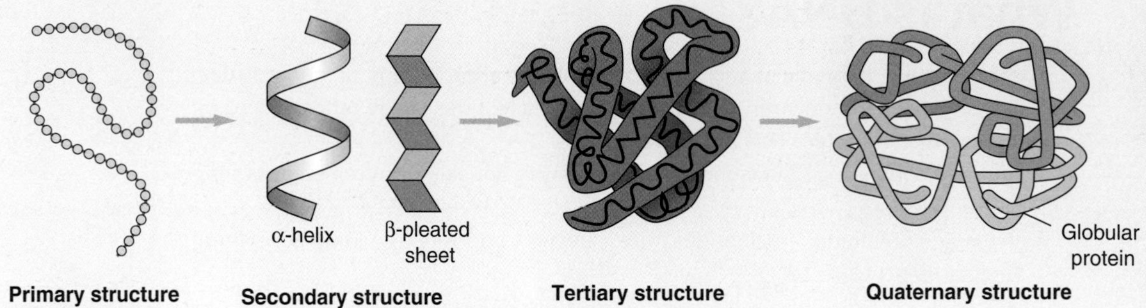

Figure 10.2 Levels of organisation within a protein molecule

group. As the phosphate and sugar are the same in all nucleotides, it is, effectively, the sequence of the bases that codes for the amino acids.

What is the code like?

A triplet of bases codes for each amino acid.

Box 10.1 The triplet code

If the genetic code were a singlet code, four bases could only specify four amino acids:

A = amino acid 1 T = amino acid 2
C = amino acid 3 G = amino acid 4

A doublet code could specify 16 amino acids. The possible codes are:

AA	AT	AC	AG
TA	TT	TC	TG
CA	CT	CC	CG
GA	GT	GC	GG

Notice that order matters. The code AT is not the same as the code TA.

In a triplet code, all 16 of the doublet codes could have any of the four bases added to the start of the code. For example, the first line of the doublet code above becomes:

AAA	AAT	AAC	AAG
TAA	TAT	TAC	TAG
CAA	CAT	CAC	CAG
GAA	GAT	GAC	GAG

This generates 16 combinations, as do the three other lines, making a total of 64 combinations.

There are 64 codes, then, but only 20 amino acids. What is the purpose of the other 44 codes? In fact, none of these is spare or redundant. Most amino acids have more than one code. Only methionine and tryptophan have just one triplet that codes for them; arginine has six. Three of the triplets (TAA, TAG and TGA) do not code for amino acids at all. They are 'stop' codes that signify the end of a coding sequence. Because there is this extra capacity in the genetic code, over and above what is essential, it is said to be a **degenerate code**.

Besides being a triplet and degenerate code, the DNA code is a **non-overlapping code** (Figure 10.3). This means that each triplet is distinct from all other triplets. The last base in one triplet cannot also be the first base (or second base) in another triplet.

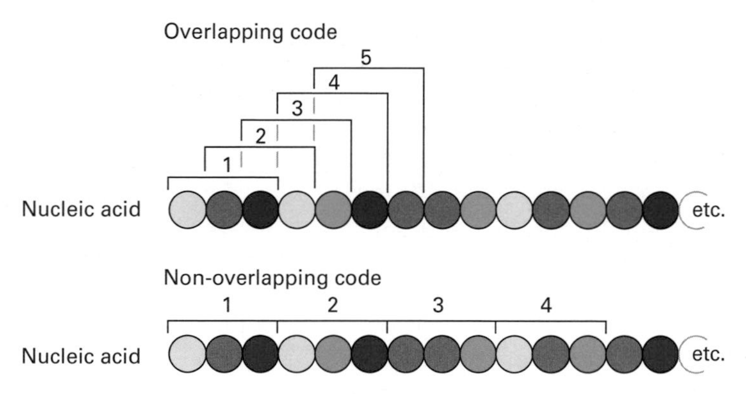

Figure 10.3 In a non-overlapping code, each triplet is distinct from all others

The genetic code is also a **universal code**. This means that the triplet TAT is the DNA code for the amino acid tyrosine in a human, a giant redwood tree, a bacterium or in any other living organism.

The 64 DNA triplets and the amino acids they code for are shown in Table 10.1. Notice the 'stop' codes.

Table 10.1

First position	**Second position**				**Third position**
	T	C	A	G	
T	Phenylalanine Phenylalanine Leucine Leucine	Serine Serine Serine Serine	Tyrosine Tyrosine stop stop	Cysteine Cysteine stop Tryptophan	T C A G
C	Leucine Leucine Leucine Leucine	Proline Proline Proline Proline	Histidine Histidine Glutamine Glutamine	Arginine Arginine Arginine Arginine	T C A G
A	Isoleucine Isoleucine Isoleucine Methionine	Threonine Threonine Threonine Threonine	Asparagine Asparagine Lysine Lysine	Serine Serine Arginine Arginine	T C A G
G	Valine Valine Valine Valine	Alanine Alanine Alanine Alanine	Aspartic acid Aspartic acid Glutamic acid Glutamic acid	Glycine Glycine Glycine Glycine	T C A G

In Table 10.1, the first letter of each triplet specifies a horizontal band. The second letter specifies a column and the third letter specifies a horizontal line.

Take, for example, the triplet CAG:

- **C** (the first position) specifies the second horizontal band across
- **A** (the second position) specifies the third column
- **G** (the third position) specifies the fourth horizontal line

CAG is the code for the amino acid glutamine. ATT is the code for isoleucine and GAG is the code for glutamic acid.

How does the DNA code control protein structure?

Proteins are built from amino acids in the ribosomes, but DNA — the code carrier — remains in the nucleus at all times. So the code must be carried out of the nucleus to the ribosomes by a messenger. That messenger is **messenger RNA** (mRNA). A molecule of mRNA is a rewritten version of the genetic code of a single gene — a section of DNA. The process by which the code is rewritten is **transcription**. The mRNA carries the code to the ribosomes where it is used to bring together the amino acids in the correct sequence. This process is **translation**. Translation requires the amino acids be brought to the ribosomes; this is carried out by **transfer RNA** (tRNA). Figure 10.4 summarises the relationship between these processes.

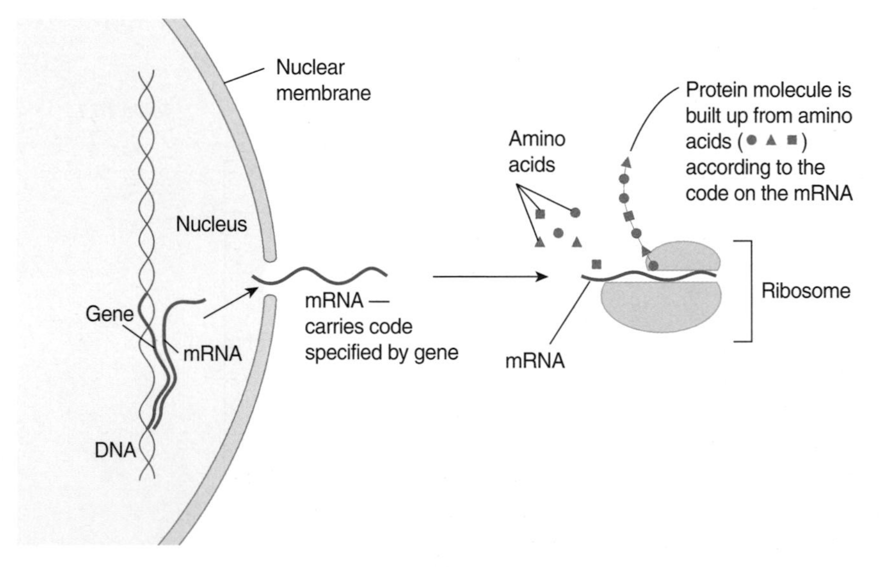

Figure 10.4 An overview of protein synthesis

How does transcription take place?

During this process, the coded information in the DNA of one gene is used to synthesise a molecule of mRNA that will carry the code to the ribosomes. Messenger RNA is similar to DNA in that it is built from nucleotides, but it is different from DNA in a number of ways:

- It is a much smaller molecule.
- It is single stranded.

If you are asked to compare the structure of DNA and mRNA, make sure that you give a genuine *comparison*. Simply saying, 'mRNA is single stranded' is not a comparison. Is the DNA also single stranded? Or maybe triple stranded? You must make your comparison clear.

- The base thymine is replaced by uracil.
- The sugar in each nucleotide is ribose, not deoxyribose.

The triplets of bases in mRNA that code for amino acids are called **codons**. The mRNA codons are identical to the DNA triplets that code for specific amino acids, except that U (uracil) is substituted for T (thymine).

First position (5′ end)	Second position: U		C		A		G		Third position (3′ end)
	U		C		A		G		
U	UUU	Phe	UCU		UAU	Tyr	UGU	Cys	U
	UUC		UCC	Ser	UAC		UGC		C
	UUA	Leu	UCA		UAA	stop	UGA	stop	A
	UUG		UCG		UAG	stop	UGG	Trp	G
C	CUU		CCU		CAU	His	CGU		U
	CUC	Leu	CCC	Pro	CAC		CGC	Arg	C
	CUA		CCA		CAA	Gln	CGA		A
	CUG		CCG		CAG		CGG		G
A	AUU	Ile	ACU		AAU	Asn	AGU	Ser	U
	AUC		ACC	Thr	AAC		AGC		C
	AUA		ACA		AAA	Lys	AGA	Arg	A
	AUG	Met	ACG		AAG		AGG		G
G	GUU		GCU		GAU	Asp	GGU		U
	GUC	Val	GCC	Ala	GAC		GGC	Gly	C
	GUA		GCA		GAA	Glu	GGA		A
	GUG		GCG		GAG		GGG		G

Table 10.2 mRNA codons. The table shows the mRNA codons for the 20 amino acids used in protein synthesis. There is more than one code for most amino acids. Some codes are 'stop' codes to tell the ribosome that this is the end of the reading sequence. The code for methionine, AUG, also acts as a 'start' code. So all mRNA molecules begin with AUG and therefore the first amino acid in all polypeptide chains is methionine, although this may be removed at a later stage.

To form the single-stranded mRNA when transcription takes place, only one of the strands of DNA is copied (Figure 10.5). Transcription takes place in the following way:

- The enzyme DNA-dependent RNA polymerase (RNA polymerase) binds with a section of DNA next to the gene that is to be transcribed.
- Transcription factors (see page 219) activate the enzyme.
- The enzyme begins to 'unwind' a section of DNA. The sense strand of this section contains the gene that codes for a protein. However, copying this would produce a complementary sequence of bases, similar to those in the antisense strand, which would *not* code for a protein.
- RNA polymerase moves along the *antisense* strand, using it as a template for synthesising the mRNA.
- The polymerase assembles free RNA nucleotides into a chain in which the base sequence is complementary to the base sequence on the antisense strand of the DNA. This, therefore, carries the same triplet code as the sense strand (except that uracil replaces thymine).
- The completed molecule leaves the DNA; the strands of DNA rejoin and re-coil.

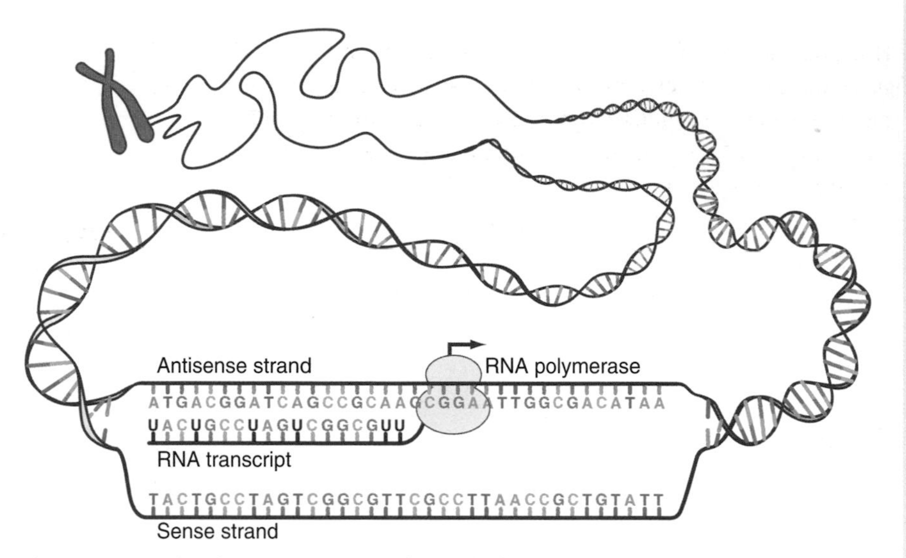

Figure 10.5 In transcription, the anti-sense strand is used to produce a single-stranded molecule of mRNA with a base sequence identical to the sense strand, except that thymine bases are replaced by uracil

The mRNA molecule now contains the code for the protein that was held in the DNA of the gene. However, it contains more than just the code. In addition to the coding regions of a gene (exons) there are:

- promoter regions in the DNA that are not transcribed but serve as recognition sites for RNA polymerase — they indicate where to start copying the DNA
- non-coding sections of DNA called introns, separating the coding regions or exons; they do not code for any amino acids but are transcribed

◀ Estimates vary, but most biologists agree that over 95% of our DNA is non-coding DNA.

At this stage, the transcribed molecule is referred to as pre-mRNA.

To convert the pre-mRNA to mRNA, the introns must be cut out and the remaining exons joined or spliced together (Figure 10.6). In addition, the end of the mRNA is 'capped' with a different nucleotide that allows the ribosome to recognise it. Ribosomes cannot recognise non-capped mRNA.

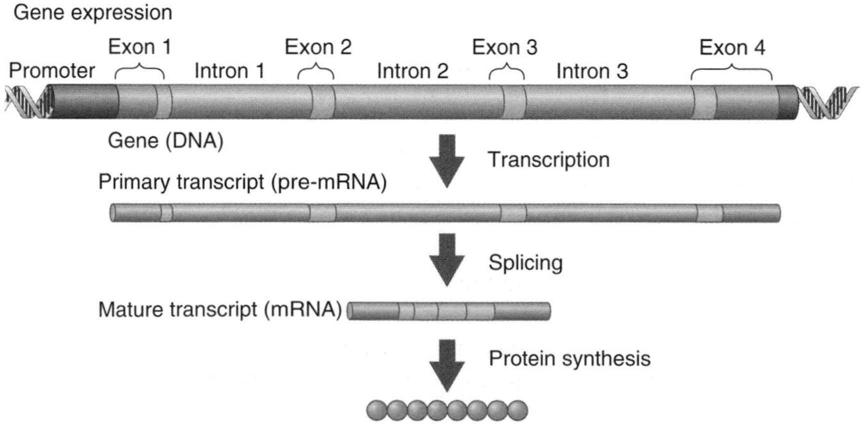

Gene expression
Exon 1
Exon 2
Exon 3
Exon 4
Promoter
Intron 1
Intron 2
Intron 3
Gene (DNA)
Transcription
Primary transcript (pre-mRNA)
Splicing
Mature transcript (mRNA)
Protein synthesis

Figure 10.6 To produce functional mRNA from pre-mRNA, the introns are cut out and the remaining exons spliced together

◀ The splicing of mRNA is carried out by yet another kind of RNA called short nuclear RNA (snRNA), which is combined with a protein molecule.

How does translation take place?

Translation of the mRNA code into a protein depends on the interaction within a ribosome between mRNA and tRNA. There are different types of tRNA, each one adapted to transfer one particular amino acid and to be recognised as carrying that amino acid. However, all tRNA molecules have the same basic structure. The 'cloverleaf' configuration of the molecule has at one end a triplet of bases called an **anticodon**. This anticodon is complementary to one of the mRNA codons. The other end of the tRNA molecule has an attachment site for the amino acid that is specified by the mRNA codon (Figure 10.7).

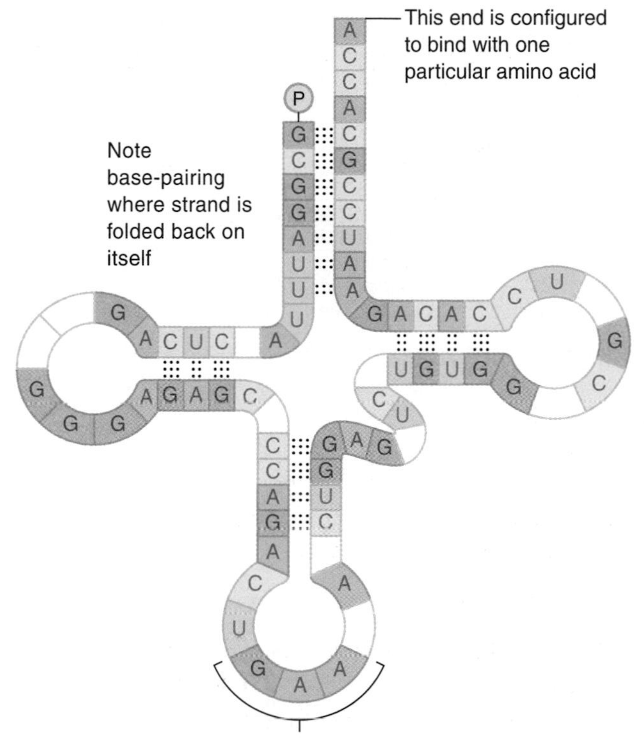

Figure 10.7 Structure of a tRNA molecule

Transfer RNA is single stranded (as is mRNA) but the single strand is conformed into a cloverleaf shape. It is held in this shape by hydrogen bonds between complementary bases.

Once mRNA arrives at a ribosome and is recognised, it begins to thread itself between the small and large ribosomal subunits. Then, the following events take place (Figure 10.8):

- The first two codons of the mRNA enter the ribosome.
- Transfer RNA molecules (with amino acids attached) that have complementary anticodons to the first two codons of the mRNA bind to those codons.
- A peptide bond forms between the amino acids carried by these two tRNA molecules.
- The ribosome moves along the mRNA by one codon, bringing the third codon into the ribosome.
- The tRNA that is freed returns to the cytoplasm.
- A tRNA with a complementary anticodon binds with the third codon, bringing its amino acid into position next to the second amino acid.

- A peptide bond forms between the second and third amino acids.
- The ribosome moves along the mRNA by one codon, bringing the fourth mRNA codon into the ribosome.
- The process is repeated until a stop codon is in position and translation ceases.

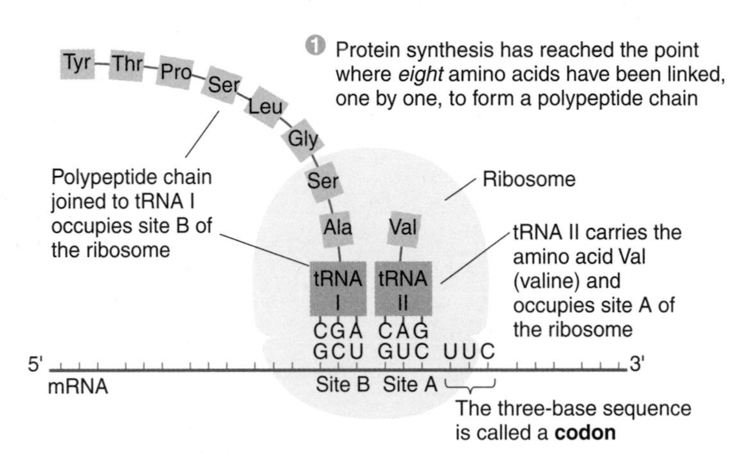

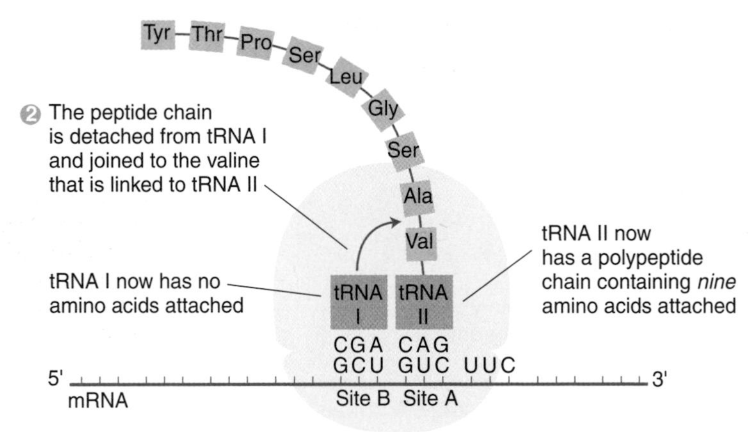

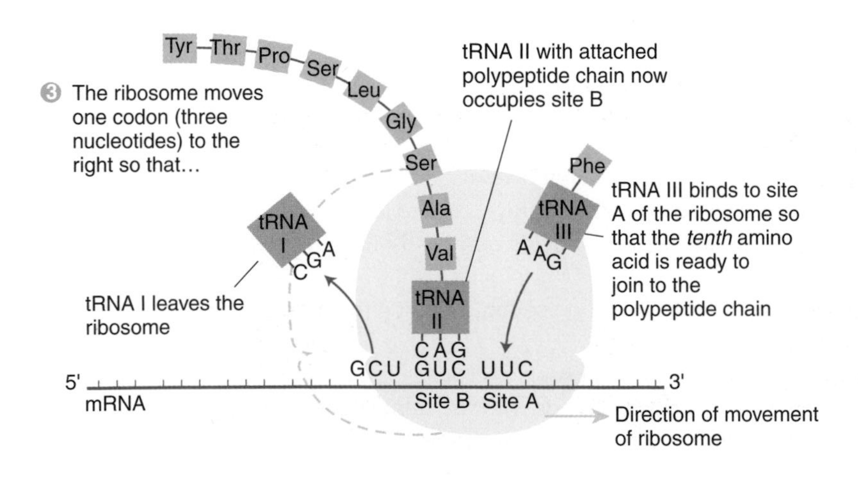

Figure 10.8 How translation takes place

What happens if genes mutate?

What types of mutations are there?

A mutation is a spontaneous change in the DNA molecule. There can be large structural changes or changes that involve only a single base. These are called **point mutations**.

There are several types of point mutation, including **substitutions** and **deletions**. In a substitution, one base in a DNA triplet is replaced by another — often when DNA is replicating. This is shown in Figure 10.9.

A-T-T -T-C-C -G-T-T -A-T-C ...
↑
Original base

A-T-G -T-C-C -G-T-T -A-T-C ...
↑
Substituted base

Figure 10.9 In a substitution mutation, one base is substituted for another

In the substitution shown in Figure 10.9, guanine replaces thymine. The triplet ATT has been changed to ATG (no other triplet is affected). The original triplet, ATT, codes for the amino acid isoleucine. However, the new triplet, ATG, codes for methionine. As a result, a different protein will be synthesised, which may or may not be significantly different from the original. One different amino acid in a protein does not always make a functional change.

If the substitution had been by any base other than guanine, because the DNA code is degenerate, the triplet would still code for isoleucine and the same protein would be synthesised. Effectively, it would still be the same gene.

Box 10.2 A substitution of one base can make a huge difference

A substitution of just one base in the sixth triplet of the gene that codes for one of the haemoglobin polypeptides alters the triplet from GAG to GTG. This results in the amino acid glutamine being replaced by valine in the polypeptide chain and causes the condition known as sickle-cell anaemia. If a person inherits two copies of the mutated gene, then all their red blood cells will contain the abnormal haemoglobin that causes the red blood cells to collapse into sickle-shaped cells under conditions of low oxygen concentration (Figure 10.10). The sickled cells often fracture and stick together in capillaries, blocking the capillaries.

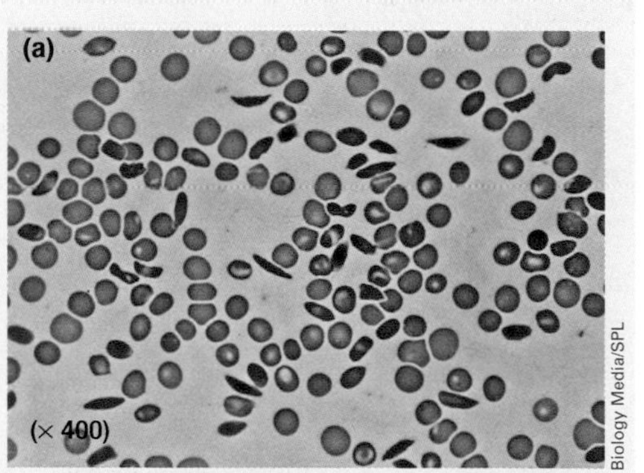

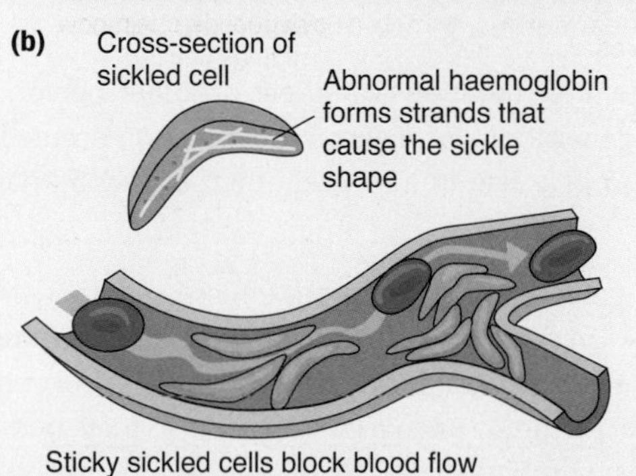

Figure 10.10
(a) A slide of blood showing normal and sickled cells
(b) Sickled red blood cells can block capillaries, preventing blood flow

In a deletion, a base is 'missed out' during replication.

In the deletion mutation shown in Figure 10.11, the third base of the first triplet is deleted (missed out) when replication occurs. As a result, the first triplet remains the same, but all triplets after the deletion are altered. The first base of each original triplet becomes the last base of the preceding triplet. This is called a **frameshift** and results in a new base sequence that codes for a different protein. It has produced a nonsense code and a non-functional protein as a result.

A-T-T -T-C-C -G-T-T -A-T-C ...
Deletion here
A-T-T -C-C-G -T-T-A -T-C ...
Replaced by first base of next triplet

Figure 10.11 In a deletion mutation, one base is omitted

What causes mutations?

Mutations occur spontaneously and randomly — they are accidents that occur when DNA is replicating. Mistakes happen. Mutations are rare events, which is quite surprising when you consider that each cell contains 6×10^9 (six billion) base pairs that might mutate!

Biologists estimate that mutations arise at the rate of 1 in 50×10^6 (one in 50 million) base pairs. This means that each new cell will have, on average, 120 mutations. This sounds rather worrying, but you should remember that:

- most of these mistakes (mutations) are detected and repaired
- 95% of our DNA is non-coding, so most mutations are unlikely to affect coding genes

The rate of mutation can be increased by a number of factors including:

- carcinogenic chemicals, e.g. those in tobacco smoke
- high-energy radiation, e.g. ultraviolet radiation, X-rays

What are the consequences of gene mutations?

There are a number of factors that influence the answer to this question. The two most important factors are in which cells, and in which genes, the mutations occur.

A mutation that occurs in a normal body cell (a non-sex cell) will have one of four possible consequences:

- It will be completely harmless.
- It will damage the cell.
- It will kill the cell.
- It will make the cell cancerous, which might kill the person.

Whatever the outcome, the mutation will affect no other person; it will not be passed on to the next generation. However, if the mutation occurs in a sex cell, or a cell that will divide to give rise to a sex cell, then it may be passed on to the next generation.

Mutations in different genes produce different effects; mutations in two types of gene are particularly important. **Proto-oncogenes** and **tumour suppressor genes** play important roles in regulating cell division and preventing the formation of tumours. When a proto-oncogene mutates, it may become an active **oncogene**. Oncogenes stimulate cells to divide in an uncontrolled manner.

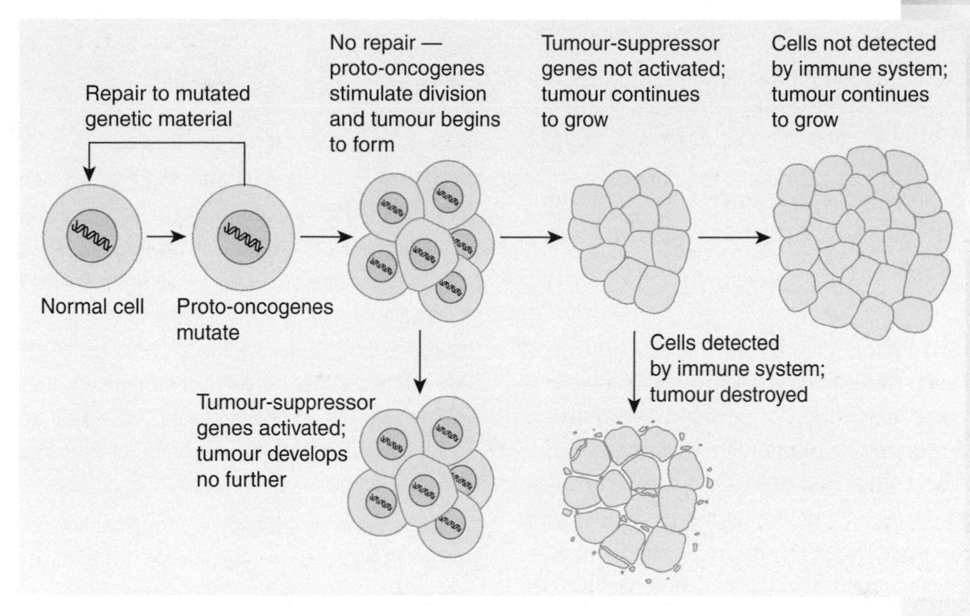

Figure 10.12 Development of cancer

Ordinarily, some growth factor is necessary to make the cell divide. However, the presence of active oncogenes stimulates repeated cell division in the absence of the external growth factor (Figure 10.12).

Tumour suppressor genes recognise uncontrolled cell division and act to suppress it. If these genes mutate and become inactive, a tumour will form as the cells continue to divide in an uncontrolled manner.

Why are cells in your big toe not the same colour as cells in your iris?

All our body cells contain all our genes — but every gene does not work in every cell. A range of regulatory factors switch on and switch off genes so that they only work under certain circumstances. For example, only muscle cells express the genes that control the synthesis of molecules of the proteins actin and myosin. Only cells in the islets of Langerhans express genes that control the synthesis of insulin and glucagon. The cells in your big toe do not express the genes that give the iris its colour.

How are genes allowed to express themselves?

Often, genes are switched on by **'transcription factors'** present in the cell. These transcription factors are usually proteins that bind to a regulatory sequence of DNA near to the gene to be expressed. They operate in the following way:

- The transcription factor binds to the promoter sequence of DNA near to the gene to be activated.

- RNA polymerase binds to the DNA–transcription factor complex.
- The RNA polymerase is 'activated' and moves away from the DNA–transcription factor complex along the gene.
- The RNA polymerase transcribes the anti-sense strand of the DNA as it moves along; the gene is now being expressed.

Transcription factors need not be synthesised within the cell. The hormone oestrogen is a steroid hormone that can diffuse through the plasma membrane of a cell. It binds with a receptor in the cytoplasm and the oestrogen–receptor complex then moves into the nucleus where it binds with and activates specific genes. In the breasts, the activated genes cause cell division and enlargement of the breast tissue. The lining of the uterus responds in a similar way to oestrogen.

Box 10.3 Where do transcription factors come from?

If the transcription factors are themselves proteins, then they must be synthesised as a result of gene expression, and some other genes must regulate their expression.

Biologists think that this goes back to the egg cell, which is able to synthesise a number of transcription factors. These are produced once the egg is fertilised to become a zygote; they are passed on to the cells formed when the zygote divides. They influence the cells that are formed and these cells produce other transcription factors, which are passed on to the next generation of cells and so on.

This 'cascade' or 'hierarchy' of transcription factors results in each cell having only certain transcription factors that activate particular genes.

Box 10.4 Oestrogen and breast cancer

Many breast cancers are said to be oestrogen-receptor positive. This means that the cancer cells have oestrogen receptors to which the hormone can bind, causing the same increase in cell division as it does in normal breast tissue. The molecule of the anti-cancer drug tamoxifen is able to bind with the oestrogen receptors. The complex then binds with the DNA as if it were the oestrogen–receptor complex. However, tamoxifen does not allow transcription factors to bind and so expression of the genes is prevented, and cell division in the cancer is slowed (Figure 10.13).

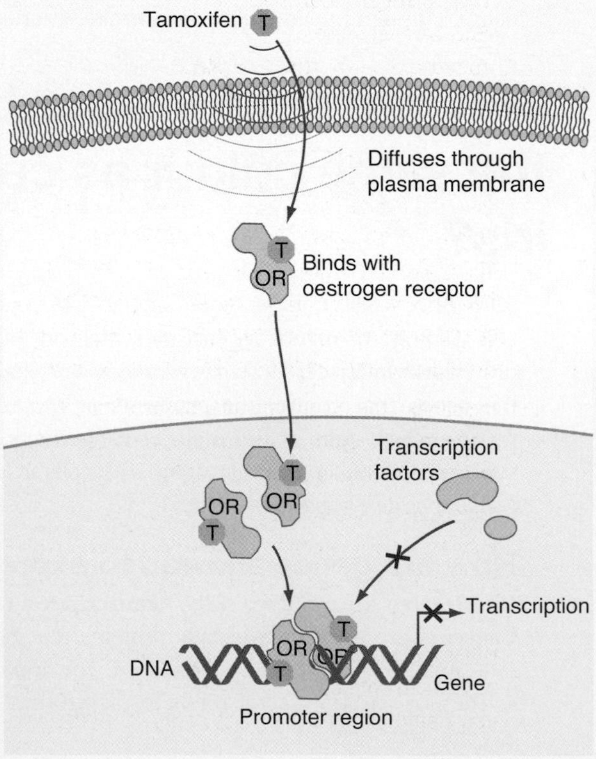

Figure 10.13 Tamoxifen prevents oestrogen from activating the DNA of cancer cells

How are genes silenced?

Besides transcription factors that promote the expression of genes, other factors can act to repress gene action. One group of substances that does this is known as **short interfering RNA (siRNA)**. These RNA molecules are unusual because they are:

- very short — only about 21–23 nucleotides long
- double stranded

Short interfering RNA molecules do not act on the gene itself. They 'interfere with' or 'silence' the mRNA once it has been transcribed from the DNA. This is called **post-transcriptional interference**. If mRNA is prevented from translating its codons into amino acids and consequently into a protein at the ribosomes, then the gene that codes for the protein is effectively 'silenced' or repressed.

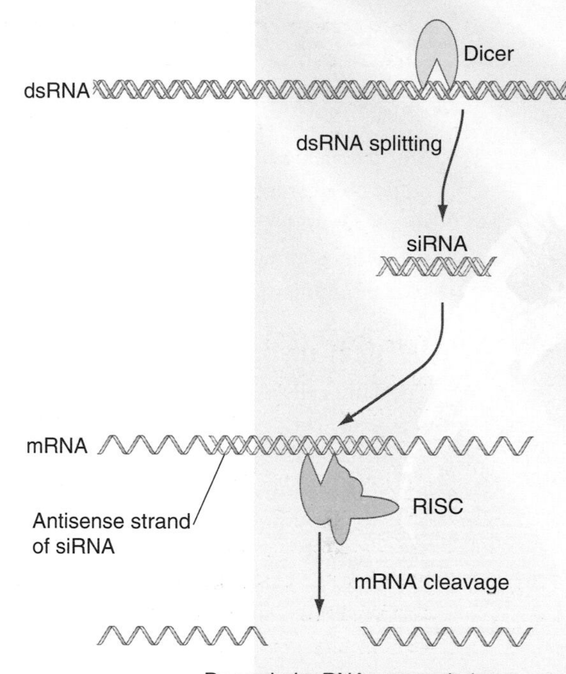

Figure 10.14 The way in which siRNAs act

Biologists think that the action of siRNA is as follows:

- Double-stranded RNA (dsRNA) is produced in the nucleus from a range of genes.
- It is then split into the very short lengths that characterise siRNA by an enzyme called 'dicer'.
- The anti-sense strand of the siRNA then binds with the mRNA it is to silence.
- This guides a complex of molecules called RNA-induced silencing complex (RISC) to the site.
- Together with the siRNA, RISC breaks down (degrades) the mRNA (Figure 10.14).

Box 10.5 How biologists can create and use siRNAs

In one method, a section of DNA is used as a template and RNA polymerase is used to build a strand of RNA that is complementary to the DNA. Then another strand of RNA, complementary to the one just created, is produced and the two are linked with a few uracil bases (Figure 10.15(a)).

In solution, the complementary bases attract each other and the molecule starts to fold. Hydrogen bonds then form between the complementary bases to form a molecule of 'hairpin siRNA' (Figure 10.15(b–c)).

Molecules of siRNAs hold out hope for the treatment of a number of conditions. Researchers have shown already that siRNA can be used to prevent the replication of HIV in cultures by silencing either some of the genes of the virus or some of the human genes on which it depends.

Other researchers have shown that siRNA molecules can silence genes associated with cancer. If oncogenes could be silenced effectively, then a new treatment for many cancers would be possible.

Figure 10.15

(a)

GUUCAGUUGCAC UU CAAGUCAACGUG

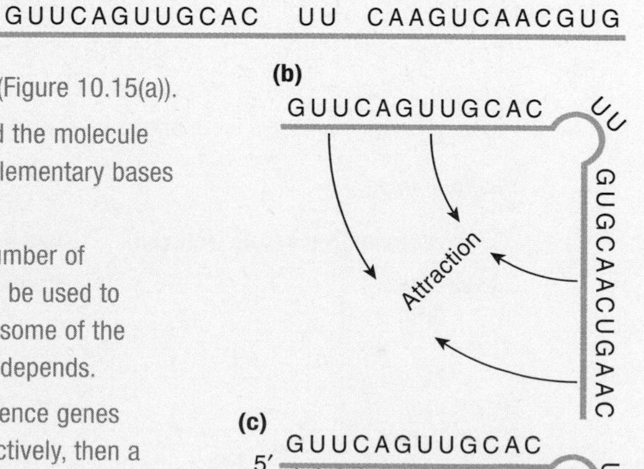

Do any cells have the potential to express all their genes?

Some cells do have this potential. They don't actually do it, but any of the genes *could* be expressed. Such cells are called **totipotent** cells. A zygote is an example of a totipotent cell. It has the ability to give rise to all the cells of an organism, by allowing all the genes to be expressed at an appropriate time in the development of the organism.

◀ Some biologists believe that a totipotent cell must be able to give rise not only to all the types of cell in the organism, but also to any extra-embryonic tissue needed for development. In humans, this refers to the placenta tissues.

Which cells are totipotent in humans?

The only cells that show totipotency are cells formed by the first few divisions of a zygote. These cells differ from other kinds of cell in the body in that they are:

- unspecialised
- capable of dividing and renewing themselves for long periods
- capable of giving rise to specialised cells

Cells with these features are called **stem cells**.

Not all stem cells are totipotent; some are **pluripotent**.

Pluripotent cells can give rise to most, but not all, of the cells needed for a fetus to develop. In humans, embryonic stem cells are pluripotent. They cannot, for example, give rise to placental tissue.

Multipotent stem cells can give rise to a few related cell types. Stem cells in the bone marrow can give rise to all the different types of blood cell, but not to muscle or nerve cells. As development progresses, the stem cells become less able to develop into a wide range of cells. The specialised cells that they form are not usually able to divide.

◀ Although this is generally true, it has now been shown that even highly specialised cells can sometimes regain their ability to divide.

Because of this ability to develop into different types of cell, stem cells have the potential to be of great value in treating degenerative diseases.

Box 10.6 Research into the use of stem cells

Researchers treated mice that had suffered heart attacks by injecting bone marrow stem cells into the heart muscle. The stem cells developed into cardiac muscle cells and the functioning of the heart became much better than when no stem cells were injected (Figure 10.16).

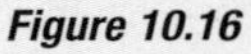

Figure 10.16

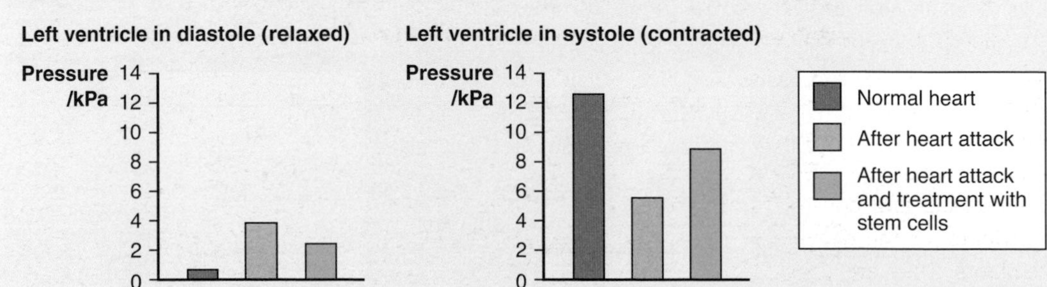

The following story appeared in several newspapers on the 20 November 2008.

Claudia Castillo, 30, who lives in Barcelona, has become the first person to be given a whole organ tailor-made for her in laboratories across Europe.

A graft from a donor was used, but because Ms Castillo's own cells have been grown over the framework, there is no sign that her body will reject the organ.

Researchers and surgeons from Britain, Italy and Spain collaborated to grow tissue from Ms Castillo's own bone marrow stem cells, using them to fashion the new bronchus — a branch of the windpipe. They believe that one day the approach will be used to create engineered replacements for other damaged organs, such as the bowel or bladder.

Researchers hope that within 5 years, using similar methods, it will be possible to make new voice boxes (larynxes). This would mean that a better treatment could be offered to patients with cancer of the larynx.

Research is underway into using stem cells to treat a range of conditions in which cells are malfunctioning or have been damaged. These include:

- burns — perhaps stem cells could differentiate into the various types of cell in the skin, rather than the scar tissue that normally forms following burns
- Alzheimer's disease — perhaps stem cells could differentiate into healthy brain cells to replace the diseased ones
- diabetes — perhaps stem cells could be used to replace the α-cells and β-cells in the islets of Langerhans, restoring the normal control of blood sugar

Which cells are totipotent in plants?

In plants, more cells are either totipotent or pluripotent than is the case in animals. This can be shown by the fact that we can take stem cuttings that then grow roots and become complete plants. This means that the cells at the base of the cutting must have started to divide again and the cells they formed then differentiated into the various tissues in the roots.

In micropropagation, the totipotency of some plant cells is exploited even further. Small sections of plant tissue called **explants** are grown in special culture media. These media are usually agar based and contain appropriate nutrients for the development of the explants. They are maintained under sterile conditions to prevent infection.

The cells at the base of this stem cutting have divided and formed roots

Andrew Lambert Photography/SPL

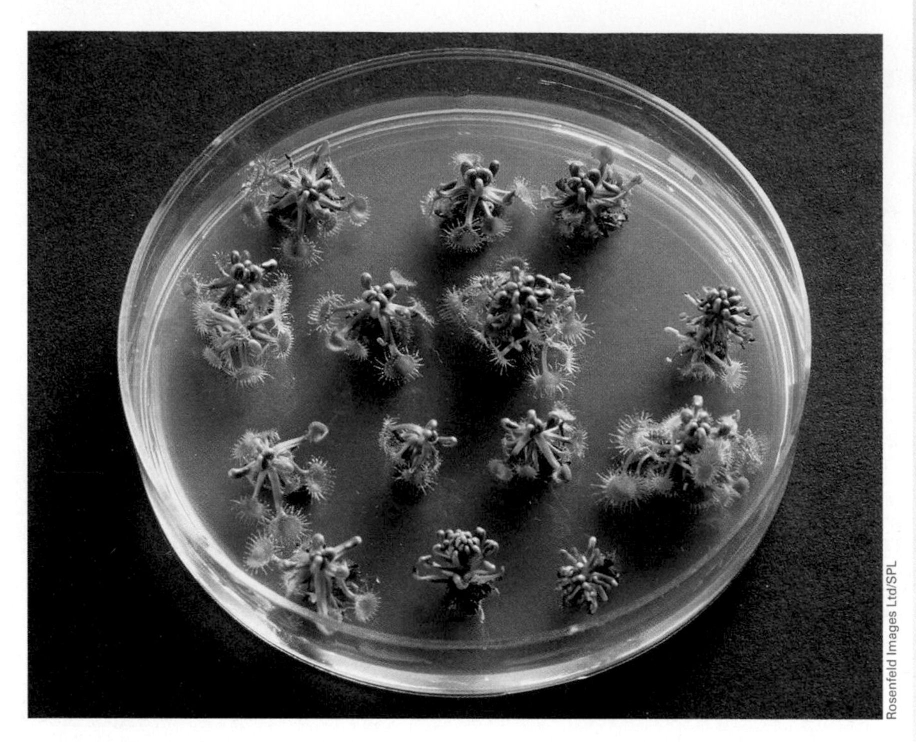

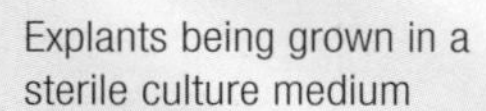
Explants being grown in a sterile culture medium

The explants are treated first with hormones that make them develop roots and later with hormones that make them develop shoots. When they have developed sufficiently, the young plantlets are transplanted into compost and grown in glass houses.

Thousands of genetically identical plants can be grown from a small piece of tissue from one parent plant. The technique is used to propagate rare plants and commercially to propagate plants to be sold in supermarkets.

Summary

Protein synthesis

- The DNA molecule is a double helix in which the base sequence on the sense strand is complementary to the base sequence on the antisense strand.
- Proteins are composed of a chain of amino acids (the primary structure) that coils or folds into an β-helix or a β-pleated sheet (secondary structure), which then folds into a unique globular or fibrous shape (tertiary structure).
- The genetic code is a:
  - triplet code
  - degenerate code
  - non-overlapping code
  - universal code
- When compared with DNA, mRNA is
  - smaller
  - single stranded (not double stranded)

- contains ribose (not deoxyribose)
- contains uracil instead of thymine
- In the transcription of the DNA in a gene:
  - RNA polymerase binds to a promoter region near to the gene
  - using RNA nucleotides, it assembles a chain of mRNA complementary to the DNA antisense strand of the gene
  - the molecule formed is pre-mRNA
  - enzymes remove the introns from the pre-mRNA and splice together the exons to form mRNA
- During translation of mRNA into a polypeptide chain, the following events occur:
  - the first two codons of the mRNA enter the ribosome
  - tRNA molecules with complementary anticodons bind with the mRNA codons
  - the amino acids carried by these two tRNA molecules are joined by a peptide bond
  - the ribosome moves along the mRNA, bringing the third codon into the ribosome
  - tRNA with a complementary anticodon binds with the third codon
  - a peptide bond forms between the second and third amino acids
  - the process is repeated until a stop codon is in position and translation ceases

Mutations

- Point mutations are spontaneous changes in a single base on the DNA molecule.
- Substitutions replace one base with another; because the code is degenerate, there may be no change in the amino acid coded for.
- Deletions remove a base from the DNA sequence causing a frameshift; these mutations alter all the DNA triplets after the point of mutation and change all the amino acids in the protein molecule after this point.
- High-energy radiation and carcinogenic chemicals increase the rate of mutation.
- If proto-oncogenes mutate to become active oncogenes, they will stimulate the cell to divide in an uncontrolled manner.
- If tumour suppressor genes mutate and fail to regulate cell division, a tumour may form.

Gene expression

- Transcription factors are necessary to activate genes:
  - these proteins bind with the promoter regions next to a gene
  - RNA polymerase binds with the DNA–transcription factor complex
  - the enzyme breaks away from the complex and begins transcribing the gene
- Oestrogen (and some other steroid hormones) can enter cells and bind with receptors that carry the hormone into the nucleus where the oestrogen–receptor complex binds with a promoter region and activates a gene.
- In the breasts and uterine lining, oestrogen binds with genes that stimulate cell division.

- Breast cancer cells may have oestrogen receptors and so be stimulated to divide more quickly by oestrogen.
- Molecules of short interfering RNA (siRNA) can 'silence' genes by degrading the mRNA transcribed from the genes.
- siRNA has the potential to treat conditions such as AIDS by preventing replication of HIV, and cancers by silencing genes that enhance cell division.
- Totipotent cells are unspecialised and retain the ability to give rise to any kind of cell in an organism.
- Pluripotent cells can give rise to most types of cell; multipotent cells can give rise to a few related types of cell, such as the different types of blood cell.
- Only very early embryonic stem cells are totipotent in humans; many adult plant cells are totipotent.
- Human stem cells have the potential (because of their ability to differentiate into any type of cell) to treat conditions in which cells are damaged and need replacing.

Questions

Multiple-choice

1 DNA consists of two polynucleotide strands in which:

A the percentage of adenine is the same in each strand

B the percentage of adenine is the same as that of thymine in each strand

C the percentage of adenine is the same as that of thymine in the whole molecule

D the percentage of adenine is 50% of that of thymine in the whole molecule

2 tRNA differs from DNA because it is:

A smaller, single stranded and shows no base-pairing

B smaller, single stranded with thymine replaced by uracil

C smaller, single stranded and with deoxyribose instead of ribose

D smaller, single stranded and linear in shape

3 The genetic code is:

A a triplet code, degenerate and overlapping

B a doublet code, degenerate and universal

C a doublet code, degenerate and non-overlapping

D a triplet code, degenerate and universal

4 The triplet AAT on the sense strand of a DNA molecule would code for an amino acid carried by tRNA with the anticodon:

A AAU

B TTA

C AAT

D UUA

5 In a ribosome, the two amino acids that are held adjacent to each other form a:

A hydrogen bond

B ionic bond

C ester bond

D peptide bond

6 Following transcription, mRNA must be modified to:
- **A** remove non-coding sections
- **B** alter its shape so that it can bind with ribosomes
- **C** alter its shape so that it can bind with an amino acid
- **D** remove unwanted amino acids

7 Transcription factors:
- **A** are proteins
- **B** bind with promoter sequences of DNA next to the gene
- **C** allow RNA polymerase to bind to the promoter sequence and be activated
- **D** all of the above

8 siRNA is:
- **A** single-stranded RNA that can 'silence' genes
- **B** single-stranded RNA that can cause genes to be expressed
- **C** double-stranded RNA that can cause genes to be expressed
- **D** double-stranded RNA that can 'silence' genes

9 Totipotent stem cells can give rise to:
- **A** all types of cells in an organism
- **B** most types of cells in an organism
- **C** a few types of cells in an organism
- **D** one or two types of cells in an organism

10 Researchers hope to be able to use stem cells to treat degenerative conditions because they:
- **A** are unspecialised
- **B** can divide to create new cells
- **C** can give rise to many types of cells
- **D** all of the above

Examination-style

1 Protein synthesis takes place in the ribosomes. The code for synthesis of a particular protein is specified by a section of the DNA molecule and is carried to the ribosomes by mRNA.

(a) (i) What do we call a section of DNA that codes for a protein? *(1 mark)*

(ii) The DNA code is sometimes called a degenerate code. What does this mean? *(2 marks)*

(b) The diagram shows protein synthesis taking place in a ribosome.

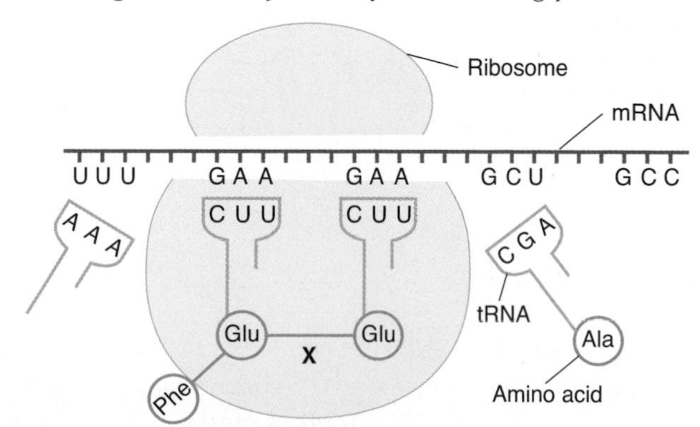

Key to amino acids
Phe = phenylalanine
Glu = glutamic acid
Ala = alanine

(i) Name the type of bond labelled **X**. (*1 mark*)

(ii) Use examples from the diagram to explain the terms codon and anticodon. (*2 marks*)

Total: 6 marks

2 Some viruses contain an enzyme called reverse transcriptase. This enzyme catalyses the reverse transcription of mRNA to DNA.

(a) Part of a molecule of mRNA has the base sequence:

AAUGCCUUAGGU

What would be the sequence of the bases on the DNA strand formed by reverse transcription of this molecule of mRNA? (*1 mark*)

(b) Give *two* ways in which a molecule of DNA formed by reverse transcription would be different from a molecule of DNA in a typical eukaryotic cell. (*2 marks*)

(c) Give *three* differences between mRNA and DNA found in a typical eukaryotic cell. (*3 marks*)

Total: 6 marks

3 Short interfering RNA (siRNA) is a type of RNA that is important in 'silencing' genes.

(a) Give *two* differences between siRNA and:

(i) DNA (*2 marks*)

(ii) mRNA (*2 marks*)

(b) Huntington's disease is a disorder in which the protein produced by a mutant gene causes progressive death of cells in the brain. The cells of sufferers from this condition frequently contain one mutant gene and one normal gene. The diagram shows how siRNA could be used in the treatment of such conditions.

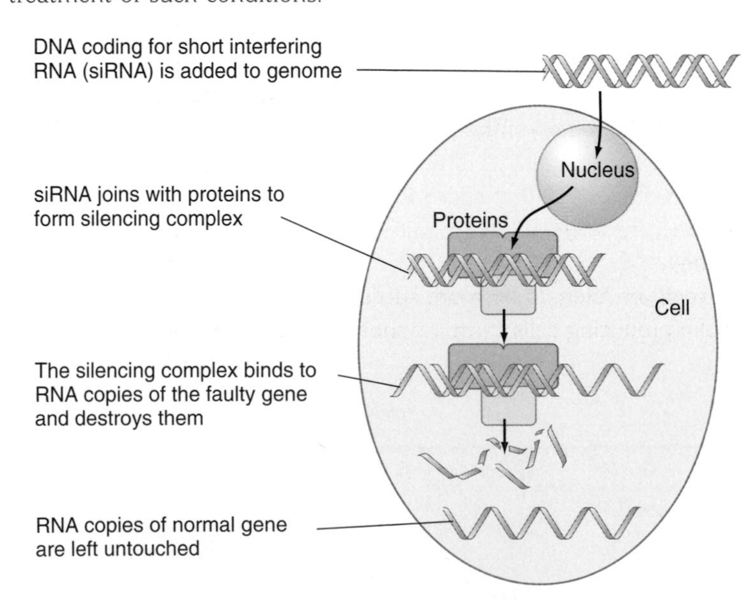

Use the diagram, and your knowledge of siRNA, to explain how siRNA might one day be used to treat Huntington's disease. (*4 marks*)

Total: 8 marks

4 Read the following passage concerning protein synthesis.

A section of DNA in the nucleus 'unzips' and molecules of free RNA nucleotides are assembled into a strand of mRNA using one of the single DNA strands as a template. The resulting molecule of mRNA contains the same genetic information as the section of DNA from which it was copied. It then leaves the nucleus through a pore in the nuclear envelope and moves into the cytoplasm, where it encounters ribosomes. Here, using tRNA as a sort of adaptor molecule, the coded information is used to synthesise a protein.

(a) (i) Name the enzyme involved in producing the mRNA molecule from a section of DNA. *(1 mark)*

(ii) Name the process by which mRNA is produced. *(1 mark)*

(iii) Why is only one of the strands of DNA used to produce mRNA? *(2 marks)*

(b) Using the DNA triplet ATC as an example, explain how the mRNA formed carries the same genetic information as the DNA from which it was copied. *(3 marks)*

(c) Describe how mRNA and tRNA are used in protein synthesis in the ribosomes. *(5 marks)*

Total: 12 marks

5 (a) Insulin is a peptide hormone containing 51 amino acids. It is produced in the islets of Langerhans in the pancreas.

(i) How many bases would there be in the coding section of the gene that controls insulin production? Explain your answer. *(2 marks)*

(ii) Describe how insulin is produced in cells in the islets of Langerhans. *(4 marks)*

(b) In a person who suffers from type I diabetes, the cells of the pancreas that normally produce insulin are destroyed by the person's own immune system. New studies indicate that it may be possible to use human embryonic stem cells in cell culture to form insulin-producing cells that eventually could be used to replace the defective insulin-producing cells in the pancreas.

(i) What are the main features of stem cells? *(3 marks)*

(ii) Suggest why this might be possible. *(3 marks)*

(iii) Suggest why this would be likely to be more successful than injecting insulin-producing cells from a donor. *(3 marks)*

Total: 15 marks

Chapter 11

How are genes cloned and how can cloned genes be used?

This chapter covers:

- the techniques involved in producing specific fragments of DNA
- the use of the polymerase chain reaction to form multiple copies of the fragments of DNA
- the use of restriction enzymes, ligases and vectors in transferring genes from one organism to another, producing a genetically modified organism
- the use of genetically modified organisms in producing desirable products
- the actual and potential use of gene therapy
- the way in which DNA probes are produced and their use in diagnosing genetic disease
- genetic fingerprinting

The elucidation of the structure of DNA in 1953 led to an explosion of related discoveries and the development of new techniques. At about the same time, biologists discovered the presence of restriction enzymes in bacteria. These enzymes cut DNA molecules at specific base sequences. The genetic code was deciphered in 1966 and in 1970 the true nature of the restriction enzymes was realised. In 1976, Fred Sanger and his team of researchers developed the technique of gene sequencing that allowed them to work out the sequence of bases in a fragment of DNA. Within a year, a genetically engineered bacterium containing the gene for human insulin was created. The first purified insulin from this source was available in 1982. In 1987, the technology of the polymerase chain reaction was developed. In 1990, the human genome project was set up and the first draft of the human genome was published in 2001.

A genome is the complete set of genetic information of an organism.

Biotechnology is used increasingly in criminal investigations. It may be used to eliminate a person from suspicion or to provide extra evidence to secure a conviction. Some forensic evidence depends on the fact that, apart from identical twins, the DNA of each person is slightly different. The genetic fingerprint of a person can be important evidence in ascertaining whether or not that individual was present at the scene of a crime.

Other developments have led to our current ability to alter the genome of some organisms by transferring genes to them from other organisms. This is known commonly as genetic engineering. It may be that our ever-increasing knowledge of how to transfer genes, switch them on and switch them off will lead to gene-therapy based treatment of some genetic diseases.

How are genes cloned?

What *is* gene cloning?

We met some clones in Chapter 10. Plant cuttings are clones and the thousands of plants produced by micropropagation also represent a clone. When talking about whole organisms, a clone is a group of individuals that are:

- produced asexually from one parent
- genetically identical to each other
- genetically identical to the parent organism

Although identical twins are genetically identical, they are not clones because they are not produced asexually and they are not genetically identical to either of their parents.

Gene cloning means making multiple copies of a gene. There are several ways in which this can be done. The principal methods are divided into two main categories:

- **in vivo cloning** — the gene is introduced into a cell and is copied as the cell divides
- **in vitro cloning** — this does not take place in living cells; the DNA is copied many times, using the **polymerase chain reaction** (PCR)

There are advantages and disadvantages to both methods of gene cloning. In vitro cloning using PCR is both quicker and cheaper than in vivo cloning. Billions of copies of a gene can be made within a few hours at low cost. However, if the gene is to be used by an organism to make a product — for example by a bacterium to allow it to make insulin — then in vivo cloning delivers the gene already in the organism.

How do biologists obtain the fragments of DNA (genes) to be cloned?

There are two main methods:

- extract the DNA from a donor cell
- create the DNA in vitro from mRNA

Extracting the gene from a donor cell

First, the DNA must be isolated from the donor cells. Figure 11.1 summarises how this is done.

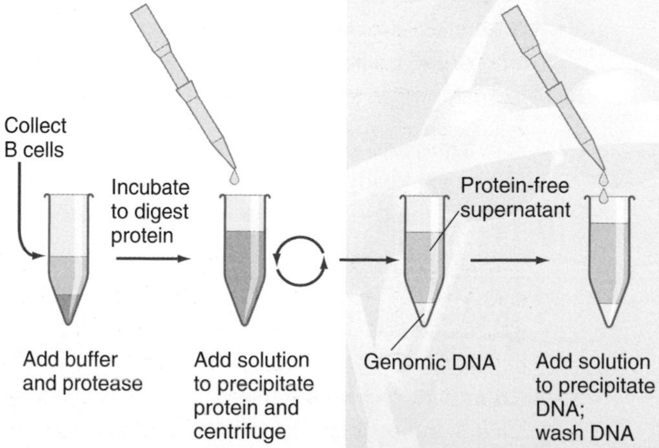

Figure 11.1 How DNA is isolated from donor cells

After isolating DNA from donor cells, it is incubated with enzymes called **restriction endonucleases**. These enzymes 'cut' the DNA molecule at places called **restriction sites**. Each restriction site is a short sequence of 4–8 nucleotides. The enzymes can be used to cut out any gene of interest. Different restriction enzymes recognise different restriction sites. One of these enzymes is called *Eco*R1. It recognises the sequence G–A–A–T–T–C, and cuts the strand between the G and the A nucleotides wherever it finds the sequence (Figure 11.2).

Restriction endonucleases are often simply called restriction enzymes. *Eco*R1 is so named because it was the first restriction enzyme (R1) to be isolated from the *E. coli* bacterium (*Eco*).

Box 11.1 Sticky ends

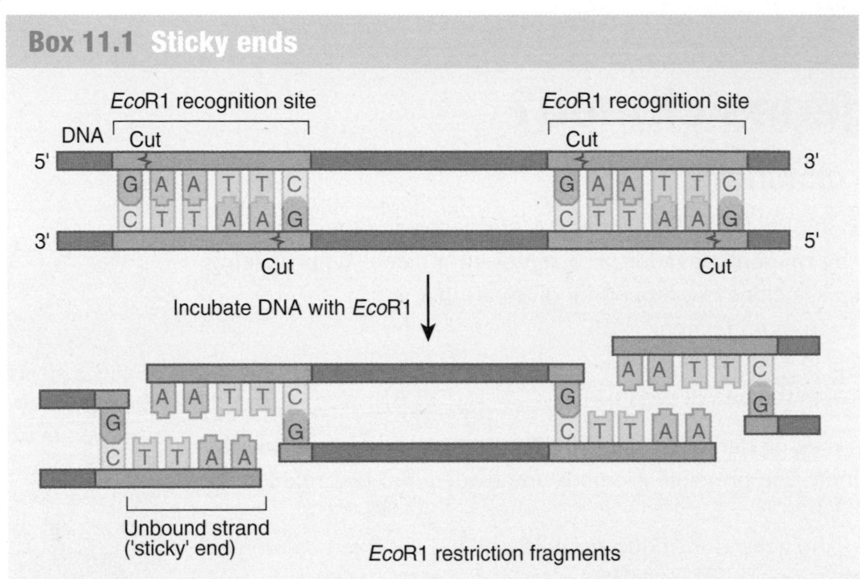

Figure 11.2 The restriction enzyme *Eco*R1 cuts DNA at specific restriction sites

Notice that complementary base pairing means that the sequence occurs on both strands, but running in opposite directions. By cutting between the G and the A on both strands, the enzyme makes a zigzag cut and the section of DNA that is cut out has overlapping ends. The single-stranded portions at each end of the section of the cut-out DNA will be able to form hydrogen bonds with a complementary DNA sequence. They are called **sticky ends**.

Life is not so simple that a restriction enzyme will cut out only the required gene. It cuts the DNA *wherever it finds its restriction site*. By chance alone, this sequence will occur once in every 4000 nucleotide pairs and there are millions of nucleotide pairs in the human genome. So, treating a sample of human DNA with a restriction enzyme cuts it into many fragments — a **gene library**. Only one of the fragments (or two if there are two copies of the gene in the cell) will contain the gene of interest.

Creating the gene from mRNA

We have already seen that not all genes are expressed in all cells. Only cells that express the gene in which we are interested will contain the mRNA transcribed from that gene. For example, the gene controlling the synthesis of insulin is only expressed in β-cells in the islets of Langerhans in the pancreas. These cells contain thousands of molecules of the mRNA that codes for insulin. This mRNA is comparatively easy to extract and gives a high yield from relatively few cells.

To initiate protein synthesis, a section of the antisense strand of a DNA molecule is transcribed into mRNA that has a complementary base sequence to the

DNA transcribed (uracil replacing thymine). By throwing this process into reverse, a section of DNA complementary to the mRNA can be created. This **complementary DNA (cDNA)** is equivalent to the antisense strand of a DNA molecule for the gene of interest.

The extracted mRNA is incubated with free DNA nucleotides and an enzyme called **reverse transcriptase**. This is the enzyme that allows transcription to be reversed and to create a strand of cDNA. However, cDNA is single-stranded, and is, effectively, the antisense strand of the gene. To be of use, the sense strand is also needed.

The mRNA is 'washed' out of the mixture and the newly synthesised cDNA is incubated with free DNA nucleotides and DNA polymerase. The cDNA strand forms the template for, and then binds with, a strand that is complementary to it. This is the sense strand of the gene of interest (Figure 11.3).

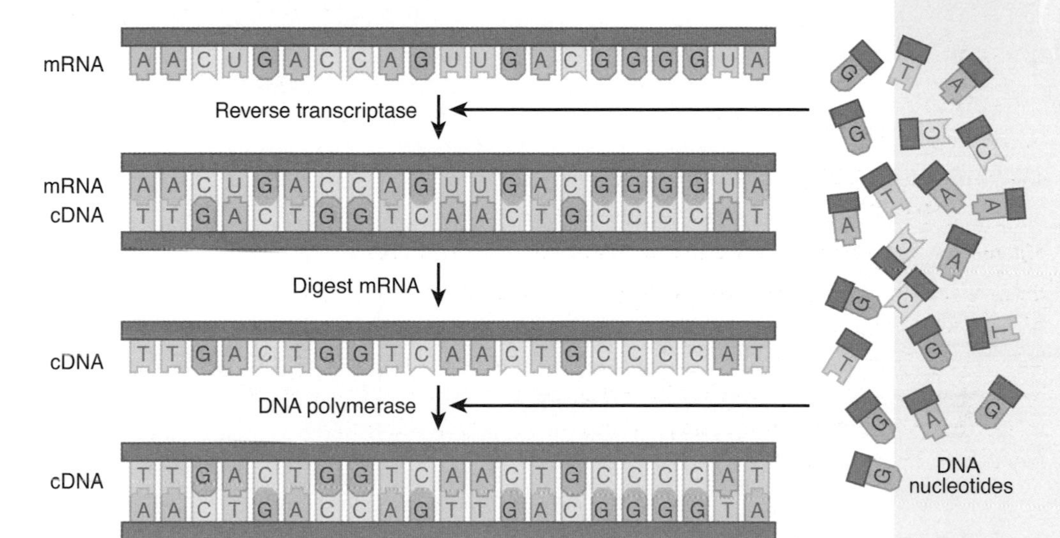

Figure 11.3 Creating the gene from mRNA

Making more DNA using the polymerase chain reaction

Whichever technique is used, it is unlikely to yield as much DNA as is needed. It has to be copied to give us enough DNA to be of use. This is carried out using the polymerase chain reaction (PCR).

The polymerase chain reaction (PCR) is an automated technique that mimics the process of DNA replication that occurs in living cells. It allows a tiny sample of DNA to be amplified many times in a short period of time — over a million copies can be made in just a few hours. Essentially, there is a repeating cycle of separation of the two DNA strands, followed by synthesis of a complementary strand for each. The amount of DNA doubles with each cycle (Figure 11.4). However, there are a few technical problems:

- There is no suitable helicase to separate the strands; instead this is achieved by heating to 95°C.

- The DNA polymerase cannot work on DNA that is completely single stranded. There must be double-stranded regions at the start of the sequence to be copied on each strand. This means that:
  - special primers (short sequences of DNA), complementary to the bases at the start of the region to be amplified, must be included; these bind to the DNA and provide a 'starting point' for DNA polymerase
  - to synthesise these primers, the base sequences at the start of the sequences to be copied must be known

There are two sequences to be copied because the sequences on the two strands are complementary, not identical. Therefore, two different primers are needed.

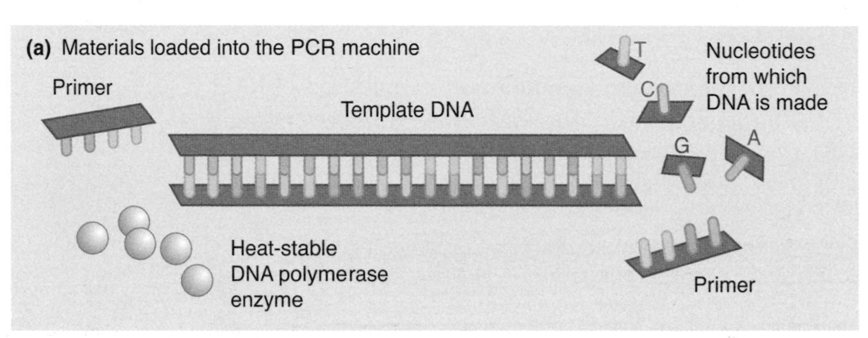

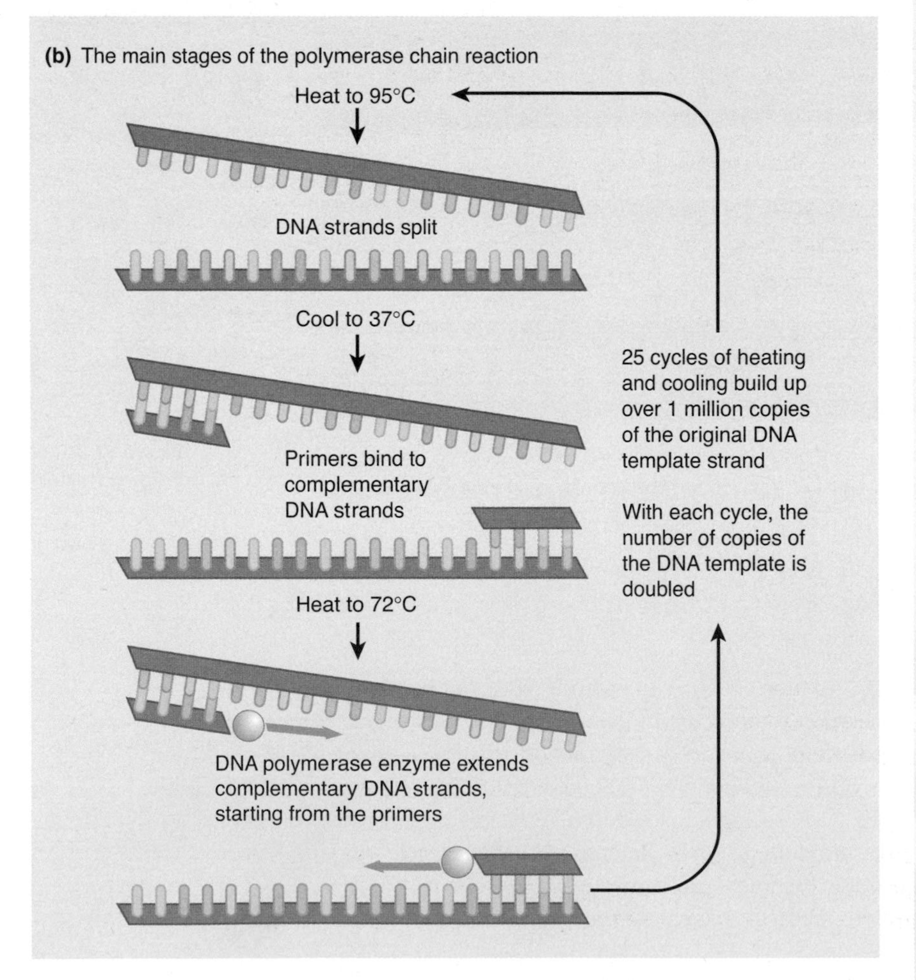

Figure 11.4 The polymerase chain reaction

- To avoid having to keep adding fresh DNA polymerase, the enzyme must be thermostable. Originally, the DNA polymerase used was obtained from bacteria that lived in hot water springs. It is now produced by genetically modified bacteria.
- The replication must be carried out at 72°C as this is the optimum temperature for the DNA polymerase.

Great care is needed to ensure that:
- the nucleotides used are of the highest purity
- the DNA is not contaminated in any way (any foreign DNA would also be copied)

Box 11.2 How much DNA do we get?

Theoretically, each cycle (taking about 7–8 minutes) doubles the amount of DNA. After two cycles, the amount of DNA should have quadrupled. After three cycles, there should be eight times as much DNA. The total number of molecules produced is given by the formula:

$$T = N \times 2^c$$

where T = total number of DNA molecules formed, N = number of DNA molecules at the start, and c = number of cycles.

In theory, after 25 cycles (taking between 3 and 4 hours) and starting with just one molecule of DNA, there should be 1×2^{25} = 33 554 432 molecules of DNA!

In practice, no process is 100% efficient and this number is not achieved. However, 25 cycles can produce well over 1 million copies of the initial DNA sample.

A PCR machine does not take up much space in a forensic laboratory

How do we transfer a gene into a bacterium?

To do this requires **vectors**. Vectors are carriers. Just as the female *Anopheles* mosquito is needed to carry the malarial parasite from person to person, a vector is needed to carry genes from organism to organism. There are two types of vector:
- **plasmids** are small circular fragments of DNA found naturally in bacteria; they replicate independently of the main bacterial DNA molecule
- **viruses** containing DNA (not RNA) as their genetic material that can infect bacterial cells

How do we transfer a gene into a vector?

Plasmids are usually preferred to viruses as viruses may cause disease. They are prepared in the following way:

- The plasmid is cut open using the same restriction enzyme used to cut out the DNA fragments. The sticky ends of the two types of DNA will contain complementary base sequences.
- The DNA fragments are incubated with the plasmids; the plasmid DNA and the gene DNA **anneal** (join) in the following way:
  - Hydrogen bonds form between the bases in the sticky ends, weakly holding the gene DNA in place in the plasmid.
  - Catalysed by the enzyme **ligase**, covalent bonds form between the sugar–phosphate backbones of the plasmid DNA and gene DNA; the gene has now been firmly **spliced** into the plasmid (Figure 11.5).

In the case of genes that have been synthesised, rather than extracted, sticky ends have to be added by a special technique.

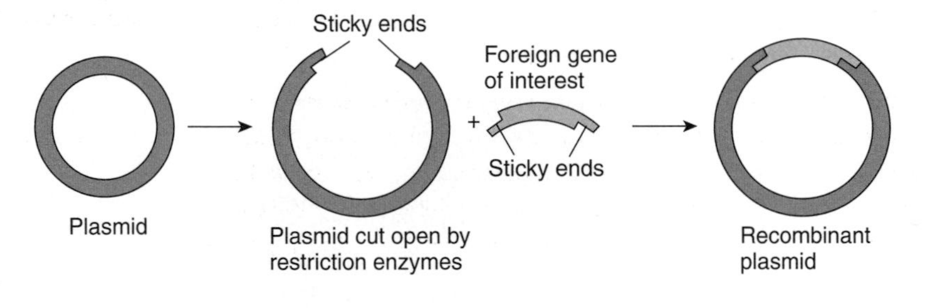

Figure 11.5 A gene is spliced into a plasmid to form a recombinant plasmid

Any DNA that has had 'foreign DNA' inserted into it is called **recombinant DNA**, so the plasmid is now a **recombinant plasmid**.

How do we get the plasmid into the host bacterium?

Obtaining plasmids from bacterial cells is relatively simple — the cells are broken open and the plasmids isolated from the other components. Putting plasmids into bacterial cells is rather more difficult because the bacteria must continue to live. The process usually involves treating the bacteria with a solution of calcium chloride. This alters the bacterial cell wall, making it permeable to plasmids. The bacteria are then incubated with the plasmids. However, there are two important points to note:

- The frequency of plasmid take-up by the bacteria may be as low as 1 in 10 000.
- Of those bacteria that do take up plasmids, some will take up recombinant plasmids and others will take up the original, non-recombinant plasmids. Those that take up the recombinant plasmids are called **transformed bacteria**.

How do we know which bacteria have taken up the gene?

Many plasmids will not take up any new DNA; they will remain exactly as they were at the start. So how do we know which plasmids have taken up the gene? There are two main ways of doing this.

Method 1: using marker genes

Bacteria contain many different plasmids, each of which contains a different combination of genes. Some plasmids contain genes that confer resistance to a particular antibiotic — for example, ampicillin or tetracycline. Some plasmids contain genes that confer resistance to two antibiotics — for example, both ampicillin *and* tetracycline (Figure 11.6).

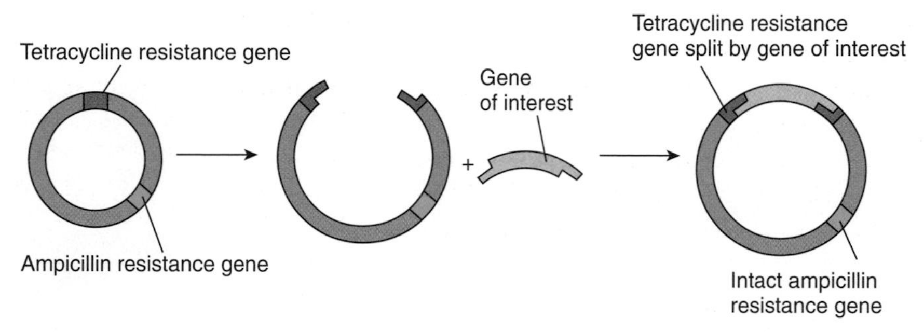

Figure 11.6 Introducing a new gene into a plasmid can damage genes already present

Introducing a gene into a plasmid that contains genes giving resistance to tetracycline *and* ampicillin allows the bacteria that have taken up the plasmid to be identified. The process splits the gene for tetracycline resistance and makes it inactive. So, bacteria that have taken up the original plasmid are resistant to both antibiotics; bacteria that have taken up the recombinant plasmid are resistant to ampicillin only. At the end of the process, there will be the following types of bacteria:

- those that have not taken up any plasmids and are not resistant to either antibiotic
- those that have taken up the original plasmids and are resistant to both antibiotics
- those that have taken up the recombinant plasmids and are resistant to ampicillin only

The bacteria are cultured on media containing either ampicillin or tetracycline. Those that survive on the ampicillin culture *only* are the transformed bacteria. These bacteria contain the gene in which we are interested.

Method 2: using DNA probes

- Incubate the DNA fragments with **plasmids** that have been previously cut open by the same restriction enzyme as was used to create the DNA fragments. Each fragment is taken up by a plasmid to form a **recombinant plasmid**.
- Incubate the recombinant plasmids with bacteria. The ratio of plasmids to bacteria is low so that it is unlikely that any one bacterium will take up more than one plasmid.
- Dilute the bacterial suspension and culture it on agar gel in Petri dishes. Each bacterium will occupy a unique location in a Petri dish.
- Incubate the Petri dish at a suitable temperature so that each bacterium multiplies to form a colony of millions (all containing copies of the recombinant plasmid taken up by the original bacterium).

- 'Blot' each Petri dish with filter paper. This will transfer a few cells from each colony to the filter paper. The cells will be in the same relative positions on the filter paper as the colonies are in the Petri dish.
- Break open the cells on the filter paper to expose the DNA and split the two strands of DNA.
- Incubate with a radioactive **gene probe** (**DNA probe**) that has a sequence that is complementary to at least part of the gene of interest. Only DNA from cells with the gene will bind with the probe and become radioactive.
- Locate the radioactive DNA by producing an X-ray photograph.
- The positions of the radioactive DNA on the X-ray photograph correspond to the colonies of bacteria containing the gene of interest in the Petri dish (Figure 11.7).

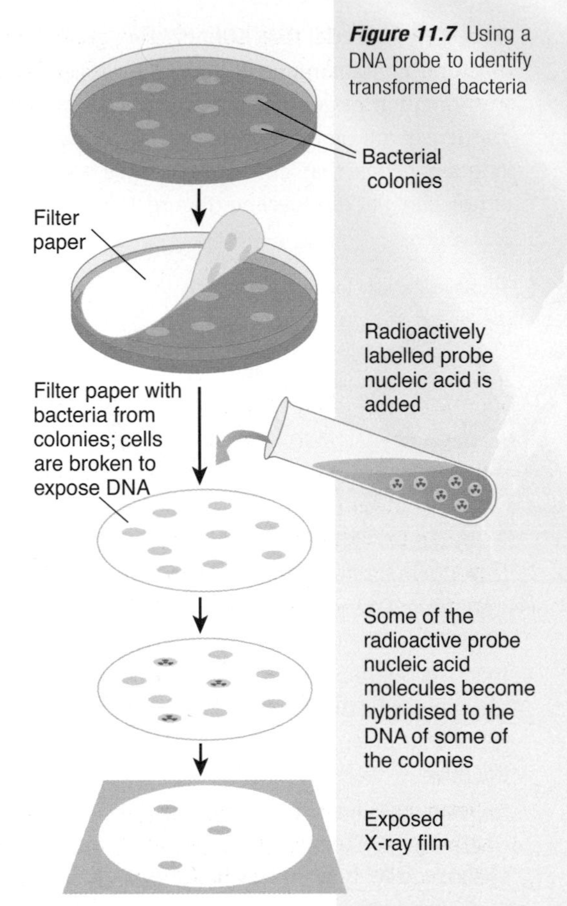

Figure 11.7 Using a DNA probe to identify transformed bacteria

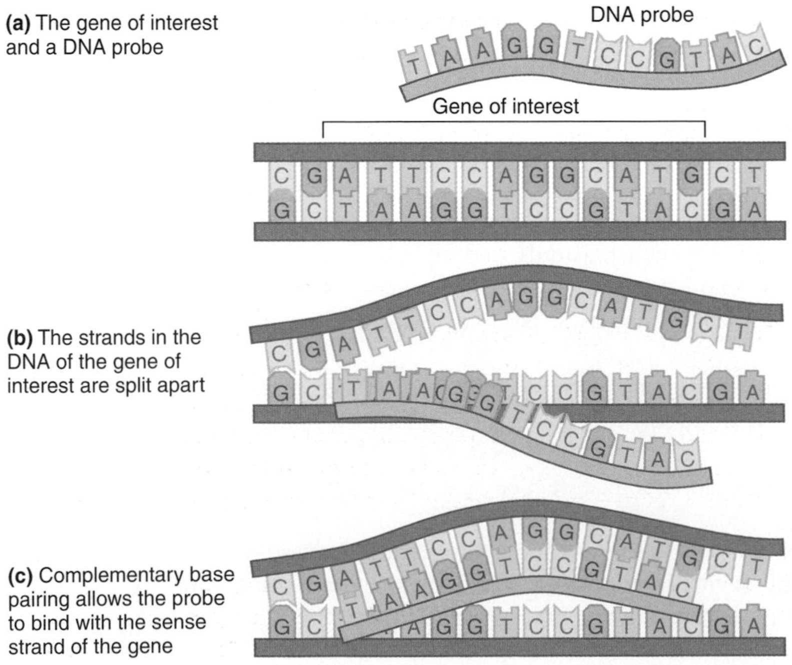

> A gene probe is a short length of single-stranded DNA that is radioactive (or sometimes fluorescent). The sequence of bases is complementary to at least part of the sequence of bases in the gene of interest. So a gene probe can pair with complementary DNA, provided the two strands of the DNA molecule have been separated (Figure 11.8).

(a) The gene of interest and a DNA probe

DNA probe

Gene of interest

(b) The strands in the DNA of the gene of interest are split apart

(c) Complementary base pairing allows the probe to bind with the sense strand of the gene

Figure 11.8 Inserting a DNA probe

Once the bacteria that contain the gene have been identified, they can be cultured to give large numbers of bacteria for a specific purpose or to act as a store of the gene.

How can gene technology be used for human benefit?

What products are made by genetically modified organisms?

Genetically modified bacteria produce a range of products, including:

- enzymes for the food industry
- thermostable enzymes for washing powders
- human insulin
- human growth hormone
- vaccines (e.g. for prevention of hepatitis B)
- bovine somatotrophin (to increase milk yield and muscle development in cattle)

Box 11.3 Two hormones produced by genetically modified bacteria

Human growth hormone is produced by the pituitary gland at the base of the brain. Before growth hormone from genetically modified bacteria was available, the only source of the hormone was human corpses. Obtaining growth hormone involved a rather gruesome procedure and carried health risks. A number of children treated with the hormone from corpses developed Creutzfeldt–Jakob disease (the human form of 'mad cow' disease). When this became apparent, the treatment was withdrawn.

Before human insulin was available from genetically modified bacteria, the only form for treatment of diabetes was non-human insulin. This had to be obtained from other animals. Insulin from genetically engineered bacteria has, therefore, not just saved the lives of people.

Other organisms have also been genetically modified. There has been more progress in creating genetically modified plants than genetically modified animals, although some genetically modified animals have been produced. Genetically modified salmon and *Tilapia* fish grow bigger and faster than the non-modified fish and could prove to be an important source of protein in some regions of Africa. Other animals have been genetically modified to produce specific products, as illustrated in Box 11.4.

Many genetically modified crop plants have been produced by using the bacterium *Agrobacterium tumefaciens*. This bacterium infects plant cells and can transfer a plasmid (the T_i plasmid) into the DNA of the host cells (Figure 11.9).

Box 11.4 'Pharming' animals

Genetic engineers have created animals that are able to produce valuable substances, such as drugs.

	Developer	Purpose
Chicken	University of Guelph	To produce the antibiotic lysozyme in its eggs; this would lower infection rates in eggs
Cow	Pharming Incorporated	Its milk contains the human protein lactoferrin; this can be used to treat infections in people
Goat	Genzyme Corporation	Its milk carries the human blood protein antithrombin III; this can prevent blood clotting in people
	Nexia Biotechnologies	To produce spider silk in its milk; this could have a use in strong, lightweight products, such as bulletproof vests
Sheep	PPL Therapeutics	Secretes a-antitrypsin in its milk; this is used to treat cystic fibrosis
Pig	University of Guelph	Produces the bacterial enzyme phytase; this aids the digestion of the pollutant phosphorus, resulting in less of it in manure

Table 11.1 The new 'pharmyard'

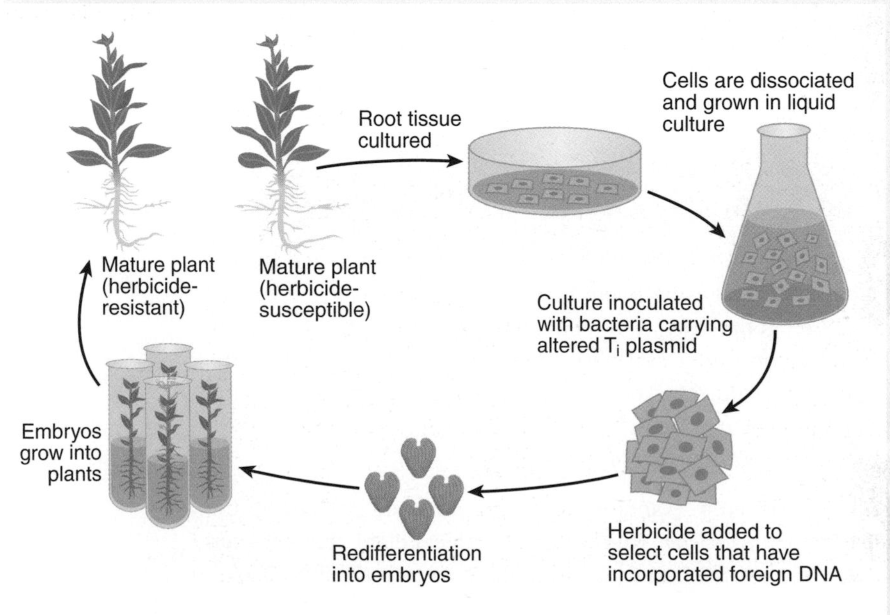

Figure 11.9 Production of herbicide-resistant plants

Modifying the T_i plasmid by adding the desired gene (e.g. one for herbicide resistance) and then allowing the bacterium to do the rest allows the new gene to be introduced. This only produces relatively few genetically modified cells, so these cells are then cultured to produce herbicide-resistant plants.

Unfortunately, *Agrobacterium* does not infect cereals. A new technique has allowed the production of a genetically modified rice called 'golden rice'. Golden rice has three foreign genes — two from daffodils and one from a bacterium — added to its normal DNA content. Together, these genes allow the rice to make beta-carotene, which colours the rice. Beta-carotene is converted into vitamin A when eaten. In less economically developed regions of the world, millions of children go blind because they have no source of vitamin A in their diet. Golden rice could save their eyesight.

Golden Rice Humanitarian Board

Golden rice

How can gene technology be used in analysis of DNA?

Gene technology is not only used to create genetically modified organisms. A recent advance with huge implications in a range of areas is the development of **DNA microarrays** (DNA chips). A gene chip contains thousands of DNA probes immobilised onto a glass slide. A sample of DNA is added to the chip and the chip is analysed to find out which genes are present and how active they are.

A DNA chip: the intensity and colour of each spot encodes information about a specific gene in the test sample

This information is provided by fluorescence analysis of the probe–gene complex.

DNA microarrays can be used to find out the base sequence of new genes and, also, which genes are active under different conditions. This can give valuable information about:

- why drugs are effective in some people, but not others
- why some drugs are toxic to some people, but not others
- the genes that are active during different cancers
- DNA profiles that make people more susceptible to certain diseases
- whether the DNA sample is contaminated by viral or bacterial DNA

How can gene technology be used to provide gene therapy for genetic disease?

Some conditions, for example cystic fibrosis and sickle-cell anaemia, are determined genetically but affect only some of the cells in the body. It might be possible to alter the genes of the affected cells only, so that they no longer produce the faulty protein that causes the symptoms. Researchers in this area might use three main techniques:

- add a normal gene to the affected cells to replace normal functioning
- repair the abnormal gene by selective reverse mutation
- alter the regulation of the gene (it could be 'silenced' or turned on as required)

Much of the research to date has centred on the first technique. People who suffer from cystic fibrosis have a mutant CFTR gene that results in the production of excessive, viscous mucus in the lungs and the pancreas. This makes breathing difficult and reduces the passage of pancreatic enzymes into the small intestine, impairing digestion.

Researchers have tried to introduce the normal gene into these cells using liposomes and viruses to carry the gene into the cells (Figure 11.10).

Liposomes are tiny spheres made from phospholipids and other chemicals. Because of their small size and high phospholipid content, they can pass easily through plasma membranes.

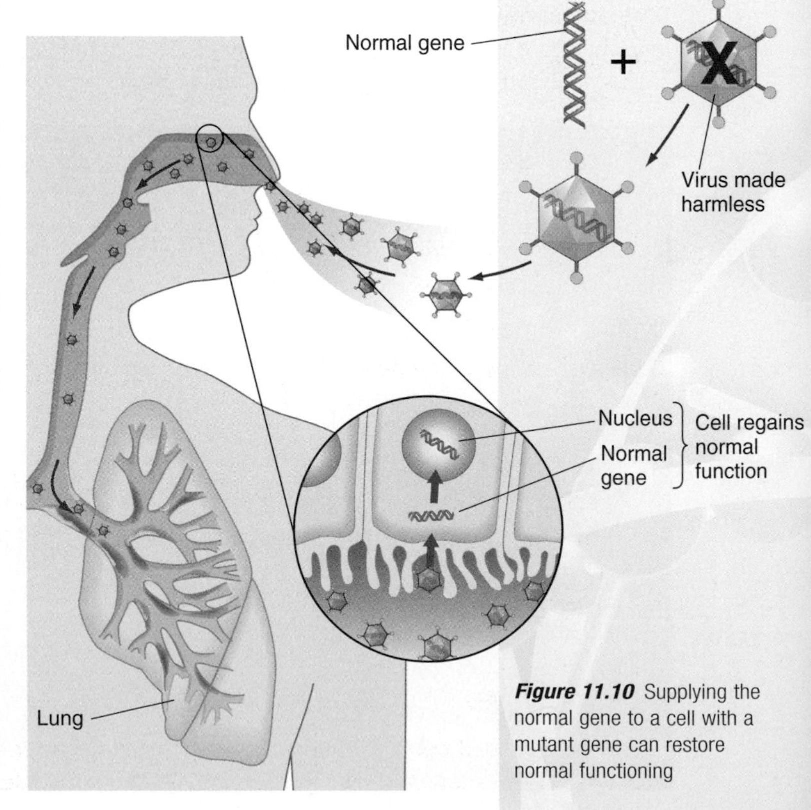

Figure 11.10 Supplying the normal gene to a cell with a mutant gene can restore normal functioning

Once in the cell, the gene is incorporated into the cell's DNA. The gene is activated and controls the production of the normal protein that results in the production of normal amounts of normal mucus. However, there are several problems:

- the transfer rate of the gene is very low
- the 'normal gene' does not integrate fully into the cell's DNA and so is not passed on to the daughter cells when it divides; this means that multiple treatments will be needed
- the virus may stimulate an immune response, which would destroy it before it can deliver the gene; this response will be stronger with each successive treatment
- the virus may recover its ability to cause disease

Memory cells from previous immune responses make subsequent responses quicker and stronger.

Research into gene therapy in the USA using viruses was halted after two children had died following gene therapy for severe combined immunodeficiency (SCID). Ironically, the gene therapy had cured the SCID, but the children had developed a leukaemia-like condition caused by the virus.

Box 11.5 Gene therapy cures sickle-cell anaemia in mice

In sickle-cell anaemia, a mutant gene causes two of the four polypeptide chains of the haemoglobin molecule to be abnormal. This results in the sickling of the cells when the concentration of oxygen is low (see p. 217).

Researchers have found a method of using a harmless virus to introduce a gene into bone marrow stem cells of mice that codes for the polypeptides found in fetal haemoglobin. This haemoglobin is very stable under low oxygen concentrations. They found that months after the gene therapy, the mice continued to produce the fetal haemoglobin and showed no signs of anaemia.

The researchers believe that once technical barriers have been overcome this technique will eventually be used to treat human sufferers.

How is gene sequencing carried out?

Gene sequencing is the determination of the sequence of bases in a section of DNA.

The first breakthrough

The first team to have real success was led by Fred Sanger working in Cambridge. In 1975, they devised the **chain-terminator** technique. This technique has some parallels with the PCR as it also requires:

- a single-stranded section of DNA
- DNA polymerase
- DNA primers
- DNA nucleotides

The single-stranded DNA is primed and mixed with the polymerase, nucleotides and one further key reactant — modified DNA nucleotides called dideoxynucleotides (ddNTP). These molecules can form bonds with only one nucleotide and so, if they enter a chain of DNA, the chain stops at that point (Figure 11.11).

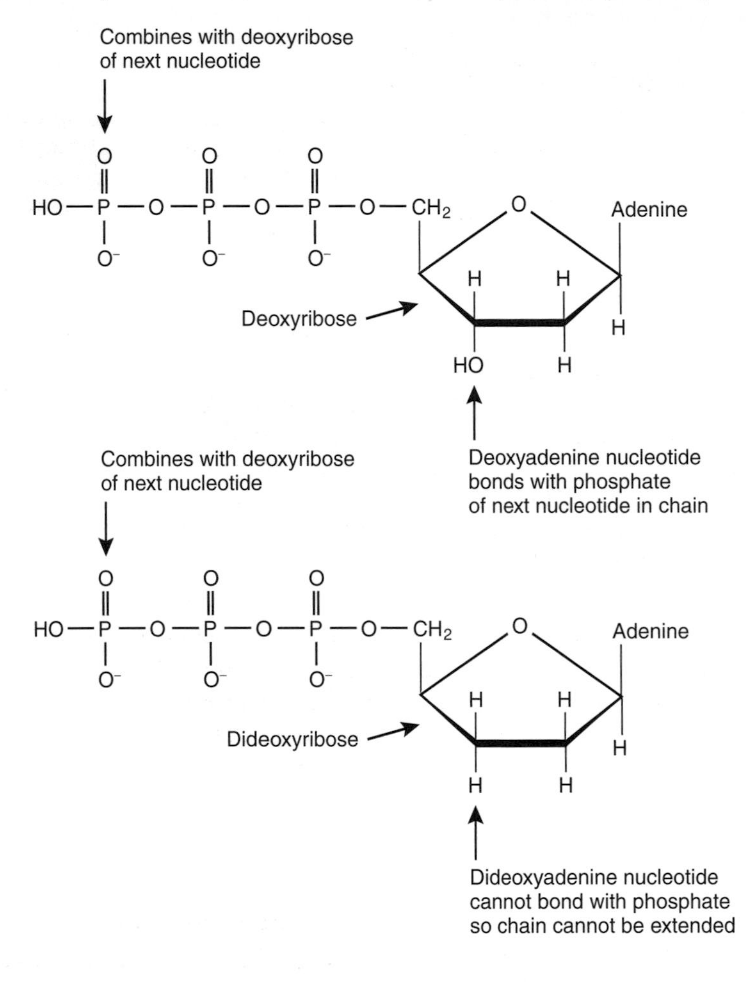

Figure 11.11 Because dideoxynucleotides lack an oxygen atom, they can combine with only one nucleotide

◀ Dideoxynucleotides lack an oxygen atom. This prevents the ribose sugar from binding with the phosphate in another nucleotide and so terminates the chain.

In the Sanger technique:

- Four samples of the single-stranded DNA are primed.
- Each sample is mixed with all four normal nucleotides and one dideoxynucleotide (e.g. dideoxycytosine or dideoxythymine), which is either radioactive or tagged with a fluorescent dye.
- DNA polymerase is added and it begins to build a chain complementary to the DNA sample.
- In the dideoxycytosine reaction tube, the polymerase will use, at random, the normal nucleotides and the dideoxycytosine nucleotides in assembling the complementary strand. Wherever a dideoxyucleotide is used, that particular strand will stop (the millions of other strands in the reaction tube will continue to build).
- At the end of the procedure, each tube will contain many fragments of different lengths. Each will have cytosine at the end of the fragment.

A similar process will occur in the other three reaction tubes (Figure 11.12).

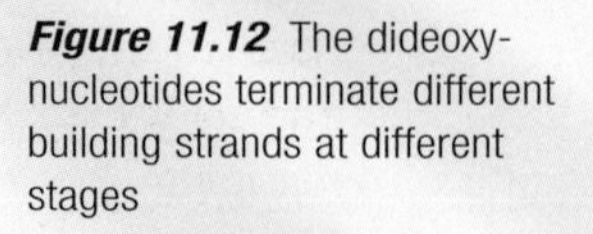

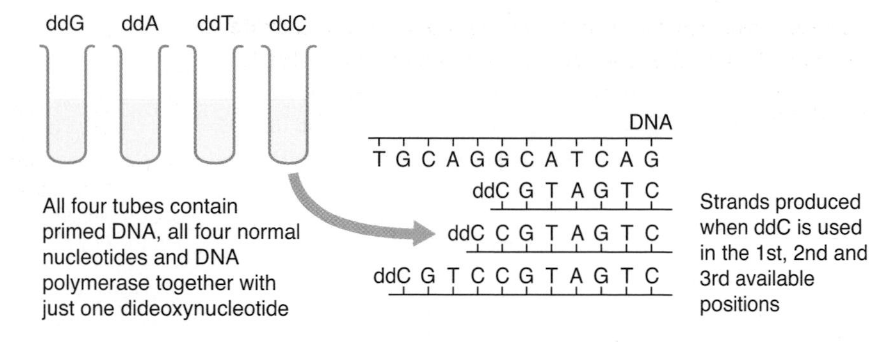

Figure 11.12 The dideoxy-nucleotides terminate different building strands at different stages

In the dideoxycytosine (ddC) tube, the polymerase uses ddC in the first available positions (opposite guanine on some of the strands). These strands then stop building; the others continue building. In some of the others, normal cytosine is used in the first position and ddC is used in the second position. These strands then stop building. In other strands, ddC is not used until the third position, in others the fourth position and so on. At the end of the run in that tube, there will be new DNA strands with ddC in each available position. The ddC nucleotide will be at the end of each of these strands.

The trick now is to find out where the ends of all the fragments from all four tubes occur in relation to the complete section of DNA that has been built. To do this, the fragments from each tube are separated at the same rate for the same length of time by gel electrophoresis. This separates the fragments to an accuracy of one base pair.

Box 11.6 Gel electrophoresis

This is a technique used to separate charged particles. It is based on the principle that in an electric field, charged particles in solution migrate towards either the positive electrode or the negative electrode. Negatively charged ions migrate towards the positive electrode; positively charged ions migrate towards the negative electrode. In electrophoresis, the electric field moves the charged particles towards the electrodes through a gel (Figure 11.13).

The gel acts as a kind of 'molecular sieve' with the result that larger particles move more slowly than smaller ones.

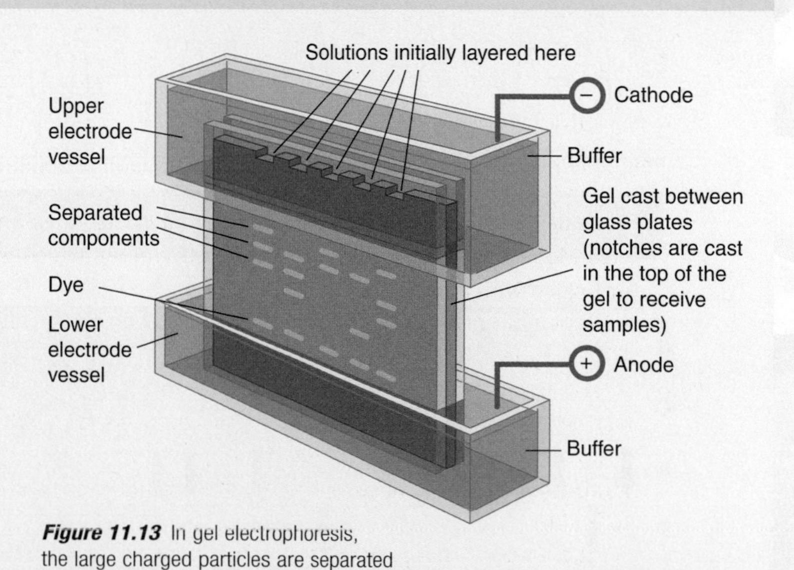

Figure 11.13 In gel electrophoresis, the large charged particles are separated according to their molecular mass

Once the separation is complete, the positions of the fragments can be located by making an X-ray image (if the dideoxynucleotides were radioactive) or by mapping the fluorescence (if they were tagged with a fluorescent dye). Figure 11.14 shows how strands of different lengths from the different tubes might be distributed and how it is then possible to work out the base sequence of the whole DNA sample.

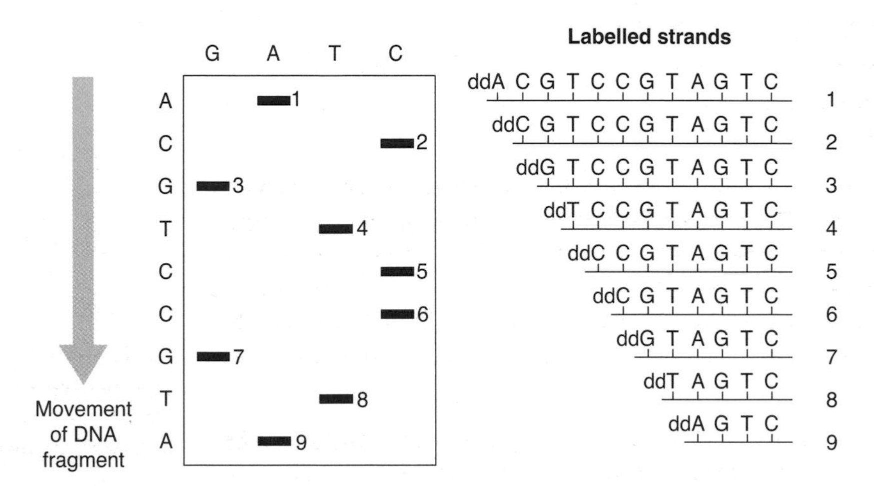

Figure 11.14 Results of gel electrophoresis of DNA fragments. The different fragments move different distances, according to their mass. The shortest fragment contains ddA and moves furthest; the longest also contains ddA and moves the least. Notice that all strands begin with CTG. This is the region that is complementary to the primer on the DNA.

More recent developments

Since the development of the chain-terminator technique, other methods of gene sequencing have been developed. One of these involves tagging the four dideoxynucleotides with different dyes (Figure 11.15). All four tagged nucleotides are contained in the same vessel and, at the end of the run, the entire mixture is

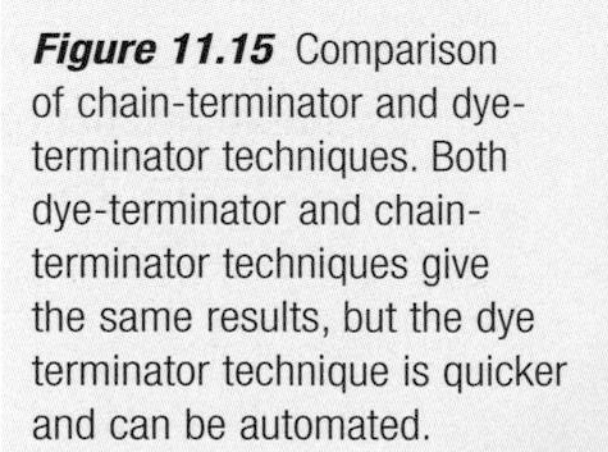

Figure 11.15 Comparison of chain-terminator and dye-terminator techniques. Both dye-terminator and chain-terminator techniques give the same results, but the dye terminator technique is quicker and can be automated.

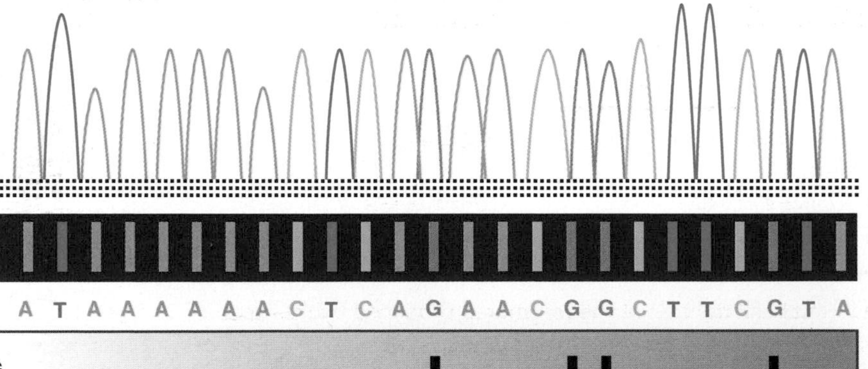

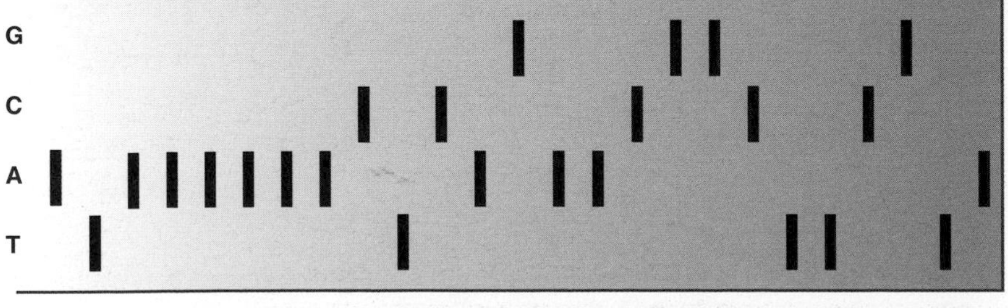

loaded into a single lane of a gel electrophoresis plate. As the nucleotides migrate, they pass a laser that detects which dye is passing and, therefore, which dideoxynucleotide.

The whole system can be automated and, if there is only a tiny amount of the initial DNA, linked to a PCR machine.

What is restriction mapping?

This is another technique that is sometimes used with sequencing procedures to find the base sequence of large sections of DNA, such as particularly big genes or whole chromosomes.

We know that restriction enzymes cut DNA at specific restriction sites (specific base sequences). If some DNA is mixed with a restriction enzyme and two fragments are produced, the DNA has been cut in only one place and so there is only one restriction site for that enzyme in the DNA. If three pieces are produced, the DNA has been cut twice and there are two restriction sites. If 25 pieces are produced, 24 cuts have been made and there are 24 restriction sites for that enzyme.

◀ The number of restriction sites is always one less than the number of fragments produced.

We can use gel electrophoresis to find out the sizes of the DNA fragments by seeing how far they move in comparison with DNA markers of known size. But where are they located in the piece of DNA? Cutting with just one enzyme will not give us that information. Figure 11.16 shows how using two enzymes can enable predictions to be made about possible locations of the restriction sites.

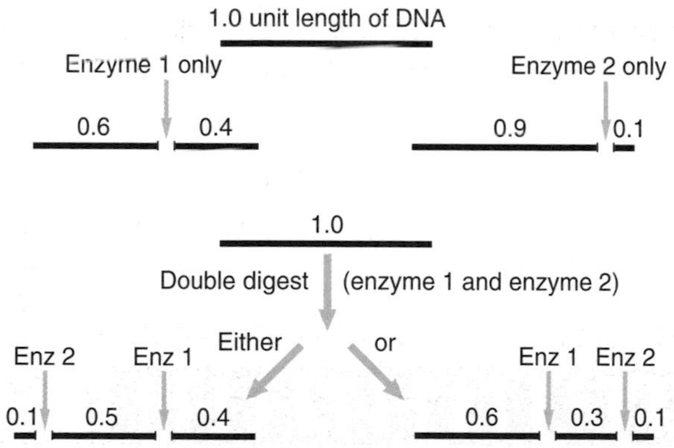

Figure 11.16 Using two restriction enzymes allows us to predict where the restriction sites might be

Using electrophoresis to check the size of the fragments produced by treatment with both enzymes together will confirm one of the predictions.

But why restriction map anyway? One reason is to enable us to combine data from gene sequences of small fragments of DNA from the same source. This is necessary because many gene sequencing procedures cannot sequence very large fragments of DNA.

Suppose a large piece of DNA is digested into several smaller pieces and a restriction map constructed for each piece. The maps could then be compared for overlap of restriction sites, as this would indicate a common piece of DNA. This is illustrated in the flow chart in Figure 11.17.

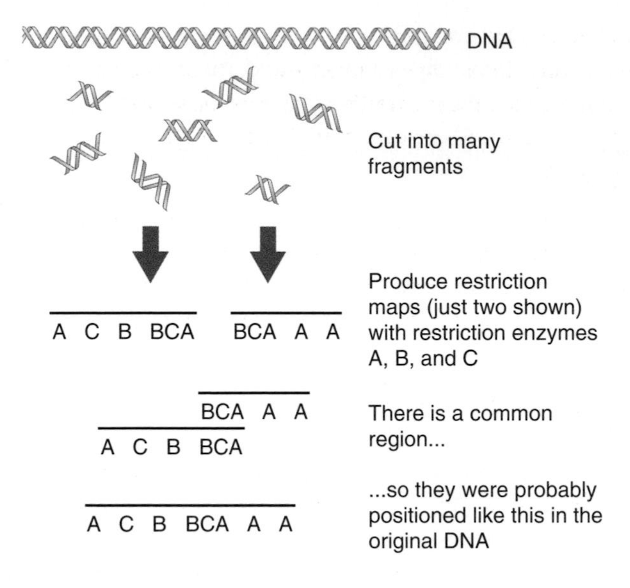

Figure 11.17 Using restriction mapping to determine base sequences of large fragments of DNA. The letters A, B and C represent the sites where the respective enzymes cut the DNA.

If the DNA fragments had been sequenced, we could now start to put together the base sequences into a larger sequence, remembering not to duplicate the overlaps.

What use are restriction mapping and sequencing?

The techniques described can be used to screen patients for mutant or disease-causing genes.

Using the information gained, patients can be counselled as to the likely outcomes of having children. Newly diagnosed patients, new parents or couples planning a pregnancy, may need support in understanding the nature of the genetic disease and its implications for future generations and what options there might be for treatment or preventing recurrence. Other family members (brothers, sisters and cousins) may be concerned that they too may carry a disorder and may also need support.

Knowledge of the sequences of specific oncogenes may lead to new methods of silencing them or methods of gene therapy to replace them.

How is gene technology used in forensic science?

Fingerprints have been used for many years to help place a suspect at the scene of a crime. They continue to provide strong evidence because, with the exception of identical twins, an individual's fingerprints are unique. They do not change throughout life.

Genetic fingerprinting has nothing to do with actual fingerprints. It is a technique for comparing the DNA of different people, without the need to know anything about the function of the DNA. Much of the DNA in the cells of the body is non-coding DNA. The non-coding DNA found between genes contains base sequences that are repeated, sometimes many times over. These repeating sequences of non-coding DNA are called **mini-satellites** and it is these that form the basis of a genetic fingerprint. The mini-satellites are inherited along with the coding DNA from one or other parent.

The coding DNA within the genes is unlikely to vary a great deal between individuals. For example, the base sequence in the gene for normal haemoglobin is the same in all of us, as is the base sequence in the gene for pepsin.

The DNA to be used for analysis can be obtained from a sample of blood (white blood cells could supply the DNA), skin or semen — in fact, from any type of cell that has a nucleus. If the sample does not contain sufficient DNA for analysis, then the amount can be amplified using the **polymerase chain reaction** (see page 233).

The main stages in preparing a genetic fingerprint are as follows:

- DNA is isolated from the cells.
- The DNA is cut into fragments using one or more restriction enzymes. These are chosen to cut, where possible, outside the repeating sequences of non-coding DNA.
- The fragments that are obtained are treated with alkali to separate the strands of each DNA fragment.
- The fragments are separated by gel electrophoresis (pp. 245). Smaller fragments (with a lower molecular mass) move further than larger fragments (Figure 11.18).

The size of DNA molecules is usually measured in kilobase pairs (thousands of pairs of bases). These fragments are single stranded, so there are no base pairs. Their size is measured in kilobases.

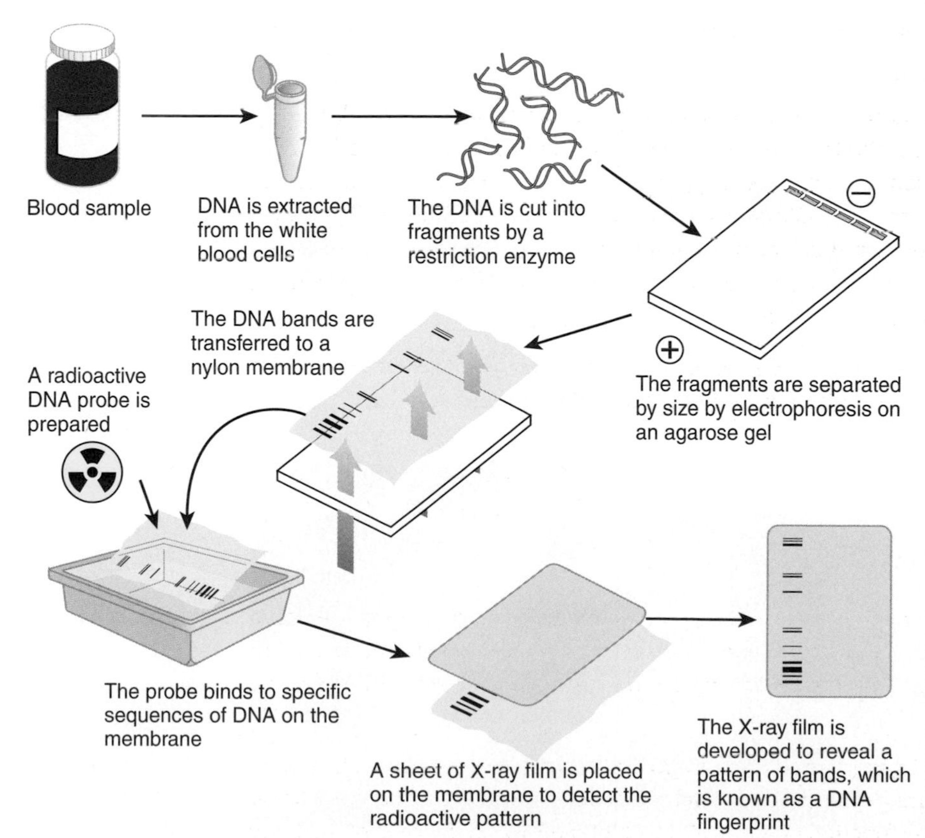

Figure 11.18 Stages in preparing a genetic fingerprint

- The (invisible) pattern of separated DNA fragments is transferred from the gel to a nylon membrane. The membrane is placed over the gel in a tray of 'flow-buffer' and is held in place by paper towels and a weight. The buffer soaks up through the gel, carrying the fragments of DNA with it. The buffer can pass through the membrane (to be absorbed by the paper towels), but the DNA cannot. It remains in the nylon membrane in the same relative position as it was in the gel.
- A radioactive gene probe is applied to the membrane. This is designed to bind with base sequences in the mini-satellite regions.
- Placing the membrane over a piece of X-ray film reveals the positions of those fragments that have base sequences complementary to the probe.

The technique of transferring DNA fragments from the gel to the nylon membrane was devised by Professor E. M. Southern and is called Southern blotting. A similar technique can be used to transfer mRNA fragments to a nylon membrane. It is known as Northern blotting — but no Professor Northern was involved!

Remember that the mini-satellites are inherited along with the coding DNA. Genetic fingerprints can, therefore, be used to help resolve disputed parentage — each fragment of DNA in the fingerprint must have come from one or other parent.

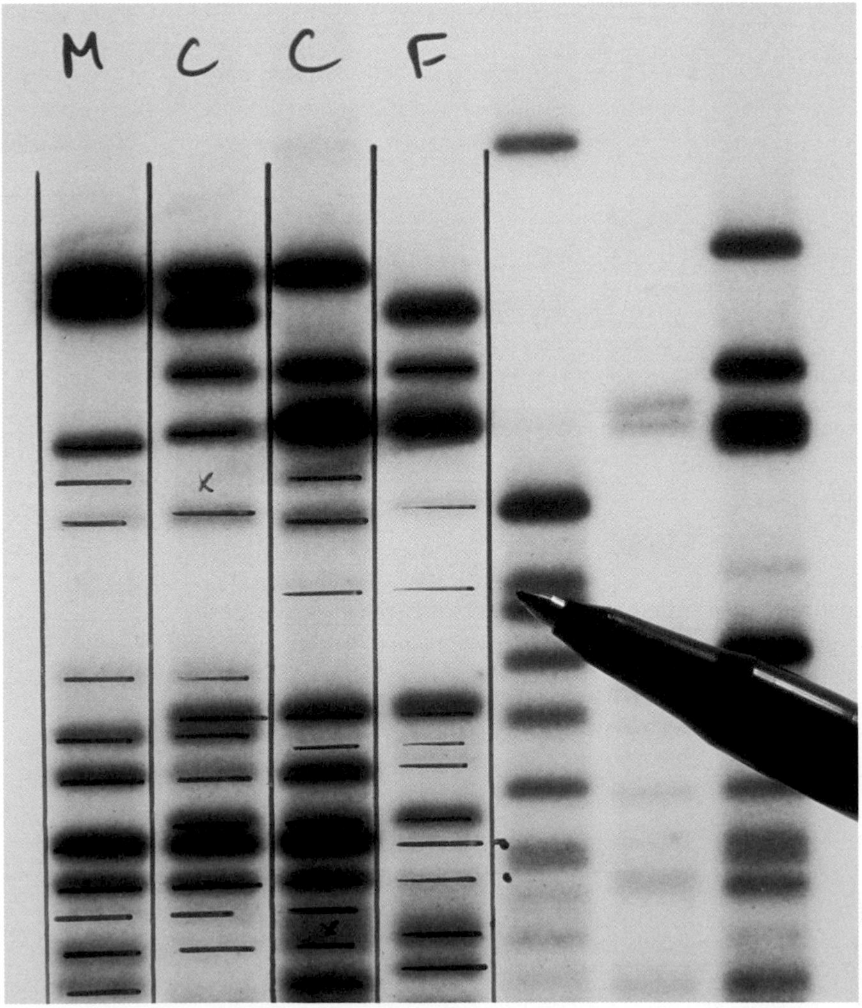

David Parker/SPL

Genetic fingerprints of two children and their parents. Each fragment of DNA (dark bands) in the children is also present in one or both of the parents.

What are the moral and ethical considerations of using gene technology?

Is genetic engineering right or wrong? A debate about right and wrong involves the principles of ethics and morality.

- Morality is our personal sense of what is right, or acceptable, and what is wrong. Morality is not necessarily linked to legality.
- Ethics also involve a sense of right and wrong. However, ethics are not individual opinions. They represent the 'code' adopted by a particular group to govern its way of life.

Many people have passionate views about genetic engineering. Some hold an unshakeable belief in the technology, which they see as something that will bring great benefits to humankind. Other people hold the equally strong belief that genetic engineering is tampering with nature and is likely to cause serious ecological and physiological problems. Some of the issues people are concerned about are discussed below.

- A species is sacrosanct and should not be altered genetically in any way.

This is a personal, moral viewpoint. People who take this moral stance usually do so on the basis that the genes from one species would not normally find their way into another species. However, genes have been 'jumping' from one species to another (albeit at a very low frequency) for millions of years.

- Not enough is known about the long-term ecological effects of introducing genetically modified organisms into the field. They may out-compete wild plants and take over an area.

This is also a moral viewpoint. The effects of any new crop cannot be determined without field trials. Ten thousand years ago, the early farmers who cross-bred wild wheat plants to produce the forerunner of today's strains could not have known what impact these would have. Does this make it wrong?

- If plants are genetically engineered to be resistant to herbicides, the gene could 'jump' into populations of weeds and other wild plants.

This is true — it could. However, non-genetically modified herbicide-resistant strains of plants already exist. The gene could just as easily jump from these.

- Gene technology might give doctors the ability to create designer babies. It could become possible to obtain a newly fertilised human egg, determine its genotype and ask the parents which genes they would like to be modified. Initially, only genes that cause disease might be replaced. Subsequently, the technology might be used to replace other genes.

The human genome project has identified all the base sequences in the human genome. However, much of this is 'junk' DNA and the exact start and end points of many genes are not yet known.

Most doctors would find this morally and ethically unacceptable. They might consider replacing genes that cause disease but not replacing genes merely to improve a child's image in the eyes of its parents. However, if such practices become possible, who will define for doctors what is ethically acceptable? What will be the dividing line between cosmetic gene therapy and medical gene therapy?

- Using **genetic fingerprinting** to combat crime will only be useful if there is a genetic database — a file of the genetic fingerprints of everyone in the country, so that a genetic fingerprint found at the scene of a crime could instantly implicate that person. Who will have access to this information?

There are concerns that a genetic database would be subject to misuse. If insurance companies had access to the genetic database, they might refuse insurance (or charge higher premiums) to people with an increased risk of, say, heart disease. Employers could (covertly) refuse employment to people because their 'genetic profiles' did not meet particular requirements. A recent ruling from the European court states that the police have no right to hold the DNA of someone unless they have been convicted of a crime.

You should also consider the fact that biotechnology (including gene technology) is sometimes merely a refinement of less controversial practices. Organic farmers use the naturally occurring soil bacterium *Bacillus thuringiensis* as a non-chemical insecticide. Genetic engineers have extracted a gene from this bacterium and transferred it to cotton plants to make them resistant to attack by insects. Is there any real difference? People have known for centuries that rubbing a certain blue mould onto cuts can stop them turning septic. In 1922, Alexander Fleming discovered penicillin in the blue mould *Penicillium*.

Summary

The genome

- The genome of an organism is the complete set of genetic information in that organism.

Gene cloning and transfer

- A transgenic organism has had genes from a different type of organism (usually a different species) added to its genome.
- The methods for obtaining a gene for transfer to another organism involve:
  - using mRNA as a template and synthesising single-stranded cDNA from free DNA nucleotides using the enzyme reverse transcriptase; the cDNA is then used (with more nucleotides and DNA polymerase) to make the complementary second strand of DNA
  - cutting the DNA into fragments using a restriction enzyme and isolating the gene
- If insufficient DNA is obtained, the amount can be amplified using the polymerase chain reaction; in this reaction:
  - the template DNA (the sample) is placed in a PCR machine with free DNA nucleotides, DNA primers and thermostable DNA polymerase

- the strands of the template DNA are separated by heating to 95°C
- the mixture is cooled to 37°C to allow the primers to bind
- the mixture is heated to 72°C to allow replication at the optimum temperature of the DNA polymerase
- the cycle repeats itself many times, each cycle taking about 7–8 minutes

- Restriction enzymes cut DNA at specific sequences, called restriction sites, to leave overlapping, sticky ends.
- A gene is inserted into a plasmid using a ligase enzyme.
- Plasmids are chosen that contain marker genes, such as resistance to an antibiotic, which allow identification of the bacteria that have taken up the gene of interest.
- Testing the DNA with fluorescent cDNA probes also allows identification of bacteria that have taken up a specific gene.
- Once the gene has been inserted successfully and the transformed bacteria identified, they are cultured and the product is harvested.
- Genetically modified organisms can be used to manufacture specific products to benefit humans (e.g. insulin, bovine somatotrophin and vaccines).
- Other organisms have been genetically modified to produce increased yields.
- Gene technology is also used to produce DNA microarrays (DNA chips) that allow the identification and assessment of activity of specific genes.

Gene therapy

- Gene therapy involves treating defective genes, usually by supplying a copy of the normal gene; progress has been limited because of:
  - low rate of gene take up by cells with the defective gene
  - attack by the immune system
  - the virus vectors regaining virulence and causing disease

Gene sequencing and restriction mapping

- In the chain terminator technique of gene sequencing:
  - four samples of the single-stranded DNA are primed
  - each is mixed with all four normal nucleotides and one radioactive dideoxynucleotide
  - DNA polymerase is used to build a chain complementary to the DNA sample
  - at the end of the procedure, each tube will contain many fragments of different lengths; each will have its specific dideoxynucleotide at the end of each fragment
  - the position of the different-length strands can be located by electrophoresis
  - the positions of all the strands allows the base sequence to be deduced
- Restriction mapping allows biologists to find the location of different restriction sites in a piece of DNA.
- Fragments of DNA from the same source can be pieced together in the correct sequence by looking for overlapping areas in their restriction maps.
- Gene sequencing and restriction mapping provide information for genetic counsellors to advise patients on what options may be open to them, knowing that they have certain defective genes.

Genetic fingerprinting

- In genetic fingerprinting:
  - a DNA sample is cut into fragments by restriction enzymes
  - the fragments are denatured by separating the two strands
  - the fragments are separated by gel electrophoresis and transferred to a nylon membrane by Southern blotting
  - a radioactive gene probe is added to the membrane and the pattern of complementary sequences is revealed using X-ray film

Moral and ethical concerns

- There are moral and ethical concerns about the use of gene technology.
- Morals are our personal sense of what is right and what is wrong.
- Ethics represent the 'code' adopted by a particular group to govern its way of life.

Questions

Multiple-choice

1 Ligase is an enzyme that:
  A cuts DNA molecules, leaving sticky ends
  B joins sticky ends of DNA fragments
  C copies DNA fragments
  D separates DNA fragments

2 Reverse transcriptase is an enzyme that makes:
  A a cDNA copy of mRNA
  B an mRNA copy of DNA
  C an mRNA copy of cDNA
  D a DNA copy of mRNA

3 A gene probe could be:
  A a short length of single-stranded DNA that has been made radioactive
  B a tRNA molecule that has been made radioactive
  C a short length of double-stranded DNA that has been made radioactive
  D all of the above

4 Viruses and plasmids are examples of:
  A vectors
  B transformed bacteria
  C restriction enzymes
  D liposomes

5 It is often preferable to create a gene from mRNA rather than extract the gene from the donor cell because:
  A it saves the tedious process of isolating the gene from many DNA fragments
  B there can be thousands of copies of mRNA in cells where the gene is active
  C it creates a copy of the gene without any introns
  D all of the above

6 Restriction mapping produces maps of:
 A where each gene is in a fragment of DNA
 B where each base is in a fragment of DNA
 C where each restriction enzyme cuts the fragment of DNA
 D where each restriction enzyme links one piece of DNA to another

7 The DNA used in producing a genetic fingerprint is:
 A coding DNA
 B non-coding DNA
 C both A and B
 D neither A nor B

8 In gel electrophoresis of DNA:
 A the DNA fragments migrate towards the positive electrode and are separated according to their molecular mass
 B the DNA fragments migrate towards the negative electrode and are separated according to their molecular mass
 C the DNA fragments migrate towards the negative electrode and are separated according to their electric charge
 D the DNA fragments migrate towards the positive electrode and are separated according to their electric charge

9 In the polymerase chain reaction, the sample of DNA is mixed with:
 A free DNA nucleotides
 B free DNA nucleotides and DNA primers
 C free DNA nucleotides, DNA primers and thermostable DNA polymerase
 D free DNA nucleotides, DNA primers and thermolabile DNA polymerase

10 The chain terminator technique of gene sequencing produces:
 A different length DNA fragments with the same dideoxynucleotides
 B different length DNA fragments with different dideoxynucleotides
 C DNA fragments of the same length with different dideoxynucleotides
 D DNA fragments of the same length with the same dideoxynucleotides

Examination-style

1 The polymerase chain reaction (PCR) is a method of obtaining large amounts of DNA from a small initial sample. The diagram shows the main stages in the polymerase chain reaction.

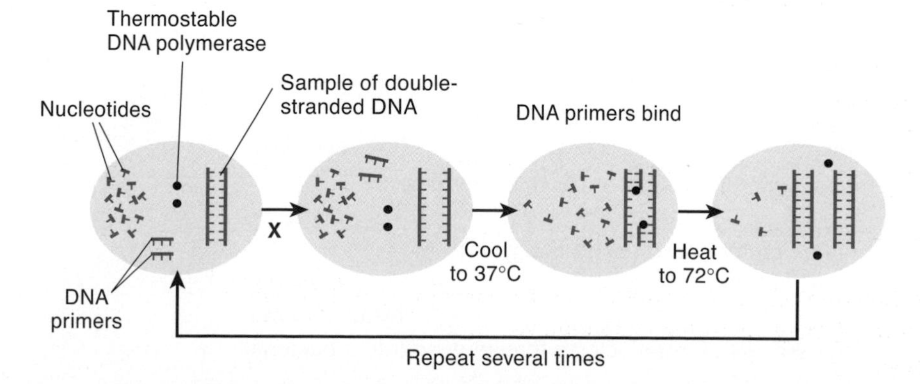

(a) (i) What must be done to separate the strands of DNA (process **X**)? *(1 mark)*

(ii) What are primers? *(2 marks)*

(b) (i) What is meant by 'thermostable DNA polymerase'? *(2 marks)*

(ii) What is the main advantage of using a thermostable DNA polymerase in this process? *(1 mark)*

Total: 6 marks

2 The flow chart shows some stages in creating a genetically modified bacterium.

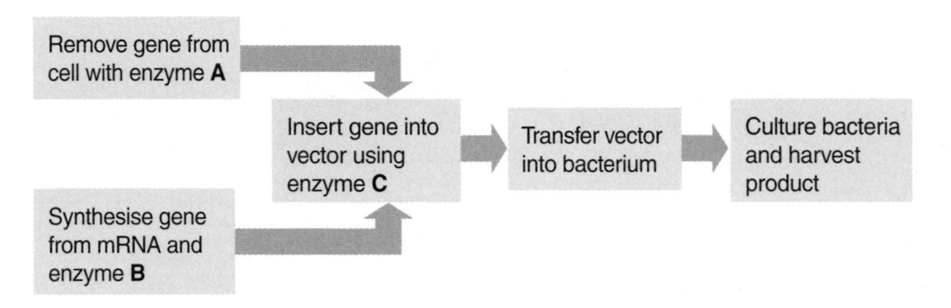

(a) (i) Name the enzymes **A**, **B** and **C**. *(3 marks)*

(ii) Name two possible vectors. *(2 marks)*

(b) Explain the benefits of using genetically engineered bacteria to produce insulin for humans. *(2 marks)*

Total: 7 marks

3 The diagram shows how the action of enzyme X cuts a molecule of DNA.

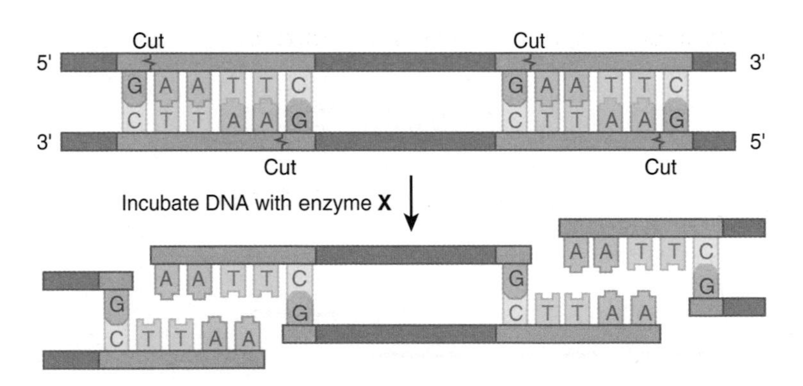

(a) (i) Name the type of enzyme that cuts DNA in this way. *(1 mark)*

(ii) What name is given to the places where the enzyme makes the cuts? *(1 mark)*

(iii) Explain the importance of the type of cut made by this enzyme. *(3 marks)*

(b) Suggest why it may be preferable to obtain the gene from mRNA, rather than from DNA. *(2 marks)*

Total: 7 marks

4 The diagram shows the process of introducing a foreign gene into a bacterial plasmid.

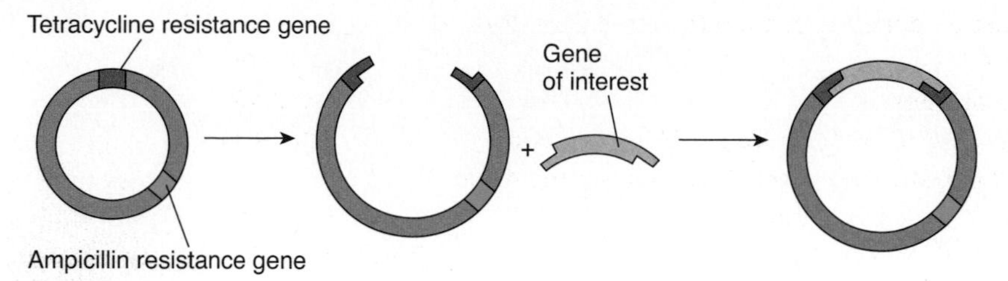

(a) Name the enzyme used to:

  (i) produce the 'sticky ends'

  (ii) insert the foreign gene into the plasmid *(2 marks)*

(b) The plasmids are then introduced into host bacteria with no resistance to antibiotics. What would be the result of culturing the transformed bacteria on Petri dishes of agar containing:

  (i) tetracycline?

  (ii) ampicillin?

  Explain your answers. *(4 marks)*

(c) Bacteria naturally 'swap' plasmids in a process called conjugation. Suggest how this might have given rise to bacteria resistant to many antibiotics. *(2 marks)*

Total: 8 marks

5 A couple have three children. However, the father believes that the third child is not his, but that of another man. Genetic fingerprints are produced from samples of blood taken from the three adults and the three children. These are shown below.

Father | Mother | Other man | Child 1 | Child 2 | Child 3

(a) (i) Which blood cells would be used to provide the DNA sample? Give a reason for your answer. *(2 marks)*

  (ii) Explain why it was not necessary to use the polymerase chain reaction. *(2 marks)*

(b) Explain how genetic fingerprinting is carried out. *(6 marks)*

(c) (i) Explain how the fingerprints confirm that child 1 is the child of the parents. *(2 marks)*

  (ii) Is the other man the father of child 3? Explain your answer. *(3 marks)*

Total: 15 marks

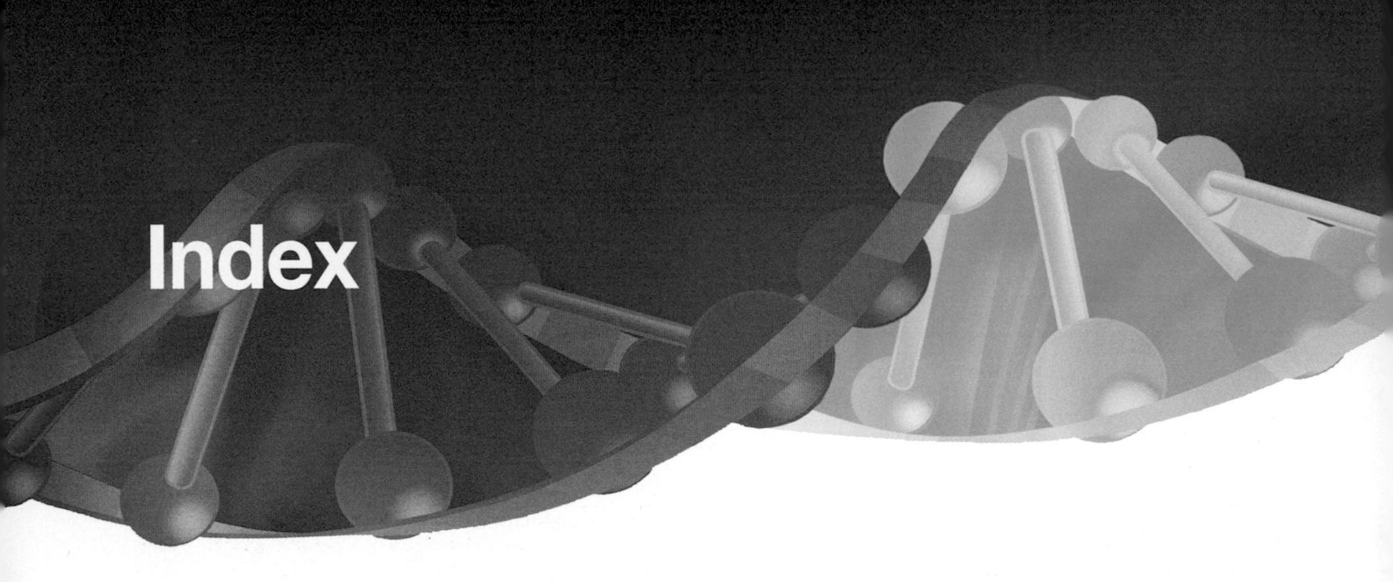

Index

A

B

C

D

G

M

Q

R

S